Electroless
Plating:
Fundamentals
And Applications

Electroless Plating:
Fundamentals
And Applications

Glenn O. Mallory
Juan B. Hajdu
Editors

Reprint Edition

Sponsored and published by
American Electroplaters and Surface Finishers Society

American Electroplaters and Surface Finishers Society
The International Technical and Educational Society for
Surface Finishing
International Headquarters in Orlando, Florida

Library of Congress Card Number 90-081578

Published in the United States of America by
Noyes Publications/William Andrew Publishing, LLC
13 Eaton Avenue, Norwich, New York 13815

Transferred to Digital Printing 2009

PREFACE

The term "electroless plating" describes the methods of depositing metals and alloys by means of electrochemical reactions. However, chemical plating is the more accurate term that can be used to denote the several means of metal deposition without the application of electric current from an external source. Hence, immersion deposition, as well as electroless deposition, covered in this book, are two forms of chemical plating. In usage, the term "electroless plating," as coined by Abner Brenner, has come to be synonymous with autocatalytic plating. In this process, the chemical reaction proceeds continuously on selected surfaces, providing the means to produce uniform coatings with unique properties on a wide variety of substrates.

The practice of electroless plating is a relatively young art, developed over the past fifty years for a large number of applications. Several major industries, such as printed circuit boards, hard memory disks and electroplated plastics were made possible by the development of electroless technology. This book describes the chemical principles of the major electroless processes and the practical applications of these techniques in industry. Of the different electroless processes available, electroless nickel and electroless copper have gained the largest industrial use and are discussed extensively. Other electroless plating processes and related subjects are discussed in individual chapters. A limited number of techniques, mentioned in the literature, that have no experimental proof or applications background were not included. It is important to note here that electroless plating is a fast-growing field and the references should be updated continuously.

Two points should be made on editorial decisions. As a result of our intention to cover both principles and applications of electroless plating, some subjects required a theoretical approach, while other subjects demanded pragmatic and descriptive treatment. For this reason, the authors had very few constraints on style, format, units and the general outlay of their chapters.

In addition to electroless plating, immersion plating is reviewed. While this process is not based strictly on chemical reduction, it is closely related to electroless plating in industrial applications.

The editors would like to express their gratitude to the many persons who have made this book a reality: First of all to the authors for their cooperation and patience; to the staff and authorities of the American Electroplaters and Surface Finishers Society for their help and support; and to the members of the Electroless Finishing Committee, especially to our colleagues, Michael Aleksinas, Dr. Moe El-Shazly, David Kunces, CEF, and Fred Pearlstein, CEF, in reviewing the manuscripts. Special thanks are also due Harry Litsch, CEF-SE, for preparing the index.

No work on the subject of electroless plating should be published without acknowledging the industry's lasting debt to the pioneering work of Dr. Abner Brenner.

We hope this book will fill the void which has existed for a complete reference on electroless deposition, and that you will find it a most useful addition to your library.

Glenn O. Mallory
Editor

Juan B. Hajdu
Editor

CONTRIBUTORS

MICHAEL J. ALEKSINAS, Fidelity Chemical Products Co., Newark, NJ

DONALD W. BAUDRAND, CEF, Allied-Kelite Division, Witco Chemical Corp., Melrose Park, IL

PETER BERKENKOTTER, Western Digital Corp., Santa Clara, CA

DR. PERMINDER BINDRA, International Business Machines Corp., Endicott, NY

ROBERT CAPACCIO, P.E., Mabbett, Capaccio & Associates, Boston, MA

DR. JOSEPH COLARUOTOLO, Occidental Chemical Corp., Berwyn, PA

JOHN G. DONALDSON, CEF, Surface Finishing Engineering, Tustin, CA

DR. E.F. DUFFEK, Adion Engineering Co., Cupertino, CA

DR. NATHAN FELDSTEIN, Surface Technology Inc., Trenton, NJ

DR. JUAN B. HAJDU, Enthone-OMI International, Inc., West Haven, CT

DR. N. KOURA, Science University of Tokyo, Machida City, Japan

JOHN J. KUCZMA, JR., Elnic, Inc., Nashville, TN

DAVID KUNCES, CEF, Fidelity Chemical Products Corp., Newark, NJ

JOHN KUZMIK, MacDermid, Inc., Waterbury, CT

HARRY J. LITSCH, CEF-SE, Litsch Consultants, Bethlehem, PA

GLENN O. MALLORY, Electroless Technologies Corporation, Los Angeles, CA

DR. YUTAKA OKINAKA, AT&T Bell Laboratories, Murray Hill, NJ

KONRAD PARKER, Consultant, Park Ridge, IL

FRED PEARLSTEIN, CEF, Temple University, Philadelphia, PA

W.H. SAFRANEK, CEF, American Electroplaters and Surface Finishers Society, Orlando, FL

PHILLIP D. STAPLETON, The Stapleton Co., Long Beach, CA

DR. DONALD STEPHENS, Consultant, Westlake Village, CA

FRANK E. STONE, Inland Specialty Chemical Corporation, Tustin, CA

DIANE M. TRAMONTANA, Occidental Chemical Corp., Grand Island, NY

DR. ROLF WEIL, Stevens Institute of Technology, Hoboken, NJ

DR. JAMES R. WHITE, International Business Machines Corp., Austin, TX

CATHERINE WOLOWODIUK, AT&T Bell Laboratories, Murray Hill, NJ

CONTENTS

Preface   vii

Contributors   viii

Chapter 1   1
The Fundamental Aspects of Electroless Nickel Plating
Glenn O. Mallory

Chapter 2   57
Composition and Kinetics of Electroless Nickel Plating
Glenn O. Mallory

Chapter 3   101
Troubleshooting Electroless Nickel Plating Solutions
Michael J. Aleksinas

Chapter 4   111
Properties of Electroless Nickel Plating
Rolf Weil and Konrad Parker

Chapter 5   139
Equipment for Electroless Nickel
John Kuczma, Jr.

Chapter 6   169
Test Methods for Electroless Nickel
Phillip Stapleton

Chapter 7   193
Surface Preparation for Electroless Nickel Plating
Juan Hajdu

Chapter 8   207
Engineering Applications of Electroless Nickel
Joseph Colaruotolo and Diane Tramontana

Chapter 9   229
Electronic Applications of Electroless Nickel
E.F. Duffek, D.W. Baudrand, CEF, and J.G. Donaldson, CEF

Chapter 10   261
Electroless Deposition of Alloys
Fred Pearlstein, CEF

Chapter 11   269
Composite Electroless Plating
Nathan Feldstein

Chapter 12   **289**
Fundamental Aspects of Electroless Copper Plating
Perminder Bindra and James R. White

Chapter 13   **331**
Electroless Copper in Printed Circuit Fabrication
Frank E. Stone

Chapter 14   **377**
Plating on Plastics
John J. Kuzmik

Chapter 15   **401**
Electroless Plating of Gold and Gold Alloys
Yutaka Okinaka

Chapter 16   **421**
Electroless Plating of Platinum Group Metals
Yutaka Okinaka and Catherine Wolowodiuk

Chapter 17   **441**
Electroless Plating of Silver
N. Koura

Chapter 18   **463**
Electroless Cobalt and Cobalt Alloys
Section I
W.H. Safranek, CEF
Section II
Peter Berkenkotter and Donald Stephens

Chapter 19   **511**
Chemical Deposition of Metallic Films from Aqueous Solutions
David Kunces, CEF

Chapter 20   **519**
Waste Treatment of Electroless Plating Solutions
Robert Capaccio, P.E.

Index   **529**

About the Editors   **539**

Chapter 1
The Fundamental Aspects
Of Electroless Nickel Plating
Glenn O. Mallory

The chemical deposition of a metal from an aqueous solution of a salt of said metal has an electrochemical mechanism, both oxidation and reduction (redox), reactions involving the transfer of electrons between reacting chemical species. The oxidation of a substance is characterized by the *loss* of electrons, while reduction is distinguished by a *gain* of electrons. Further, oxidation describes an anodic process, whereas reduction indicates a cathodic action. The simplest form of chemical plating is the so-called *metal displacement reaction*. For example, when zinc metal is immersed in a copper sulfate solution, the zinc metal atoms (less noble) dissolve and are spontaneously replaced by copper atoms from the solution. The two reactions can be represented as follows:

Oxidation: $Zn^0 \rightarrow Zn^{+2} + 2e^-$, anodic, $E^0 = 0.76$ V
Reduction: $Cu^{+2} + 2e^- \rightarrow Cu^0$, cathodic, $E^0 = 0.34$ V

Overall reaction: $Zn^0 + Cu^{+2} \rightarrow Zn^{+2} + Cu^0$, $E^0 = 1.1$ V

As soon as the displacement reaction begins, the surface of the zinc substrate becomes a mosaic of anodic (zinc) and cathodic (copper) sites. The displacement process continues until almost the entire substrate is covered with copper. At this point, oxidation (dissolution) of the zinc anode virtually stops and copper deposition ceases. Chemical plating by displacement yields deposits limited to only a few microns in thickness, usually 1 to 3 μm. Hence, chemical plating via the displacement process has few applications.

In order to continuously build thick deposits by chemical means without consuming the substrate, it is essential that a sustainable oxidation reaction be employed as an alternative to the dissolution of the substrate. The deposition reaction must occur initially and exclusively on the substrate and subsequently continue to deposit on the initial deposit. The redox potential for this chemical process is usually more positive than that for a metal being deposited by immersion. The chemical deposition of nickel metal by hypophosphite meets both the oxidation and redox potential criteria without changing the mass of the substrate:

Reduction:	$Ni^{+2} + 2e^- \rightarrow Ni^0$	$E^0 = -25$ mV
Oxidation:	$H_2PO_2^- + H_2O \rightarrow H_2PO_3^- + 2H^+ + 2e^-$	$E^0 = +50$ mV

$$Ni^{+2} + H_2PO_2^- + H_2O \rightarrow Ni^0 + H_2PO_3^- + 2H^+ \qquad E^0 = +25 \text{ mV}$$

which is the sum of the oxidation and reduction equations. This reaction does not represent the true electroless plating reaction, since EN deposition is accompanied by hydrogen evolution. Figure 1.1 shows the difference between immersion and electroless deposition by comparing deposit thickness vs. time.

The term *electroless plating* was originally adopted by Brenner and Riddell (1) to describe a method of plating metallic substrates with nickel or cobalt alloys without the benefit of an external source of electric current. Over the years, the term has been subsequently broadened to encompass any process that continuously deposits metal from an aqueous medium.

In general, electroless plating is characterized by the selective reduction of metal ions only at the surface of a catalytic substrate immersed into an aqueous solution of said metal ions, with continued deposition on the substrate through the catalytic action of the deposit itself. Since the deposit catalyzes the reduction reaction, the term *autocatalytic* is also used to describe the plating process.

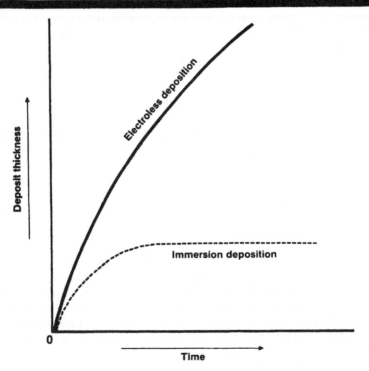

Fig. 1.1—Thickness vs. time—comparison between electroless and immersion deposition.

In 1844, Wurtz (2) observed that nickel cations were reduced by hypophosphite anions. However, Wurtz only obtained a black powder. The first bright metallic deposits of nickel-phosphorus alloys were obtained in 1911 by Breteau (3). In 1916, Roux (4) was issued the first patent on an electroless nickel plating bath. However, these baths decomposed spontaneously and formed deposits on any surface that was in contact with the solution, even the walls of the container. Other investigators studied the process, but their interest was in the chemical reaction and not the plating process. In 1946, Brenner and Riddell (1) published a paper that described the proper conditions for obtaining electroless deposition as defined above. Over the years, the process has been investigated further and expanded by many workers to its present state of development.

THE ELECTROLESS NICKEL PLATING BATH: COMPONENTS

Electroless nickel (EN) plating is undoubtedly the most important catalytic plating process in use today. The principal reasons for its widespread commercial and industrial use are to be found in the unique properties of the EN deposits. The chemical and physical properties of an EN coating depend on its composition, which, in turn, depends on the formulation and operating conditions of the EN plating bath. Typically, the constituents of an EN solution are:

- A source of nickel ions
- A reducing agent
- Suitable complexing agents
- Stabilizers/inhibitors
- Energy

The Nickel Source
The preferred source of nickel cations is nickel sulfate. Other nickel salts, such as nickel chloride and nickel acetate, are used for very limited applications. The chloride anion can act deleteriously when the EN plating bath is used to plate aluminum, or when the EN deposit is used as a protective coating over ferrous alloys in corrosion applications. The use of nickel acetate does not yield any significant improvement in bath performance or deposit quality when compared to nickel sulfate. Any minor advantages gained by nickel acetate are offset by its higher cost vs. the cost of nickel sulfate. The ideal source of nickel ions is the nickel salt of hypophosphorus acid, $Ni(H_2PO_2)_2$. The use of nickel hypophosphite would eliminate the addition of sulfate anions and keep to a minimum the buildup of alkali metal ions while replenishing the reactants consumed during metal deposition. The concentration of nickel ions and its relationship to the reducing agent and complexing agent concentrations will be discussed in a succeeding chapter.

Reducing Agents

Four reducing agents are used in the chemical reduction of nickel from aqueous solutions:

Sodium hypophosphite $NaH_2PO_2 \cdot H_2O$

Sodium borohydride $NaBH_4$

Dimethylamine borane (DMAB) $(CH_3)_2NHBH_3$

Hydrazine $N_2H_4 \cdot H_2O$

The four reducing agents are structurally similar in that each contains two or more reactive hydrogens, and nickel reduction is said to result from the catalytic dehydrogenation of the reducing agent. Table 1.1 gives a summary of nickel reducing agents.

Electroless nickel deposition can be viewed, in a very elementary manner, as the sum of two chemical reactions occurring in an electrochemical cell—a chemical oxidation reaction that liberates electrons and a nickel reduction reaction that consumes electrons:

Oxidation of reducing agent
$$Red \rightarrow Ox + ne$$

Reduction of nickel ion
$$mNi^{+2} + 2me^- \rightarrow mNi^0, \; 2m = n$$

Overall or sum reaction

$$mNi^{+2} + Red \rightarrow mNi^0 + Ox$$

The sum equation is a schematic illustration of the type of stoichiometric reactions usually written to describe the chemical reduction of nickel by a reducing agent. However, these overall reactions do not account for all of the phenomena that are observed during plating. Experimentally observed reaction characteristics indicate that the course of the reaction is considerably more complex than described by simple stoichiometric equations. Hence, it is necessary to attempt to ascertain the mechanism of the nickel reduction by the various reducing species.

Table 1.1
Nickel Reducing Agents

Reducing agent	Mol. wt.	Equiv. wt.	pH range	E^0, volts
Sodium hypophosphite			4-6	0.499
$NaH_2PO_2 \cdot H_2O$	106	53	7-10	1.57
Sodium borohydride				
$NaBH_4$	38	4.75	12-14	1.24
Dimethylamine borane				
$(CH_3)_2NHBH_3$	59	9.8	6-10	—
Hydrazine				
H_2NNH_2	32	8.0	8-11	1.16

An explicit understanding of the reaction mechanisms that govern electroless nickel deposition is necessary from both theoretical and practical viewpoints. Knowledge of the mechanisms of the reaction of a reducing agent with nickel ions can lead to the solution of a series of problems—development of methods to increase the plating rate, for enhancing hypophosphite efficiency, and for regulating the phosphorus or boron content of the deposit. It must be emphasized that an understanding of the course of the reaction, especially as it relates to the reduction of phosphorus or boron, is extremely important. It is the inclusion of P or B in the respective nickel alloys (Ni-P and Ni-B) that determines the properties of each alloy.

Before discussing the individual reducing agents and the proposed mechanisms of their reactions with nickel, it might be informative to recall certain characteristics of the process that the mechanism must explain:

- The reduction of nickel is always accompanied by the evolution of hydrogen gas.
- The deposit is not pure nickel but contains either phosphorus, boron or nitrogen, depending on the reducing medium used.
- The reduction reaction takes place only on the surface of certain metals, but must also take place on the depositing metal.
- Hydrogen ions are generated as a by-product of the reduction reaction.
- The utilization of the reducing agent for depositing metal is considerably less than 100 percent.
- The molar ratio of nickel deposited to reducing agent consumed is usually equal to or less than 1.

Hypophosphite

Nickel deposition by hypophosphite was sometimes represented in the literature by the following equations:

$$Ni^{+2} + H_2PO_2^- + H_2O \rightarrow Ni^0 + H_2PO_3^- + 2H^+ \tag{1}$$

$$H_2PO_2^- + H_2O \xrightarrow{\text{cat}} H_2PO_3^- + H_2 \tag{2}$$

overall
$$Ni^{+2} + 2H_2PO_2^- + 2H_2O \rightarrow Ni^0 + 2H_2PO_3^- + 2H^+ + H_2 \tag{3}$$

The reduction of nickel ions with hypophosphite yields alloys of nickel and phosphorus; however, Eqs. 1, 2, and 3 completely fail to account for the phosphorus component of the alloy. Further, if the plating reaction took place in accordance with the above equations, the rate of deposition would be proportional to the concentrations of the reactants. Gutzeit (5) has shown that in acid plating solutions (pH >3.0), the plating rate has a first order dependence upon the hypophosphite concentration. That is, plating rate is dependent on the hypophosphite concentration, over a very wide concentration range. Gutzeit further showed the rate to be independent of the nickel ion concentration beyond about 0.02M Ni^{++}; the rate is said to have a zero order dependence on nickel concentration. In alkaline solution, the rate is dependent only on the hypophosphite concentration.

Since the publication in 1946 of the paper by Brenner and Riddell (1), four principal reaction mechanisms have been proposed to explain electroless nickel deposition, which is incompletely represented by the stoichiometric reactions in Eqs. 1 to 3. These reaction schemes attempt to explain nickel reduction by hypophosphite in both acid and alkaline media. To account for the phosphorus in the deposit, the proposed mechanisms involve a secondary reaction of hypophosphite to elemental phosphorus.

Each of the heterogeneous reaction mechanisms outlined below requires a catalytic surface on which the reaction sequence will proceed. Hence, electroless nickel plating occurs only on specific surfaces. The reduction reaction begins spontaneously on certain metals—almost all metals of Group VIII of the periodic table (Fe, Co, Ni, Rh, Pd, Pt) in active form. The active metals of Group VIII are well known as hydrogenation-dehydrogenation catalysts (5). Nickel, cobalt, palladium, and rhodium are considered to be catalytically active. Metals that are more electropositive than nickel, such as iron and aluminum, will first displace nickel from a solution of its ions as follows:

$$Fe + Ni^{+2} \rightarrow Fe^{+2} + Ni^{0}_{cat} \qquad\qquad [4]$$

or

$$2Al^{0} + 3Ni^{+2} \rightarrow 2Al^{+3} + 3Ni^{0}_{cat} \qquad\qquad [5]$$

forming the catalytic surface, e.g., $Ni^{0}cat$. It is interesting to note that when steel or aluminum, the most commonly plated substrates, are electroless nickel plated, the initial phase in the deposition process is the displacement reaction.

If the substrate metal is more electronegative than nickel, it can be made catalytic by electrolytically depositing a thin nickel deposit on its surface. This can also be accomplished by providing contact, in the solution, between the substrate and a more electropositive metal, thereby forming an internal galvanic cell. For example, copper and its alloys are usually rendered catalytic to EN plating by either of these techniques.

A surface reaction, such as electroless nickel deposition, can be divided into the following elementary steps:

1. Diffusion of reactants (Ni^{+2}, $H_2PO_2^-$) to the surface;
2. Adsorption of reactants at the surface;
3. Chemical reaction on the surface;
4. Desorption of products (HPO_3^-, H_2, H^+) from the surface;
5. Diffusion of products away from the surface.

These are consecutive steps, and if any one has a much slower rate constant than all the others, it will become rate determining.

Appropriately, the first electroless nickel reaction mechanism proposed was advanced by Brenner and Riddell. They postulated that the actual nickel reductant is atomic hydrogen, which acts by heterogeneous catalysis at the catalytic nickel surface. The atomic hydrogen, generated by the reaction of water with hypophosphite, is absorbed at the catalytic surface:

$$H_2PO_2^- + H_2O \rightarrow H_2PO_3^- + 2H_{ad} \qquad\qquad [6]$$

The absorbed atomic hydrogen reduces nickel ions at the catalytic surface:

$$Ni^{+2} + 2H_{ad} \rightarrow (Ni^{+2} + 2H^+ + 2e) \rightarrow Ni^0 + 2H^+ \qquad\qquad [7]$$

The evolution of hydrogen gas, which always accompanies catalytic nickel reduction, was ascribed to the recombination of two atomic hydrogen atoms:

$$2H_{ad} \rightarrow (H + H) \rightarrow H_2 \tag{8}$$

Gutzeit (5) essentially agrees with the Brenner-Riddell atomic hydrogen concept of nickel reduction. However, Gutzeit attributes the formation of atomic hydrogen to the dehydrogenation of the hypophosphite ion during formation of the metaphosphite ion:

$$H_2PO_2^- \xrightarrow{cat} PO_2^- + 2H \tag{9}$$

followed by the formation of an orthophosphite molecule and an hydrogen ion according to:

$$PO_2^- + H_2O \rightarrow HPO_3^{-2} + H^+ \tag{10}$$

A secondary reaction between hypophosphite and atomic hydrogen results in the formation of elemental phosphorus:

$$H_2PO_2^- + H \rightarrow P + OH^- + H_2O \tag{11}$$

Although the atomic hydrogen mechanism, which has received the support of several authors, sustains the observed results, it fails to explain certain other phenomena. The scheme does not account for the simultaneous reduction of nickel and hydrogen, nor does it explain why the stoichiometric utilization of hypophosphite is always less than 50 percent.

The second mechanism, known as the *hydride transfer* mechanism, was first suggested by Hersch (6), who claimed that the behavior of hypophosphite is analogous to the reduction of nickel ions by borohydride ions. That is, Hersch assumed that hypophosphite acts as the donor of hydride ions (H^-). Hersch's proposed mechanism was later modified by Lukes (7).

In acid solutions, the primary step in the mechanism involves the reaction of water with hypophosphite at the catalytic surface, and may be described by the following equation:

$$2H_2PO_2^- + 2H_2O \xrightarrow{cat} 2H_2PO_3^- + 2H^+ + 2H^- \tag{12}$$

The corresponding reaction in alkaline solution is given by:

$$2H_2PO_2^- + 6OH^- \xrightarrow{cat} 2H_2PO_3^- + 2H_2O + 2H^- \tag{13}$$

The reduction of nickel ion in this mechanism proceeds as follows:

$$Ni^{+2} + 2H^- \rightarrow (Ni^{+2} + 2e^- + 2H) \rightarrow Ni^0 + H_2 \tag{14}$$

The hydride ion can also react with water or a hydrogen ion:

Acid
$$H^+ + H^- \rightarrow H_2 \qquad \text{[15a]}$$

Alkaline
$$H_2O + H^- \rightarrow H_2 + OH^- \qquad \text{[15b]}$$

According to Lukes, the hydrogen that appears as hydride ion was originally bonded to phosphorus in the hypophosphite. If Eq. 11 is included in this scheme, the codeposition of phosphorus is also accounted for. The hydride mechanism presents a satisfactory explanation for the coupled reduction of nickel and hydrogen.

The third mechanism is the so-called *electrochemical* mechanism, originally proposed by Brenner and Riddell, and later modified by others. This theory can be represented as follows:

An anodic reaction where electrons are formed by the reaction between water and hypophosphite:

$$H_2PO_2^- + H_2O \rightarrow H_2PO_3^- + 2H^+ + 2e^-, \; E^0 = 0.50 \text{ V} \qquad \text{[16]}$$

Cathodic reactions that utilize the electrons generated in Eq. 16:

$$Ni^{+2} + 2e \rightarrow Ni^0, \; E^0 = -0.25 \text{ V} \qquad \text{[17]}$$

$$2H^+ + 2e \rightarrow H_2, \; E^0 = 0.000 \text{ V} \qquad \text{[18]}$$

$$H_2PO_2^- + 2H^+ + e \rightarrow P + 2H_2O, \; E^0 = 0.50 \text{ V} \qquad \text{[19]}$$

According to this mechanism, the evolution of hydrogen gas that takes place during nickel deposition is a result of the secondary reaction represented in Eq. 18. The electrochemical mechanism implies that the nickel ion concentration should have a significant effect on the rate of deposition; however, the converse is true.

The fourth mechanism involves the coordination of hydroxyl ions with hexaquonickel ion. This mechanism was proposed by Cavallotti and Salvago (8), and later supported by the results of calorimetric studies on electroless nickel plating by Randin and Hintermann (9). The chemical reduction of nickel at a catalytic surface can be represented by the following reactions:

Ionization of water at catalytic nickel surface:

$$2H_2O \rightarrow 2H^+ + 2OH^- \qquad \text{[20]}$$

Coordination of hydroxyl ions to solvated nickel ion:

$$Ni(H_2O)_6^{+2} + 2OH^- \rightarrow \left[Ni(aq) \begin{smallmatrix} {}^{\cdots}OH \\ {}^{\cdots}OH \end{smallmatrix} \right] + 2H_2O \qquad [21]$$

Reactions of hydrolized nickel species with hypophosphite:

$$\left[Ni(aq) \begin{smallmatrix} {}^{\cdots}OH \\ {}^{\cdots}OH \end{smallmatrix} \right] + H_2PO_2^- \rightarrow NiOH_{ads}^+ + H_2PO_3^- + H \qquad [22]$$

$$NiOH_{ads} + H_2PO_2^- \rightarrow Ni^0 + H_2PO_3^- + H \qquad [23]$$

where $NiOH_{ads}$ represents a hydrolyzed Ni^+ species adsorbed at the catalytic surface. The hydrogen atoms formed by reactions 22 and 23 result from P-H bonds. The two hydrogen atoms can react and evolve as hydrogen gas:

$$H + H \rightarrow H_2 \qquad [24]$$

Salvago and Cavallotti (10) proposed the direct interaction of the catalytic nickel surface with hypophosphite to give phosphorus codeposition:

$$Ni_{cat} + H_2PO_2^- \rightarrow P + NiOH_{ad}^+ + OH^- \qquad [25]$$

The authors point out that copper, silver, and palladium can be reduced by $H_2PO_2^-$ without P codeposition, and hence, show the direct intervention of the chemical nature of the metal in the codeposition reaction.

According to the reaction scheme proposed by Salvago and Cavallotti (10), the hydrolyzed Ni species can react with water as follows:

$$NiOH_{ads}^+ + H_2O \rightarrow [Ni(OH)_2]_{aq} + H \qquad [26]$$

Reactions 23 and 26 are seen to be competing reactions, that is:

$$NiOH_{ads} \begin{array}{c} \xrightarrow{H_2PO_2^-} Ni^0 + H_2PO_3^- + H \qquad [27a] \\ \xrightarrow{H_2O} [Ni(OH)_2]_{aq} + H \qquad [27b] \end{array}$$

Salvago and Cavallotti further argue that the adsorbed NiOH species plays a role in the lamellar morphology observed in electroless nickel deposits. Gutzeit (5) attributes the lamellae of the EN deposit to a periodic variation of the phosphorus content in the coating. As long as there is a constant supply of adsorbed NiOH species on the catalytic surface and reaction 23 takes place, the deposition of phosphorus via reaction 25 cannot occur. The metallic Ni_{cat} surface must be available for a direct interaction with $H_2PO_2^-$ to deposit phosphorus. However, if reaction 26 occurs, the metallic catalytic nickel surface that was previously covered by adsorbed NiOH species is now free to interact with $H_2PO_2^-$. It is evident that any periodicity between reactions 23 and 26 will lead to a lamellar morphology.

The reaction of hypophosphite ions with water must also be included in the reaction scheme:

$$H_2PO_2^- + H_2O \rightarrow H_2PO_3^- + H_2 \tag{28}$$

According to Randin and Hintermann (9), the overall reaction can be written as:

$$Ni^{+2} + 4H_2PO_2^- + H_2O \rightarrow Ni^0 + 3H_2PO_3^- + H^+ + P + 3/2H_2 \tag{29}$$

From Eq. 29, the mole ratio $[Ni^{+2}]/[H_2PO_2]$ is observed to be ¼ (0.25).

$$\left[Ni_{aq} \begin{array}{c} \text{---OH} \\ \text{---OH} \end{array} \right]$$

is a schematic representation of the reacting species formed by loosely bonded OH^- ion in the coordination sphere of partially solvated $Ni(H_2O)_4^{+2}$ ions, and not nickel hydroxide. Cavallotti and Salvago, in fact, point out that when nickel hydroxide precipitates, inhibition phenomena are evident. If the hydroxyl ions are bonded too tightly to the nickel, they will not react with the hypophosphite ion. The adsorption of the hydrolyzed species on the active surface probably increases the lability of the coordination bonds.

The results of the studies by Franke and Moench (11) and later by Sutyagina, Gorbunora and Glasunov (12) on the origin of the hydrogen gas evolved from the interaction between water and hypophosphite ion can be explained by the reaction mechanism proposed by Cavallotti and Salvago.

Reducing Agents Containing Boron

In the forty or so years since the discovery of "electroless" nickel plating by Brenner and Riddell, hundreds of papers describing the process and the resulting deposits have been published. Although other electroless systems depositing metals such as palladium, gold, and copper are covered, the vast majority of these publications (papers and patents) are concerned with nickel and cobalt-phosphorus alloys and the plating solutions that produce them.

Attempts to develop alternative reducing agents led several workers to investigate the boron-containing reducing agents, in particular, the boro-hydrides and amine boranes. Subsequently, several patents were issued covering electroless plating processes and the resulting deposits.

The deposits obtained from electroless systems using boron-containing reducing agents are nickel-boron alloys. Depending on the solution operating conditions, the composition of the deposit can vary in the range of 90 to 99.9

percent nickel, with varying amounts of reaction products. In some cases, a metallic stabilizer will be incorporated in the deposit during the plating reaction.

As is the case with nickel-phosphorus alloys, nickel-boron deposits are characterized by their unique chemical and physical properties.

The Borohydride (BH_4^-) Ion

The borohydride reducing agent may consist of any water soluble borohydride compound. Sodium borohydride is generally preferred because of its availability. Substituted borohydrides in which not more than three of the hydrogen atoms of the borohydride ion have been replaced can also be used; sodium tri-methoxyborohydride ($NaB(OCH_3)_3H$) is an example of this type of compound.

The borohydride ion is a powerful reducing agent. The redox potential of BH_4^- is calculated to $E_{ca} = 1.24$ V. In basic solutions, the decomposition of the BH_4 unit yields 8 electrons for the reduction reaction:

$$BH_4^- + 8OH^- \rightarrow B(OH)_4^- + 4H_2O + 8e^- \qquad [30]$$

Then, each borohydride ion can theoretically reduce four nickel ions. That is, adding Eq. 30 to:

$$4Ni^{+2} + 8e^- \rightarrow 4Ni^0 \qquad [31]$$

gives the overall reaction:

$$4Ni^{+2} + BH_4^- + 8OH^- \rightarrow B(OH)_4^- + 4Ni^0 + 4H_2O \qquad [32]$$

However, it has been found experimentally that one mole of borohydride reduces approximately one mole of nickel.

There are only a few published articles that are concerned with the mechanism of nickel deposition with borohydride, and most of the proposed schemes are not supported by experimental data. Experimental evidence to the contrary, several authors still assume that each borohydride ion reduces four nickel ions.

Although the authors of three separate mechanisms agree that nickel reduction proceeds as expressed in Eq. 31, the reduction to boron is approached differently in each case.

Case 1 (13)

Here the author assumes that only three hydride ions are oxidized to protons and that the fourth hydride is oxidized to a hydrogen atom, which leads to the formation of a molecule of evolved hydrogen gas:

$$4Ni^{+2} + 2BH_4^- + 6OH^- \rightarrow 2Ni_2B + 6H_2O + H_2 \qquad [33]$$

Case 2 (14)

In this instance, it is assumed that all hydride ions are oxidized to protons, so that:

$$5Ni^{+2} + 2BH_4^- + 8OH^- \rightarrow 5Ni^0 + 2B + 8H_2O \qquad [34]$$

Case 3 (15)
Boron reduction is, as assumed by these authors, the catalytic decomposition of borohydride to elemental boron that takes place independently of nickel reduction:

$$2BH_4^- + 2H_2O \rightarrow 2B + 2OH^- + 5H_2O \qquad [35]$$

Gorbunova, Ivanov and Moissev (16) raised an objection to the three above hypotheses. They argue that, based on data relating to the reduction reactions by hypophosphite, it is doubtful that the hydrogen atoms, formed during the oxidation of the hydride ions of BH_4^-, are intermediate products that can take part in either nickel or boron reduction:

$$BH_4^- + 4H_2O \rightarrow B(OH)_4^- + 4H + 4H^+ + 4e \qquad [36]$$

Data obtained from mass-spectrometric measurements of the isotope composition of hydrogen gas evolved during electroless Ni-B deposition experiments using heavy water (D_2O), led Gorbunova et al. to propose a mechanism that more nearly fit the results of their studies. Their experiments were carried out using plating solutions prepared with D_2O, and also containing NaOD. It should be noted that calculations based on previously proposed schemes yielded results that deviated by almost 200 percent from the experimental data obtained in isotope experiments.
The proposed scheme of Gorbunova et al. for the reaction mechanism of nickel-boron plating consists of three main steps:

Reduction of nickel
$$BH_4^- + 4H_2O \rightarrow B(OH)_4^- + 4H + 4H^+ + 4e^- \qquad [37]$$
$$2Ni^{+2} + 4e^- \rightarrow 2Ni^0 \qquad [38]$$

$$BH_4^- + 2Ni^{+2} + 4H_2O \rightarrow 2Ni^0 + B(OH)_4^- + 2H_2 + 4H^+ \qquad [39]$$

Reduction of boron
$$BH_4^- + H^+ \rightarrow BH_3 + H_2 \rightarrow B + 5/2H_2 \qquad [40]$$

Hydrolysis of borohydride
$$BH_4^- + 4H_2O \rightarrow B(OH)_4^- + 4H + 4H^+ + 4e^- \rightarrow B(OH)_4^- + 4H_2 \qquad [41]$$

Mallory (17) has also investigated the reduction of nickel ions by amine boranes. On the basis of experimental data (16,17), he suggests that the hydrolyzed nickel mechanism of Cavallotti and Salvago, proposed to explain nickel reduction with hypophosphite, can be adapted to explain nickel reduction with both borohydride and amine boranes. The modified hydrolyzed nickel mechanism with borohydride can be represented by the following sequence of reactions:

Ionization of water

$$4H_2O \rightarrow 4H^+ + 4OH^-$$ [42]

Coordination of hydroxyl ions to solvated nickel ion

$$2Ni(H_2O)_6^{+2} + 4OH^- \rightarrow 2 \left[Ni_{aq} \overset{-OH}{\underset{-OH}{\diagdown}} \right] + 4H_2O$$ [43]

Reaction of hydrolized nickel species with borohydride ion

$$\left[Ni \overset{-OH}{\underset{-OH}{\diagdown}} \right] + BH_4^- \rightarrow NiOH_{ads} + BH_3OH + H$$ [44]

$$NiOH_{ads} + BH_3OH^- \rightarrow Ni^0 + BH_2(OH)_2^- + H$$ [45]

The $BH_2(OH)_2^-$ species reacts with the second hydrolyzed nickel ion in a similar manner:

$$\left[Ni \overset{-OH}{\underset{-OH}{\diagdown}} \right] + BH_2(OH)_2 \rightarrow NiOH_{ads} + BH(OH)_3^- + H$$ [46]

$$NiOH_{ads} + BH(OH)_3 \rightarrow Ni^0 + BO_2 + 2H_2O + H$$ [47]

The four atoms of atomic hydrogen react to form hydrogen gas

$$4H \rightarrow 2H_2$$ [48]

Thus, Eqs. 42 through 48 can be represented by the overall reaction:

$$2Ni^{+2} + 4H_2O + BH_4^- \rightarrow 2Ni^0 + B(OH)_4^- + 2H_2 + 4H^+$$ [49]

The reduction of boron is accounted for in the reaction of BH_4^- with a hydrogen ion:

$$BH_4^- + H^+ \rightarrow BH_3 + H_2 \rightarrow B + 5/2H_2$$ [50]

Equations 49 and 50 can be combined to obtain:

$$2Ni^{+2} + 2BH_4^- + 4H_2O \rightarrow 2Ni^0 + B + B(OH)_4^- + 3H^+ + 9/2H_2$$ [51]

The mechanism proposed by Gorbunova et al. and the modified Cavallotti and Salvago proposal lead to the same overall reaction. Also, in both schemes, the increase in acidity observed in the process is a result of hydrogen ions that originate from water molecules only. Finally, the two mechanisms indicate that the mole ratio of nickel reduced to borohydride consumed is 1:1, which is supported by experimental evidence.

The Amine Boranes

In the BH_3 molecule, the boron octet is incomplete, that is, boron has a low-lying orbital that it does not use in bonding, owing to a shortage of electrons. As a consequence of the incomplete octet, BH_3 can behave as an electron acceptor (Lewis acid). Thus, electron pair donors (Lewis bases), such as amines form 1:1 complexes with BH_3 and thereby satisfy the incomplete octet of boron. The linkage between BH_3 and dimethylamine is illustrated by the following:

$$H \overset{H}{\underset{H}{:B:}} \; + \; :N \overset{CH_3}{\underset{CH_3}{\diagup}} H \quad \longrightarrow \quad H \overset{H}{\underset{H}{\diagdown}} B : N \overset{CH_3}{\underset{CH_3}{\diagup}} H$$

The amine boranes are covalent compounds whereas borohydrides such as $Na^+BH_4^-$ are completely ionic, that is, $Na^+ BH_4^- = Na^+ + BH_4^-$. Although the amine boranes do not ionize, one of the atoms has a greater affinity for the electrons than the other and the bond will therefore be polar:

$$\overset{CH_3}{\underset{CH_3}{\diagdown}} H \overset{\delta^+}{-} N \quad \overset{\delta^-}{:}B \overset{H}{\underset{H}{\diagup}} H$$

In this case, the electrons are displaced toward the boron atom, giving the boron atom excess negative character, whereas the nitrogen atom displays excess positive charge. The electrical polarity of a molecule, expressed as its dipole moment, plays an important role in the reactions of covalent compounds.

The commercial use of amine boranes in electroless nickel plating has, in general, been limited to dimethylamine borane (DMAB), $(CH_3)NHBH_3$. DMAB has only three active hydrogens bonded to the boron atom and, therefore, should theoretically reduce three nickel ions for each DMAB molecule consumed (each borohydride will theoretically reduce four nickel ions). The reduction of nickel ions with DMAB is described by the following equations:

$$3Ni^{+2} + (CH_3)_2NHBH_3 + 3H_2O \rightarrow 3Ni^0 + (CH_3)_2NH_2^+ + H_3BO_3 + 5H^+ \qquad [52]$$

$$2[(CH_3)_2NHBH_3] + 4Ni^{+2} \; 3H_2O \rightarrow Ni_2B + 2Ni^0 + 2[(CH_3)_2NH_2^+] +$$
$$H_3BO_3 + 6H^+ + \tfrac{1}{2}H_2 \qquad [53]$$

In addition to the above useful reaction, DMAB can be consumed by wasteful hydrolysis:

Acid
$$(CH_3)_2NHBH_3 + 3H_2O + H^+ \rightarrow (CH_3)_2NH_2^+ + H_3BO_3 + 3H_2 \qquad [54]$$

Alkaline

$$(CH_3)_2NHBH_3 + OH^- \xrightarrow{H_2O} (CH_3)_2NH + BO_2^- + 3H_2 \qquad [55]$$

The theoretical expressions for nickel reduction are not supported by experimental findings, however. The results of studies on chemical nickel plating with DMAB indicates that the molar ratio of nickel ions reduced to DMAB molecules consumed is approximately 1:1 (17). A modified hydrolyzed nickel mechanism satisfactorily accommodates the experimental data.

Based on his studies, Lelental (18) suggests that nickel deposition with DMAB is dependent on the adsorption of the reducing media on the catalyst surface, followed by cleavage of the N-B bond of the adsorbed amine borane. The adsorption step is consistent with the polar nature of the DMAB molecule. The mechanism can be illustrated as follows:

N-B bond cleavage

$$2R_2NHBH_3 \xrightarrow{cat} 2R_2NH + 2BH_{3ads} \qquad [56]$$

Reduction of hydrolized nickel with BH$_{3ads}$

$$\begin{bmatrix} Ni \begin{matrix} {}^{\nearrow}OH \\ {}_{\searrow}OH \end{matrix} \end{bmatrix} + BH_{3ads} \rightarrow Ni^0 + BH(OH)_{2ads} + 2H \qquad [57]$$

$$\begin{bmatrix} Ni \begin{matrix} {}^{\nearrow}OH \\ {}_{\searrow}OH \end{matrix} \end{bmatrix} + BH(OH)_{2ads} \rightarrow NiOH_{ads} + B(OH)_3 + H \qquad [58]$$

$$NiOH_{ads} + BH_{3ads} \rightarrow Ni^0 + BH_2OH + H \qquad [59]$$

$$\begin{bmatrix} Ni \begin{matrix} {}^{\nearrow}OH \\ {}_{\searrow}OH \end{matrix} \end{bmatrix} + BH_2(OH) \rightarrow Ni^0 + B(OH)_3 + 2H \qquad [60]$$

The sums of the above equations, including the ionization of water, is:

$$3Ni^{+2} + 2R_2NHBH_3 + 6H_2O \rightarrow 3Ni^0 + 2R_2NH_2^+ + 2B(OH)_3 + 3H_2 + 4H^+ \qquad [61]$$

Boron reduction

$$R_2NHBH_3 \xrightarrow{cat} R_2NH + BH_3 + H_2 + H^+ \rightarrow R_2NH_2^+ + B + 5/2\,H_2 \qquad [62]$$

Equations 61 and 62 can be combined to give:

$$3Ni^{+2} + 3R_2NHBH_3 + 6H_2O \rightarrow 3Ni^0 + B + 3R_2NH_2^+ + 2B(OH)_3 + 9/2H_2 + 3H^+ \quad [63]$$

Hydrazine

Soon after Brenner and Riddell published their findings on nickel reduction with hypophosphite, Pessel (19) was issued a patent (1947) claiming the use of hydrazine as a metal reductant. During the succeeding 16 years, many papers and patents were published detailing electroless nickel-phosphorus deposition. It was not until 1963, however, that Levy (20) reported the results of his investigation of electroless plating with hydrazine. Later, Dini and Coronado (21) described several electroless nickel-hydrazine plating baths and the properties of the deposits obtained from their solutions, which contained >99 percent nickel.

Hydrazine is a powerful reducing agent in aqueous alkaline solutions:

$$N_2H_4 + 4OH^- \rightarrow N_2 + 4H_2O + 4e^-, E_b = 1.16 \text{ V} \quad [64a]$$

$$2Ni^2 + 2e \rightarrow 2Ni^0, E^0 = -0.25 \text{ V} \quad [64b]$$

Levy (20) proposed the following reduction reaction for nickel ions with hydrazine in an alkaline solution:

$$2Ni^{+2} + N_2H_4 + 4OH^- \rightarrow 2Ni^0 + N_2 + 4H_2O, E^0 = 0.91 \text{ V} \quad [64c]$$

which is the sum of Eqs. 64a and 64b.

This reaction implies a reducing efficiency of 100 percent for hydrazine, since the hydrazine is involved in the reduction of nickel ions only. Equation 64c does not account for the hydrogen evolved during the nickel plating reaction with hydrazine.

The hydrolyzed nickel ion mechanism can be modified to represent the experimental observations made during nickel reduction with hydrazine:

$$Ni^{+2} + 2OH^- \rightarrow Ni(OH)_2^{+2}$$

$$Ni(OH)_2^{+2} + N_2H_4 \rightarrow Ni(OH)_{ad}^{+1} + N_2H_3OH + H$$

$$Ni(OH)_{ad}^{+1} + N_2H_3OH \rightarrow Ni + N_2H_2(OH)_2 + H$$

$$2H \rightarrow H_2$$

The overall reaction can be written as:

$$Ni^{+2} + N_2H_4 + 2OH^- \rightarrow Ni^0 + N_2 + 2H_2O + H_2 \quad [65a]$$

The above mechanism does not account for the formation of hydrogen ions (H^+) during the course of the deposition reaction. In the reaction sequence given above, the hydroxyl ions (OH^-) in the first step are present in the solution through the addition of alkali metal or ammonium hydroxides. However, if the hydroxyl ions coordinated to nickel are generated by the dissolution of water molecules, a slightly different reaction mechanism results:

$$2H_2O = 2H^+ + 2OH^-$$

$$Ni^{+2} + 2OH^- = Ni(OH)_2$$

$$Ni(OH)_2 + N_2H_4 = Ni^0 + N_2H_2(OH)_2 + 2H$$

$$N_2H_2(OH)_2 + 2H = N_2 + 2H_2O + H_2$$

Now the overall reaction is given by:

$$Ni^{+2} + N_2H_4 \xrightarrow{H_2O} Ni^0 + N_2 + H_2 + 2H^+ \qquad\qquad [65b]$$

The two plausible reaction mechanisms proposed for the reduction of nickel with hydrazine lead us to speculate that there are separate mechanisms for the reduction of nickel with hypophosphite (and possibly DMAB) in acidic and alkaline solutions. This assumption is based on the acceptance of the nickel hydroxide mechanism, or some modification of it.

In acidic solutions, the first stage of the process is the dissociation of water ($H_2O = H^+ + OH^-$) at the catalytic surface. The hydroxyl ions (OH^-) replace the hydrogen in the P-H bond of hypophosphite and as a result, an electron and a hydrogen atom are produced. The consumption of OH^- ions results in the accumulation of hydrogen ions (H^+) in the solution with a concurrent decrease in pH of the solution.

In alkaline solutions, the sources of hydroxyl ions are the basic compounds ($NaOH$, NH_4OH, etc.) that are added to the plating solution to adjust the pH into the alkaline range of >7.0 to 14.0. As a result of the reaction of OH^- with the P-H bond, the pH also decreases in the alkaline solution. In this case, however, the pH decrease is due to the consumption of OH^- ions rather than the formation and accumulation of H^+ ions.

Van Den Meerakker (22) claims that electroless deposition processes can be explained by a so-called universal electrochemical mechanism, regardless of the nature of the reducing agent. Each process can be divided into a series of elementary anodic and cathodic reactions, where the first anodic step is the chemical dehydrogenation of the reductant. Thus in alkaline media:

Anodic

(1) Dehydrogenation $RH \xrightarrow{cat} R + H$

(2) Oxidation $R + OH^- \rightarrow ROH + e$

(3) Recombination $H + H \rightarrow H_2$

(4) Oxidation $H + OH^- \rightarrow H_2O + e$

Cathodic

(5) Metal deposition $M^{+n} + ne \rightarrow M^0$

(6) Hydrogen evolution $2H_2O + 2e \rightarrow H_2 + 2OH^-$

In acid media, reactions 4 and 6 in this set of equations become:

(4a) Oxidation $H \rightarrow H^+ + e$

(6a) Hydrogen evolution $2H^+ + 2e \rightarrow H_2$

Subsequent discussion in this section will show that a universal mechanism for electroless metal deposition is not feasible.

Each of the mechanisms discussed for the various reducing agents, and in particular for hypophosphite, explain most of the characteristics of electroless nickel deposition. However, each mechanism fails to account for some experimentally observed characteristic of the plating reaction. According to the hydride ion theory, the utilization of hypophosphite is always less than 50 percent, which is contrary to an observation of Pearlstein and Weightman (23). Experimental data indicates that the hydrogen gas evolved during the plating reaction originates, in the main, from the reducing agent; this fact is not supported by the electrochemical mechanism.

Electrochemically speaking, an electroless deposition reaction can be considered the combined result of two independent electrode reactions:

• The cathodic partial reactions (e.g., the reduction of metal ions)
• The anodic partial reactions (e.g., the oxidation of the reductant)

The electrons required for the reduction of the metal ions are supplied by the reducing agent. Mixed potential theory interprets many electrochemical processes in terms of the electromechanical parameters of the partial electrode reactions. Paunovic (24) was the first to identify electroless metal deposition in terms of mixed potential theory. He suggested that electroless deposition mechanisms could be predicted from the polarization curves of the partial anodic and cathodic processes. In simple terms, mixed potential theory leads to the assumption that electroless nickel plating can be considered as the superposition of anodic and cathodic reaction at the mixed (deposition) potential, E_M. Accordingly, the rates of the anodic reactions are independent of the cathodic reactions occurring simultaneously at the catalytic surface, and the rates of the separate partial reactions depend only on the electrode potential, the mixed potential. If these assumptions are correct, it should be possible, experimentally, to separate the anodic and cathodic reactions at different electrodes.

The partial anodic and cathodic reaction for electroless nickel deposition with hypophosphite are usually written as follows:

Anodic

Oxidation of H_2PO_2
$H_2PO_2 + H_2O \rightarrow H_2PO_3 + 2H^+ + 2e$; i_{Red} = hypophosphite oxidation current

Cathodic

$Ni^{+2} + 2e \rightarrow Ni^0$; i_{Ni} = nickel deposition current

$2H^+ + 2e \rightarrow H_2$; i_H = hydrogen evolution current

$H_2PO_2 + 2H^+ + e \rightarrow P + 2H_2O$; i_p = phosphorus deposition current

At steady state equilibrium potential (mixed potential), the rate of deposition is equal to the rate of oxidation of hypophosphite (anodic current, i_{Red}), and to the rate of the cathodic reactions (cathodic current, $i_{Ni} + i_H + i_p$). That is:

$$i_{deposition} = i_{Red} = i_{Ni} + i_M + i_p$$

Using Faraday's law, the rate of nickel deposition can be expressed as:

$$Rate_{(mg/cm/hr)} = 1.09 \times i_{dep(mA/cm2)}$$

The mixed potential, E_M, and the deposition current, i_{dep}, are obtained by the intersection of the partial anodic and cathodic polarization curves. Figure 1.2 is a schematic representation of the superposition of anodic and cathodic polarization curves, showing E_M and i_{dep}.

deMinjer (25) experimentally tested the validity of the mixed potential theory as applied to electroless nickel plating with hypophosphite. She found that it was possible for the anodic and cathodic reactions to occur at separate electrodes. The anodic oxidation of hypophosphite was observed to take place at a nickel or Ni-P electrode placed in a nickel-free hypophosphite solution, while the reduction of nickel metal takes place at a separate nickel or nickel-phosphorus electrode in a hypophosphite-free nickel-containing catholyte. It should be noted that deMinjer's experiments were run under acidic conditions (pH 4.2 to 4.3). We refer the reader to Ref. 25 for the details of her experiments.

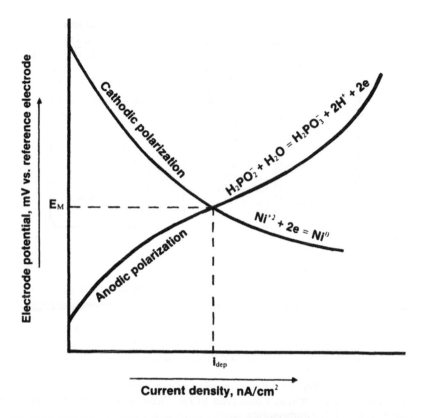

Fig. 1.2—Schematic representation of anodic and cathodic polarization curves combined in an Evan's diagram for deposition potential (E_M) and deposition current (I_{dep}).

Hydrogen evolution is usually considered to be a cathodic process (see above). However, studies by El-Raghy and Salama (26) on electroless copper plating lead to a different conclusion. They investigated electroless copper deposition electrochemically using a two-chamber galvanic cell and observed that hydrogen gas formation is an anodic process. That is, hydrogen evolution in electroless copper plating is the result of the anodic oxidation of the reducing agent formaldehyde.

Pearlstein (27) used the two-chamber technique to study electroless silver deposition with DMAB. A schematic representation of the two-chamber system used by Pearlstein is reproduced in Fig. 1.3. At open circuit, an extremely small amount of gas evolution was noted at the silver electrode in the silver-ion-free DMAB solution; there was no visible gassing on the silver electrode in the DMAB-free solution containing silver ions. When the silver electrodes in the two chambers were connected, he observed "a very evident increase in gas evolution

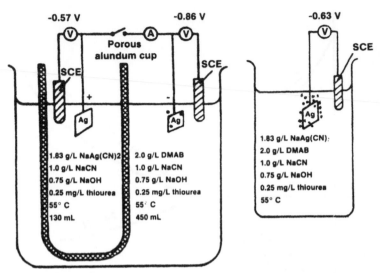

Fig. 1.3—Schematic representation of two-chamber system used to study electroless silver.

on the electrode in the DMAB-containing solution". No gassing was produced on the electrode in the silver-ion solution. Therefore, hydrogen evolution in electroless silver plating with DMAB is the product of the anodic oxidation of the DMAB.

Unfortunately, deMinjer did not report at which electrode (anode or cathode or both) hydrogen evolution occurred in her two-chamber electroless nickel experiments with hypophosphite.

Using polarization techniques, Ohno, Wakabayashi and Haruyama (28) studied the anodic oxidation of the reducing agents used in electroless plating—$H_2PO_2^-$, BH_4^-, DMAB, N_2H_4, and HCHO. The experiments were carried out in alkaline solutions on different metal electrodes (Ni, Co, Pd, Pt, Cu, Ag, Au), with special attention being paid to the catalytic aspect of electroless deposition. The authors found that the rate of anodic oxidation of a particular reductant depended on the pH of the solution, the concentration of the reductant, and the nature of the metal electrodes. Anodic polarization curves for hypophosphite on nickel electrodes exhibit a peak current that increases with increasing hypophosphite concentration. Similarly, the anodic current was also observed to increase with increasing pH (at constant $H_2PO_2^-$), with the current maximum shifting to less noble potentials. The polarization curves for the anodic oxidation of the other nickel-reducing agents (BH_4^-, DMAB, and N_2H_4) also exhibited current maximum behavior on nickel electrodes. The effects that concentration and pH have on the anodic oxidation current for $H_2PO_2^-$ on nickel electrodes are illustrated in Figs. 1.4 and 1.5, respectively.

Ohno et al. did not detect an anodic current attributable to the oxidation of hypophosphite or DMAB on copper electrodes. They found that the shape of the

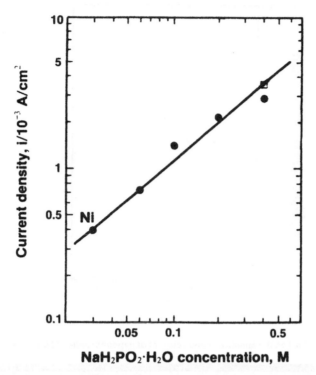

Fig. 1.4—Effect of NaH$_2$PO$_2$·H$_2$O concentration on the anodic oxidation current on nickel electrodes at constant potential.

polarization curves of the nickel-reducing agents on Pd, Pt, Ag, and Au was similar to the shapes of Tafel curves. These two facts support the contention that the nature of the metal electrode strongly affects the polarization behavior of the reductants.

Prior to the start of the polarization experiments, Ohno et al. observed that hydrogen gas evolution occurred on some electrodes at open-circuit condition, especially in solutions containing H$_2$PO$_2^-$, BH$_4^-$, or DMAB. During the experiment, shifting the potential to more noble values resulted in an increase in hydrogen gas evolution on Cu, Ag, or Au electrodes. Thus, the anodic oxidation of a reducing agent on a Group IB metal is accompanied by the formation of hydrogen gas that comes from hydrogen bonded directly to the reductant molecule. These observations coincide with and support the experimental findings of Pearlstein (27).

On the other hand, shifting the potential of the Group VIII (Co, Ni, Pd, Pt) metal electrodes to more noble potentials resulted in a decrease in hydrogen gas

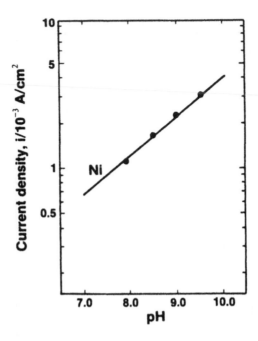

Fig. 1.5—Effect of pH on the anodic oxidation current of hypophosphite (0.2M) on nickel electrode.

evolution, and finally, its cessation during the anodic oxidation of hypophosphite, borohydride, and dimethylamine borane.

At this point, it would be premature to draw any conclusions or make assumptions as to the mechanism of electroless deposition based solely on the results reported in Refs. 26 through 28. This is especially true since the experiments undertaken in these studies were limited to alkaline solutions. What has been learned, however, is that hydrogen gas formation and evolution cannot be considered as strictly a cathodic process. A more cogent insight as to a possible mechanism or mechanisms can be gained by also reviewing the literature that is concerned with the anodic oxidation of the acid-compatible reducing agents in acid media. Therefore, of particular interest are the hypophosphite anion and DMAB.

Paunovic (29,30) reported on the electrochemical aspects of electroless nickel plating with DMAB in neutral (pH 7.0) solutions. Using polarization and other techniques, the author investigated the partial electrochemical processes. Most germane to this discussion on mechanisms, however, is the development of the kinetic diagnostic criteria $i_p/\sqrt{v}$ vs. v, where i_p is the peak current and v is the potential scan rate, mV/sec. Paunovic determined that if the ratio $i_p/\sqrt{v}$ decreases with an increase in scan rate v, then the charge-transfer reaction is preceded by a chemical reaction(s). He gave as the general kinetic scheme for this case:

(i) $Y \rightleftharpoons Red$

(ii) $Red \rightarrow Ox + ne$

where the substance Y is not directly oxidized at the applied potential, but is converted to the electroactive form, Red, by the preceding chemical reaction (i).

The direct oxidation of DMAB is usually written as:

$$(CH_3)_2NHBH_3 \xrightarrow{3H_2O} (CH_3)_2NH_2 + H_3BO_3 + 5H^+ + 6e \qquad [52a]$$

However, reactions involving DMAB are probably preceded by cleavage of the N-B bond:

$$(CH_3)_2NHBH_3 \rightarrow (CH_3)_2NH + BH_3 \qquad [56]$$

Hence, Eq. 56 is a homologue of reaction i and BH_3 is the electroactive species formed by a chemical reaction, which is subsequently oxidized and which supplies electrons for the reduction of nickel. The BH_3 unit also supplies the atomic hydrogen that is involved in H_2 evolution. The following scheme illustrates the cleavage of the N-B bond, the oxidation of BH_3, generation of electrons and evolution of hydrogen gas in acid or neutral solutions:

(1) $(CH_3)_2NHBH_3 \rightarrow (CH_3)_2NH + BH_3$

(2) $BH_3 \xrightarrow{3H_2O} BH_2(OH) + 3H^+ + 2OH^- + H + e$
    (water dissociation $3H_2O = 3H^+ + 3OH^-$)

(3) $BH_2(OH) + 2OH^- \rightarrow BH(OH)_2 + OH^- + H + e$

(4) $BH(OH)_2 + OH^- \rightarrow B(OH)_3 + H + e$

Adding steps 1 through 4:

$$(CH_3)_2NHBH_3 + H^+ \xrightarrow{3H_2O} (CH_3)NH_2 + B(OH)_3 + 3H^+ + 3/2 \, H_2 + 3e$$

A similar scheme can be written for these elementary steps in neutral or alkaline solutions.

More recently, Mital, Shrivastava and Dhaneshwar (31) sought to ascertain the validity of mixed potential theory as applied to electroless nickel deposition. Their studies involved the use of potentiostatic and voltammetric techniques on electroless nickel baths containing hypophosphite or DMAB in acid solutions (pH 6.0).

The current potential curves obtained by Mital et al. for the anodic oxidation of hypophosphite are similar in shape to those reported by Ohno et al. (28). The anodic curve attains a current maximum at which point there is a sudden rise in the anodic potential, indicating the possibility of formation of a passivating film on the electrode surface. In the case of DMAB, however, the anodic curves show a smooth monotonic increase in current with increasing anodic potential.

The deposition rate calculated from i_{dep} was compared to the rate obtained by the gravimetric method. Mital et al. (31) found the calculated rate to be significantly lower than the rate determined gravimetrically, the values being 1.63 and 10 mg/cm^2/hr, respectively. This fact indicated that an electrochemical mechanism under mixed potential control is not applicable in the case of Ni-P deposition.

The results reported (31) for plating with DMAB show the predominance of an electrochemical mechanism; i.e., the rates obtained from i_{dep} and the gravimetric method were 7.0 and 8.0 mg/cm^2/hr at 70° C, respectively.

Mital et al. state that mixed potential theory is applicable only to those electroless systems where the reduction of metal ions and the oxidation of reductant is electrochemically feasible at E_{Mix}, which they claim is the case with DMAB. On the other hand, when the reducing agent is not oxidizable electrochemically at E_{Mix}, as it is in the case of $H_2PO_2^-$, they assert that a chemical mechanism predominates. It is not clear if Mital et al. are proposing that each elementary step is a chemical reaction in an overall chemical reaction, or if at least one step is a charge-transfer reaction and the remaining steps in the overall reaction are chemical reactions. (A chemical reaction, as distinguished from an electrochemical reaction, is a reaction whose rate depends only on the concentrations of the reactants and is independent of the potential; i.e., a chemical reaction has a rate constant that is independent of the potential. If one of a number of elementary reactions is a charge-transfer reaction, then the overall reaction is an electrochemical reaction.)

It is obvious from the lengthy discussion above that considerably more experimental data will be required in order to formulate a reaction mechanism for EN deposition that satisfies all observed results for each reducing agent.

Complexing Agents

The additives referred to as complexing agents in electroless nickel plating solutions are, with two exceptions, organic acids or their salts. The two exceptions are the inorganic pyrophosphate anion, which is used exclusively in alkaline EN solutions, and the ammonium ion, which is usually added to the plating bath for pH control or maintenance. Examples of the complexing agents commonly used in electroless nickel solutions are given in Table 1.2.

There are three principal functions that complexing agents perform in the EN plating bath:

- They exert a buffering action that prevents the pH of the solution from decreasing too fast.
- They prevent the precipitation of nickel salts, e.g., basic salts or phosphites.
- They reduce the concentration of free nickel ions.

In addition to these functions, complexing agents also affect the deposition reaction and hence the resultant nickel deposit. These factors will be discussed throughout this chapter.

Nickel ions in aqueous solution interact with and are bound to a specific number of water molecules. The water molecule is oriented so that the negative end of the dipole, oxygen, is directed toward the positive nickel ion. The number of water molecules that can attach to the nickel ion is called the *coordination number*. Divalent nickel has two coordination numbers, 4 and 6. Aqueous solutions of simple inorganic nickel salts, e.g., $NiSO_4 \cdot 6H_2O$, contain the green, octahedral hexaquonickel ion $[Ni(H_2O)_6]^{+2}$. A schematic representation of a 6-coordinate nickel ion in aqueous solution is shown below:

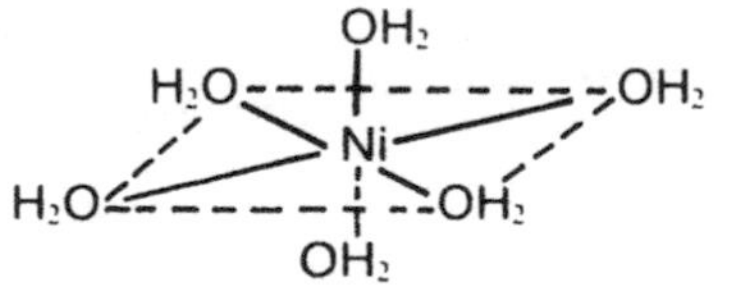

When water molecules coordinated to the nickel ion are replaced by other ions or molecules, the resulting compound is called a *nickel complex* and the combining, or donor, group is called a *complexing agent* or *ligand*. The formation of the blue ammonium complex by displacement of the 6 water molecules by ammonia is illustrated below:

$$[Ni(H_2O)_6]^{+2} + 6NH_3 = [Ni(NH_3)_6]^{+2} + 6H_2O$$

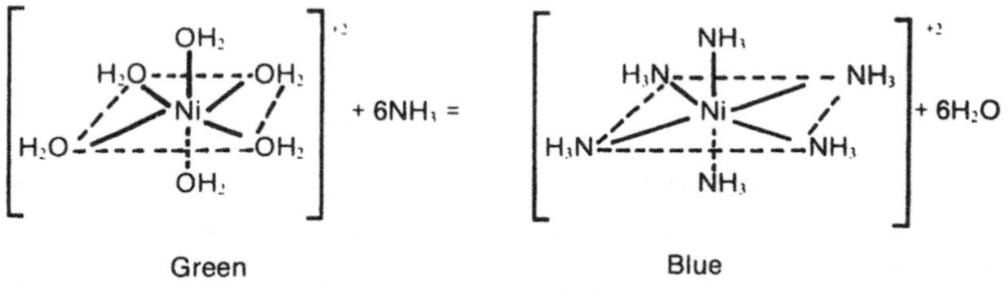

Green   Blue

The chemical properties of nickel ions in aqueous solution are altered when they are combined with complexing agents. Some of the common properties of solvated nickel ions that can be affected are color, reduction potential, and solubility. The effects complexing agents have on some of the common

properties of nickel ions, such as color, reduction potential, and solubility are
illustrated below:

$$[Ni(H_2O)_6]^{+2} + 6NH_3 = [Ni(NH_3)_6]^{+2} + 6H_2O \qquad\qquad [66]$$
Green   Blue

$$[Ni(H_2O)_6]^{+2} + 2e^- = Ni^0 + 6H_2O,\ E^0 = -0.25\ V \qquad\qquad [67a]$$

$$[Ni(CN)_4]^{+2} + 2e^- = Ni^0 + 4CN^-,\ E_{ca} = -0.90\ V \qquad\qquad [67b]$$

$$[Ni(H_2O)_6]^{+2}C_4H_8N_2O_2 = [Ni(C_4H_7N_2O_7)_2]^0 + 6H_2O + 2H^+ \qquad [68]$$
Dimethyl-
glyoxime

Equation 68 represents the classical method for separating nickel from
aqueous solutions, and shows that the nature of the complexing agent
determines whether the nickel complex remains in solution or precipitates.

Table 1.2
Ligands Commonly Used
In Electroless Nickel Plating Baths

Anion	Structure (acid)	No. of chelate rings	Ring size	pK[a]
Monodentate				
Acetate	CH_3COOH	0	—	1.5
Propionate	CH_3CH_2COOH	0	—	—
Succinate	$HOOCCH_2CH_2COOH$	0	—	2.2
Bidentate				
Hydroxyacetate	$HOCH_2COOH$	1	5	—
α-Hydroxypropionate	$CH_3CH(OH)COOH$	1	5	2.5
Aminoacetate	NH_2CH_2COOH	1	5	6.1
Ethylenediamine	$H_2NCH_2CH_2NH_2$	1	5	13.5
β-Aminopropionate	$NH_2CH_2CH_2COOH$	1	6	5.6
Malonate	$HOOCCH_2COOH$	1	6	4.2
Pyrophosphate	$H_2O_3POPO_3H_2$	1	6	5.3
Tridentate				
Malate	$HOOCCH_2CH(OH)COOH$	2	5,6	3.4
Quadridentate				
Citrate	$HOOCCH_2(OH)C(COOH)COOH$	2[b]	5,6	6.9

[a]pK = -log K, where K is the stability constant for the nickel-ligand complex.
[b]See text for explanation.

Thermodynamics is of great help in predicting equilibria in electrochemical reactions. It is the Gibbs free energy that is the criterion as to whether or not an electrochemical cell reaction will occur spontaneously at a constant temperature and pressure. There is an important relation between the change in free energy, ΔG, and the amount of electrical work done, nFE, in a reversible cell:

$$\Delta G = -nFE \qquad [69]$$

where n is the number of electrons transferred in the reaction, F is the Faraday number (96,500 coulombs), and E is the voltage or EMF of the cell. If ΔG has a negative value, the reaction is spontaneous; if $\Delta G > 0$, the reaction will not go spontaneously.

Consider Eq. 67b combined with the half-cell reaction of hypophosphite with water (Eq. 16):

$$[Ni(CN)_4]^{-2} + 2e^- = Ni^0 + 4CN^-, \ E^0 = -0.90 \text{ V} \qquad [67b]$$
$$H_2PO_2^- + H_2O \rightarrow H_2PO_3 + 2H^+ + 2e^-, \ E^0 = 0.50 \text{ V} \qquad [16]$$

$$[Ni(CN)_4]^{-2} + H_2PO_2^- + H_2O = Ni^0 + 4CN^- + H_2PO_3^- + 2H^+, \ E^0 = -0.40 \text{ V} \qquad [70]$$

From Eq. 70:

$$\Delta G^0 = -nFE^0 = (-) \frac{(2)\ (96,500)\ (-0.40)}{(4.184)} = +18 \text{ Kcal}$$

Since $\Delta G^0 > 0$, the reaction is not spontaneous and hypophosphite will not reduce nickel from the nickel tetracyanide complex. On the other hand, consider the half-cell reaction of the nickel hexamine ion, $[Ni(NH_3)_6]^{+2}$, combined with the half-cell reaction of hypophosphite in alkaline solution:

$$[Ni(NH_3)_6]^{+2} + 2e^- = Ni^0 + 6NH_3(aq), \ E^0 = -0.49 \text{ V} \qquad [71]$$
$$H_2PO_2^- + 3OH^- \rightarrow HPO_3^{-2} + 2H_2O + 2e^-, \qquad E^0 = 1.57 \text{ V} \qquad [72]$$

$$[Ni(NH_3)_6]^{+2} + H_2PO_2^- + 3OH^- =$$
$$Ni^0 + HPO_3^{-2} + 2H_2O + 6NH_3(aq), \ E^0 = 1.08 \text{ V} \qquad [73]$$

Then:

$$\Delta G^0 = -nFE^0 = \frac{-(2)\ (96,500)\ (1.08)}{(4.184)} = -49.8 \text{ kcal}$$

Since $\Delta G^0 < 0$, the cell reaction is spontaneous and hypophosphite will reduce nickel from ammoniacal solutions, as is well known. The above examples illustrate that complexing agents can have a profound effect on the reduction reaction.

An EN complexing agent can coordinate to a nickel ion through an oxygen or nitrogen atom, or both. The blue ammonium (amine) complex is an example of the general class of complexing agents (ligands) known as *monodentate* (one-tooth) ligands (see Table 1.2). Monodentate ligands use only one donor atom per ligand molecule or ion to occupy a single coordination position of the nickel ion. The monocarboxylate anions, acetate and propionate, are examples of monodentate ligands.

Nickel chelate compounds are defined as complexes in which the donor atoms. of the complexing agent are attached to each other as well as to the nickel ion. Thus, the nickel ion becomes part of a heterocyclic ring. Complexing agents having two or more donor atoms that can simultaneously fill two or more nickel coordinate positions are called *polydentate* complexing agents (ligands). Polydentate ligands whose structures are favorable for the attachment of two atoms are called *bidentate*. Those with three, four, five, and six donor sites are called *tri-, tetra-, penta-,* and *hexadentate* ligands, respectively. A polydentate ligand might not necessarily utilize all of its donor atoms when coordinating to a nickel ion, due to steric hinderance. The citrate ion, although tetradentate, probably coordinates to the nickel ion with the formation of two chelate rings, a 5-membered ring and a 6-membered ring. For pictorial simplicity, only one citrate molecule is shown coordinated to the nickel ion in the following representation:

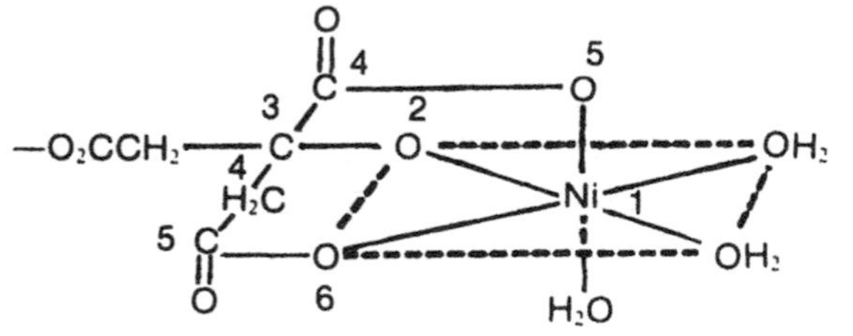

Note that the succinate anion is listed in Table 1.2 as monodentate even though it has two donor atoms per ligand molecule. The succinate ion, functioning as a bidentate ligand, would form a 7-membered ring. When the chelate ring is 7-membered or more, it becomes more probable that the one end of the ligand molecule will coordinate to another metal ion rather than form the 7-membered ring (32,33).

If the complexing agent does not have a sufficient number of donor atoms to satisfy the coordination number (sites) of the nickel ion, the remaining sites may be occupied by other ligands and/or water molecules. For example, partial chelation of the hexaquonickel ion by hydroxyacetate anion can be represented schematically by:

$$[Ni(H_2O)_6]^{+2} + HOCH_2COO^- = \left[\begin{array}{c} \text{structure} \end{array} \right] + 2H_2O$$

It is important to note that some of the chemical properties of the initial aquometal ion are retained when the nickel chelate contains one or more free aquo positions.

Aquometal ions tend to be acidic, that is, they hydrolyze in aqueous solution in a manner represented by:

$$[M(H_2O)_a]^{n+} = [M(H_2O)_{a-1}(OH)]^{(n-1)+} + H^+ \qquad [74]$$

The complete hydrolysis of hexaquonickel ion can be represented by:

$$[Ni(H_2O)_6]^{+2} = [Ni(H_2O)_5OH]^{+1} + H^+ = [Ni(H_2O)_4(OH)_2]^0 + 2H^+ \qquad [75]$$

at pH $\geqslant 5.0$.

The superscript 0 in this case means that the hydrated basic nickel salt has precipitated. Thus, the equilibrium between the aquonickel ion (Lewis acid) and the hydroxylated nickel ion can be considered to involve the displacement of the weak Lewis base, the water molecule, by the strongly coordinating hydroxyl ion. The chelation of an aquonickel ion occurs only when the ligand is a much stronger base than water, and analogously, may be considered an acid-base reaction. It should be noted that hydrolysis here occurs in the bulk of the solution and should not be confused with the hydrolyzed nickel species that originates at the catalyst surface.

Partial chelation of the nickel ion increases its resistance to hydrolysis, and it is possible to keep the nickel ion in solution at a higher pH than would otherwise be possible. However, as the pH increases, the disassociation of protons from the remaining coordinated water molecules frequently occurs and further hydrolysis of the nickel ion will result. For maximum stability against hydrolysis, the nickel ion must be completely coordinated.

In aqueous solutions, chemical reactions with nickel ions are said to take place at "free" coordination sites, that is, those sites that are weakly bound to coordinated water molecules, so that the kinetics of the electroless nickel plating reaction is a function of free nickel ion concentration. Of the several functions of complexing agents in electroless nickel plating solutions, one of the most important is the regulation and maintenance of the free nickel ion concentration, or activity; i.e., the nickel complex dissociates to form some small equilibrium amount of free nickel ions:

$$NiL_3^{2-3n} \underset{k_{-1}}{\overset{k_1}{\rightleftharpoons}} NiL_2^{2-2n} + L^{-n}, \quad k_1 = \frac{[NiL_2^{2-2n}][L^{-n}]}{[NiL_3^{2-3n}]} \tag{77}$$

$$NiL_2^{2-2n} \underset{k_{-2}}{\overset{k_2}{\rightleftharpoons}} NiL^{2-n} + L^{-n}, \quad k_2 = \frac{[NiL^{2-n}][L^{-n}]}{[NiL_2^{2-2n}]} \tag{78}$$

$$NiL^{2-n} \underset{k_{-3}}{\overset{k_3}{\rightleftharpoons}} Ni^{+2} + L^{-n}, \quad k_3 = \frac{[Ni^{+2}][L^{-n}]}{[NiL^{2-n}]} \tag{79}$$

The overall equilibrium constant for three independent equilibria is defined by:

$$K_3 = k_1 k_2 k_3 = \frac{[Ni^{+2}][L^{-n}]^3}{[NiL_3^{2-3n}]} \tag{80}$$

In general:

$$K_m = k_1 k_2 \ldots k_m = \frac{[Ni^{+2}][L^{-n}]^m}{[NiL_m^{2-mn}]} \tag{81}$$

The constant K_m is also called the *instability constant*. The term *stability constant* is more commonly used in the literature, and its value is the reciprocal of the instability constant:

$$K_{NiL} = \beta = \frac{1}{K_m} \tag{82}$$

Although not shown in the scheme, as each coordination site on the nickel ion is exposed by a leaving ligand atom, a water molecule becomes attached at that site. As one bidentate ligand molecule dissociates, for example, two water molecules associate at the sites opened by the leaving bidentate ligand. It is obvious that the activity of a free nickel ion decreases as the number of ligand molecules bound to the nickel ion approaches m; m = 6 for monodentate ligands, m = 3 for bidentate ligands, and so forth.

The rate of nickel deposition is proportional to the rate at which the nickel complex dissociates to form free nickel ion. Thus, the plating rate is inversely related to the stability constant, i.e., the larger the stability constant, the lower the rate of complex dissociation and concomitantly, the lower the rate of deposition. As an example of this relationship, the stability constant (see Table 1.2) for the nickel-citrate complex is more than an order of magnitude greater than the stability constant for the nickel-lactate complex; and as predicted, the plating rate in the citrate bath is considerably slower than the plating rate in the lactate bath under similar conditions. Hence, the choice of complexing agent for use in the electroless nickel plating bath has a profound effect on the plating reaction.

Complexing agents, being electron donors, also have a considerable affinity for hydrogen ions. Complexing agents can be considered metal buffers in a manner analogous to the function of hydrogen ion buffers. When a complexing agent is added to a solution of free metal ions, M^{+z}, equilibrium is established as shown schematically:

$$M^{+z} + mL^{-n} = ML_m^{-(n-z)} \dots \tag{83}$$

Since:

$$K_{ML} = \frac{1}{K_m} \tag{84}$$

therefore:

$$K_{ML} = \frac{[ML_m^{-(n-z)}]}{[M^{+z}] [L^{-n}]^m} \tag{85}$$

or:

$$M^{+z} = \left(\frac{1}{K_m} \right) \frac{[ML_m^{-(n-z)}]}{[L^{-n}]^m} \tag{86}$$

Analogous to the definition of pH, the negative logaritnm of the metal ion concentration, pM, is defined by:

$$pM = K_x + \log \frac{[L^{-n}]^m}{[ML_m^{(n-z)}]}, \ K_x = + \log K_{ML} \tag{87}$$

As pointed out earlier, the complexing agent shows an affinity for hydrogen ions, which can be expressed by the following equilibrium:

$$L^{-n} + H^+ = HL^{-(n-1)} \tag{88}$$

with equilibrium constant:

$$K_{HL} = \frac{[HL^{-(n-1)}]}{[H^+] [L^{-n}]} \qquad [89]$$

and:

$$[L^{-n}] = \left(\frac{1}{K_{HL}}\right) \frac{[HL^{-(n-1)}]}{H^+} \qquad [90]$$

Substituting the expression for $[L^{-n}]$ into Eq. 87 yields:

$$pM = K_x' + m\log \frac{[HL^{=(n-1)}]}{[ML_m^{(n-z)}]} + mpH = \frac{1}{\log[M]} \qquad [91]$$

where:

$$K_x' = K_x - m\log K_{HL} \qquad [92]$$

Since pM and pH are analogous concepts, an increase in the value of pM or pH results in a decrease in the concentration of the metal ion under consideration. Equation 91 indicates that as the pH increases (at constant m), pM increases and the free metal ion concentration decreases. The majority of commercial electroless nickel plating installations use acid plating baths that are operated in the pH range of 4.5 to 6.0. The complexing agents commonly used are most effective in this pH range.

Equation 91 also supports the statement made earlier that the nickel ion activity decreases as each coordinate position of the nickel ion becomes bound to a ligand atom. Using the example of the bidentate ligand (at constant pH), pM increases and the free nickel ion concentration decreases when m is increased from 1 to a maximum of 3.

Stabilizers

An electroless nickel plating solution can be operated under normal operating conditions over extended periods without adding stabilizers; however, it may decompose spontaneously at any time. Bath decomposition is usually preceded by an increase in the volume of hydrogen gas evolved and the appearance of a finely-divided black precipitate throughout the bulk of the solution. This precipitate consists of nickel particles, and either nickel phosphide or nickel boride, depending on the reducing agent being used.

Fortunately, chemical agents called *stabilizers* are available to prevent the homogeneous reaction that triggers the subsequent random decomposition of the entire plating bath. To use stabilizers effectively, the chemist must, first of all, be able to identify those problems that can be solved by the use of stabilizers. Second, the compatibility of the stabilizer with the process being used must be ascertained to avoid any adverse loss in catalytic activity due to a synergistic action with any other additive present in the bath. Third, when two or more stabilizers are employed, it is important to be sure that the action of one does not inhibit or lessen the effectiveness of the other stabilizers. Finally, the stabilizers must be selected on the basis that they only affect the plating process in a manner that the resultant deposit will be able to meet any required performance criteria.

Researchers (5) discovered the existence of hydroxyl ions at the surface of solid particles of colloidal or near-colloidal dimensions present in the solution. The hydroxyl ions are said to cause the localized reduction of nickel ions to the metal by the homogeneous reaction:

$$Ni^{+2} + 2H_2PO_2^- + 2OH^- \rightarrow Ni^0 + 2HPO_3^{-2} + 2H^+ + H_2 \qquad [93]$$

The finely divided nickel precipitate, in turn, acts as a highly efficient catalyst, because of its large surface area, thus triggering a "self-accelerating chain reaction."

The solid particles mentioned above are either formed in the solution or introduced into the solution by drag-in, dust, fumes, or other contaminants. The particles of extraneous origin can be the result of poor rinsing, poor cleaning, and/or poor housekeeping; good plating practices will eliminate their introduction.

Particles that form in the solution are insoluble nickel-phosphides and/or basic salts or metal hydroxides that precipitate at operating pH; e.g., the ions of iron and aluminum (Fe^{+2} and Al^{+3}) precipitate as gelatinous hydroxides when the bath pH is >5.0. These ions are introduced into the plating bath during the initial displacement reaction (discussed above) when steel or aluminum substrates are immersed in the plating bath. Steel and aluminum are by far the metals most likely to be plated in electroless nickel baths. Thus, the metal ions will accumulate in any production facility until their concentration exceeds their respective metal hydroxide solubility product limits and begin to precipitate.

Plating bath decomposition can be virtually eliminated by the addition of only trace amounts of stabilizer (sometimes referred to as *catalytic inhibitors*) to the plating bath. Gutzeit (5) pointed out that most of these so-called "anti-catalysts" (stabilizers/inhibitors) are identical with the materials that prevent hydrogenation/dehydrogenation catalysis.

The most effective stabilizers can be divided into the following classes:

(I)   Compounds of Group VI elements: S, Se, Te
(II) Compounds containing oxygen: AsO_2^-, IO_3^-, MoO_4^{-2}
(III) Heavy metal cations: Sn^{+2}, Pb^{+2}, Hg^+, Sb^{+3}
(IV) Unsaturated organic acids: Maleic, itaconic

The stabilizer concentration in the plating bath can be very critical. The concentration of the stabilizer depends most importantly on its structural class.

Substances that fall into Class I or Class II can function effectively as stabilizers at concentrations as low as 0.10 ppm. In many cases, when the concentration of stabilizers from either of these two classes is increased much beyond 2 ppm, the plating reaction can be completely inhibited. On the other hand, certain Class I stabilizers such as thiourea, at optimum concentration will increase the rate of deposition substantially over that of a bath without any stabilizer (see Fig. 1.5).

The usage of stabilizers found in Classes III and IV is conveniently defined in terms of molar concentrations; e.g., the concentration range of Class III stabilizers is usually 10^{-5} M to 10^{-3} M, whereas the concentration of unsaturated organic acids (Class IV) is usually in the range of 10^{-3} to 10^{-1} M.

The activity of the catalytic substrate is altered appreciably by extremely minute concentrations of stabilizers. Substances that tend to inhibit catalytic activity are known as *poisons*; some substances actually enhance activity and increase deposition rate.

Several methods are available to determine the effectiveness of these compounds as stabilizers and to also ascertain their optimum concentration in the plating bath. One method consists of plotting the mixed or deposition potential of an electroless nickel solution against the concentration of stabilizer that is incrementally added to the solution (34). A curve similar to the one shown in Fig. 1.6 is usually obtained. The initial small addition of stabilizer causes an abrupt change in potential; subsequent additions cause only minor changes in the mixed potential. When the concentration of stabilizer reaches a critical value, the plating reaction stops, as is usually evidenced by the cessation of hydrogen evolution. The mixed potential abruptly changes to more noble values by 100 to 300 mV. The potential remains constant with further additions of inhibitor beyond the critical concentration.

Another method consists of measuring the deposition rate vs. stabilizer concentration, from which curves represented by the curve of Fig. 1.7 are obtained. Some curves will cover a smaller concentration range than others and show a very sharp maximum, but the one feature common to all is the critical concentration point. (Some stabilizers will increase the rate of deposition as indicated by a maximum in the rate vs. concentration curve.)

Stability can also be determined by adding 1 to 2 mL of a 100 ppm palladium chloride solution to a sample of the warm plating solution and measuring the time before visible black precipitate is formed. The plating bath is considered stable if the time required for precipitate formation is in excess of 60 seconds. Using one of the preceding methods, it will be found that the effective stabilizer concentration varies from 0.1 to about 300 ppm for Classes I to III. The concentration for the organic stabilizers can range up to several grams per liter.

Air agitation of electroless nickel solutions has been reported to significantly enhance the stability of the plating bath. Agitation using the inert gas argon was also found to be ineffective in increasing the stability of the electroless nickel plating solution (35). When pure oxygen was bubbled through an otherwise unstabilized experimental electroless nickel plating solution, the mixed (deposition) potential shifted from -625 mV vs. SCE to -550 mV vs. SCE. In other words, the oxygen-agitated solution showed increased stability vs. the quiescent solution.

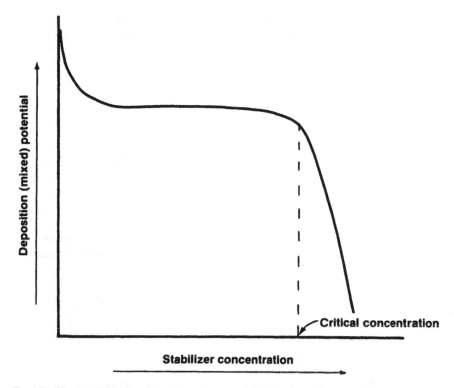

Fig. 1.6—Effect of stabilizer concentration on electroless nickel deposition potential.

Altura (36), using polarization techniques, studied the effect of stabilizers on the cathodic and anodic reactions. He first studied the partial processes in the absence of stabilizers. The initial cathodic polarization studies were made on solutions containing a complexing agent and nickel only. The initial anodic polarization studies were made in solutions containing a complexing agent (lactic acid) and sodium hypophosphite only.

Figure 1.8 presents the cathodic polarization results in solutions without hypophosphite or stabilizers. The solutions used in the cathodic studies, that is, those containing nickel, do not appear to reach a limiting current density down to -850 mV vs. SCE at the various nickel concentrations used.

The effect of hypophosphite concentration on anodic polarization curves is given in Fig. 1.9. Increasing the hypophosphite concentration at constant potential shows an increase in current density, which is to be expected. However, the shape of anodic polarization curves, especially curve D, shows limiting current behavior that indicates mass transfer as the most likely rate-controlling step for hypophosphite oxidation.

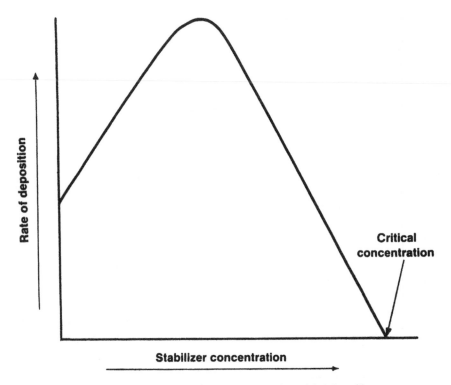

Fig. 1.7—Effect of stabilizer concentration on rate of electroless nickel deposition.

To determine the effect of stabilizers on the partial reactions, stabilizers were added to each solution used in the initial studies above. The cathodic polarization and the anodic polarization results in the presence of lead are presented in Figs. 1.10 and 1.11, respectively. Increasing the lead concentration causes a shift in cathodic curves in the cathodic direction, as shown in Fig. 1.10; curve G (Fig. 1.11) does not follow this behavior. The shift of the cathodic curves indicates an increasing inhibitive effect on the reduction reaction with increasing lead concentration. The anodic curves in Fig. 1.11 show a decrease in current density at constant potential for lead concentrations up to 2 ppm. When the lead concentration is 3 ppm or greater, the current density starts out lower than the zero lead case, but at about -600 mV. The current density increases, then decreases, causing an inflection in the curve. The inflection may be indicative of some other reaction, possibly involving lead, occurring at the surface of the electrode at lead concentrations of 3 ppm or greater.

Altura (36) also studied the effect of the widely used stabilizer, thiourea. The effect of the concentration of the sulfur-bearing substance on the cathodic

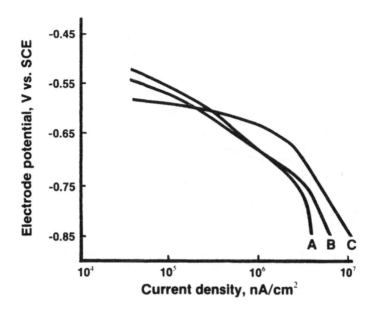

Fig. 1.8—Effect of nickel concentration on the cathodic polarization curves (no $NaH_2PO_2 \cdot H_2O$ or stabilizer). A = 0M Ni^{+2}; B = 0.02M; C = 0.15M.

polarization curves is shown in Fig. 1.12. The cathodic curves show an inconsistent shift in the location of the curve vs. thiourea concentration. This may be due to the combined or preferential inhibition of the hydrogen evolution reaction, nickel reduction, and/or the phosphorus reduction reaction. Assuming that thiourea inhibits phosphorus reduction, it could account for the decrease in the phosphorus content of deposits obtained from deposits containing thiourea. On the other hand, the family of anodic curves in Fig. 1.13 show a very consistent decrease in anodic current density with increasing thiourea concentration at constant potential. The shape of the anodic polarization curves suggests an enhanced effect on limiting current behavior vs. the stabilizer-free case, which indicates the likelihood of an anodic oxidation reaction involving thiourea to sulfur since it is well known that the potential of nickel deposits containing sulfur is different from sulfur-free Ni deposits.

Fontenals (37) suggested a theory that alternatively may explain the results obtained by Altura in his polarization studies on thiourea. The theory was developed to explain the characteristic relationship between the electroless nickel plating rate (which passes through a maximum) and the thiourea concentration in the plating bath. According to the theory, the rate increase as a

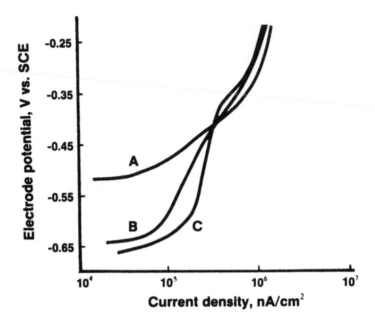

Fig. 1.9—Effect of hypophosphite concentration on the anodic polarization curves (no NH₂ or stabilizer). A = 0M NaH₂PO₂·H₂O; B = 0.1M; C = 0.6M.

result of the transfer of electrons from thiourea (anodic reaction) to the nickel ions, yielding elemental nickel and a dimer of thiourea:

$$2(H_2N - \overset{\overset{S}{\|}}{C} - NH_2) + Ni^{+2} \rightarrow Ni^0 + 2H^+ + (\overset{\overset{NH}{\|}}{\underset{NH_2}{C}} - S -)_2 \qquad [94]$$

The dimer is then reduced by the transfer of electrons from hypophosphite (anodic reaction) to the dimer to form thiourea:

$$\overset{NH}{\underset{NH_2}{C}} - S - S - \overset{NH_2}{\underset{NH_2}{C}} + H_2PO_2^- + H_2O \rightarrow \; 2(NH_2 - \overset{\overset{S}{\|}}{C} - NH_2) + H_2PO_3^- \qquad [95]$$

The actions of KIO₃ (potassium iodate) on the partial cathodic and anodic reactions is quite different from those described for lead and thiourea. Figure 1.14 shows cathodic polarization curves for the nickel-only solution containing the oxy-anion IO₃⁻. At constant potential of about -650 mV or greater, there is a

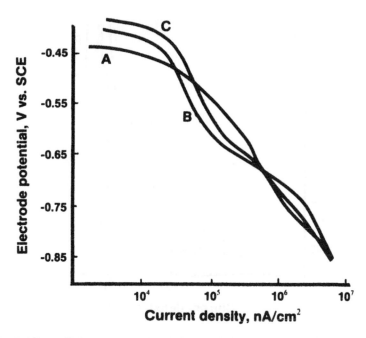

Fig. 1.10—Effect of Pb^{+2} concentration on the cathodic polarization curves. A = 0 ppm Pb^{+2}; B = 1 ppm; C = 10 ppm.

uniform shift in the curves to higher current densities as the concentration of KIO_3 is increased from 0 to 400 ppm. Also at potentials approximately $\geq$-650 mV, curves C and D reach a limiting current density where mass transfer is the controlling step. The cathodic polarization curves for the stabilizer-free case, and in the cases where thiourea or lead were present did exhibit limiting current behavior, over the range of potentials studied. If the reduction of IO_3^- is assumed to occur in the region of limiting current density, then a possible mechanism for bath stabilization by KIO_3 may be the competition of IO_3^- with Ni^{+2} for cathodic surface sites (catalysts), which would result in the reduction in the overall rate of the nickel reduction reaction. The cathodic curves at potentials below -650 mV (-650 to -850) are more or less linear and are similar to curves obtained for the stabilizer-free, lead and thiourea cases over the entire range of potentials. In this so-called linear region, the predominant reaction is most likely activation-controlled.

The results of anodic polarization in the presence of IO_3^- are shown in Fig. 1.15. There is a constant drop in current density (at constant potential) when the KIO_3 concentration is increased; curves E and F of Fig. 1.15 exhibit a dramatic shift in open circuit potential from about -650 mV to -387 mV and -354 mV, respectively. The theoretical E_{Mix} value for KIO_3 free case.

ELECTROLESS PLATING

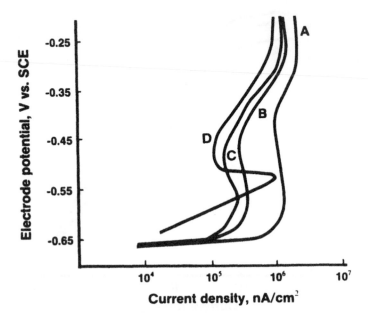

Fig. 1.11—Effect of Pb^{+2} concentration on the anodic polarization curves. A = 0 ppm Pb^{+2}; B = 1 ppm; C = 2 ppm; D = 10 ppm.

Feldstein and Amodio (38), in their study of the inhibition of EN deposition by oxy-anions, report that the steady-state potential decreases sharply when the concentration of IO_3^- is increased, resulting in a decrease in the plating rate. They propose that the inhibition effect associated with IO_3^- and other inhibiting oxy-anions (e.g., AsO_2^-, NO_2^-, BrO_3^-) is primarily by a surface adsorption mechanism. They further propose that the observed abrupt increase of inhibition (decrease in E_{Mix}) with increasing inhibitor concentration can be accounted for by one or any combination of the following mechanisms:

• A chemical reaction between the adsorbed IO_3^- (oxy-anions) and $H_2PO_2^-$ leading to a strengthening of the P-H bond in hypophosphite.
• The adsorbed IO_3^- (oxy-anions) alters the structure of the double layer, modifying (or increasing) the degree of surface adsorption and thereby affecting the kinetics of the redox reactions.
• The number of adjacent pair catalytic sites available for the catalytic dehydrogenation of the hypophosphite ion decreases rapidly once a critical concentration of adsorbed inhibitor is reached.

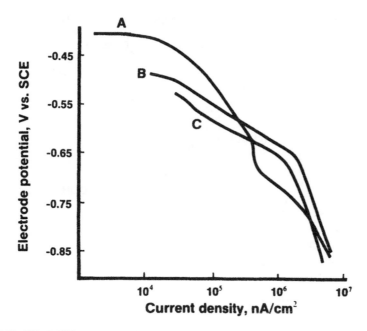

Fig. 1.12—Effect of thiourea concentration on the cathodic polarization curves. A = 0 ppm thiourea; B = 1 ppm; C = 10 ppm.

In 1966, Velemitzina and Riabchenkov (39) reported that the introduction of 1.5 to 2 g/L of maleic anhydride into an electroless nickel solution "sharply increased the resistance of the latter to self-decomposition . . .". A large variety of unsaturated short chain aliphatic compounds are claimed to function as stabilizers or at least enhance the efficacy of other stabilizers, such as lead and thiourea. The unsaturated group of the stabilizer molecule may be either the carbon-carbon double bond (-HC = CH-) alkenes, or the carbon triple bond (—C≡C—) acetylenes. Obvious choices of this class of stabilizers are the short chain dicarboxylic acids, as represented by maleic and acetylene dicarboxylic acids. However, Linka and Riedel (40) observed that the unsaturated carboxylic acids are subject to chemical reaction. The catalytic conditions (during deposition) at the electrolyte–metal substrate interface are ideal for hydrogenation of the unsaturated acid. They demonstrated analytically that maleic acid was converted to succinic and fumaric acids by hydrogenation and isomerism, respectively. These reactions can be shown schematically as follows:

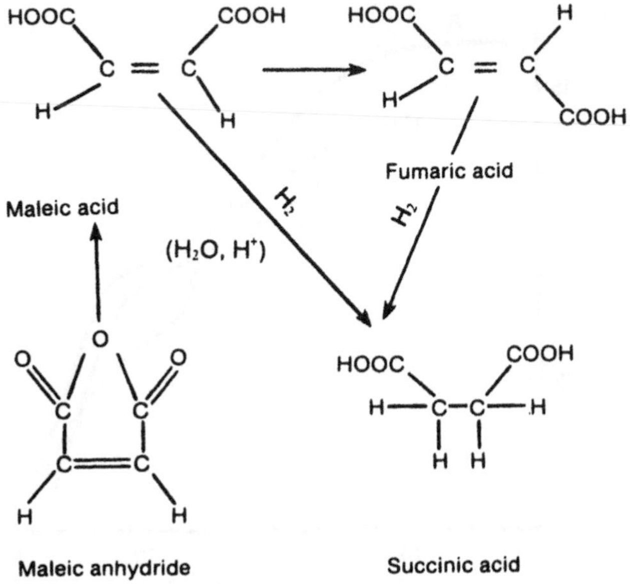

Maleic acid

(H$_2$O, H$^+$)

Fumaric acid

Maleic anhydride

Succinic acid

(For clarity, the hydrolysis of maleic anhydride to maleic acid has been included.) Based on the work of Linka and Riedel, it is assumed that the following conversions would take place:

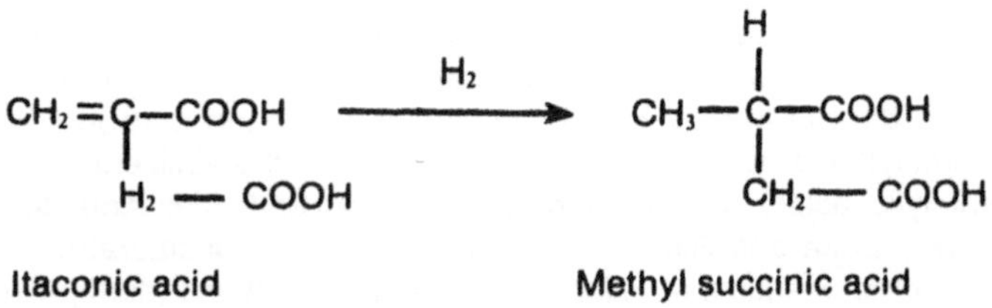

Itaconic acid

Methyl succinic acid

Aconitic acid $\xrightarrow{H_2}$ Tricarballylic acid

Acetylene dicarboxylic acid $\xrightarrow{2H_2}$ Succinic acid

Propiolic acid $\xrightarrow{2H_2}$ Propionic acid

The above examples illustrate that in some cases, the by-product of the hydrogenation reaction may be an innocuous addition to the plating solution. The formation of succinic acid can be considered a plus since its salts will help buffer the bath. On the other hand, the effect of tricarballylic acid on the plating reaction, as well as the deposit, must be determined. To reiterate, it is important to determine if the reaction by-product will have a deleterious effect on the deposition reaction and/or the final deposit.

The section on the unsaturated organic stabilizers illustrates the fact that all stabilizers in their active form are consumed during the deposition reaction, and therefore must be replenished periodically in order to maintain plating bath stability. The replenishment and maintenance of the optimum stabilizer concentration in the EN plating bath will be discussed further in the chapter on "Practical Electroless Nickel Plating."

The stabilization of an electroless nickel plating solution takes place at or near the catalytic substrate/double-layer interface. The mechanism of stabilization probably depends upon the degree of adsorption of the stabilizer onto the catalytic surface.

Class I and Class II stabilizers show a strong adsorption tendency (36,38); for example, Altura (36) reported that the steady-state potential (E_{Mix}) increases with increasing stabilizer concentration, indicating that the stabilizer is being adsorbed onto the working electrode (catalyst). Thiourea (Class I) is so strongly

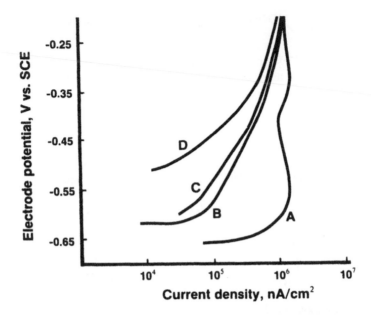

Fig. 1.13—Effect of thiourea concentration on the anodic polarization curves. A = 0 ppm thiourea; B = 1 ppm; C = 2 ppm; D = 10 ppm.

adsorbed that its use causes microporosity, as well as the codeposition of sulfur in the deposit. Feldstein and Amodio (38) show a proposed model for the interaction of adsorbed Class II anions with hypophosphite.

Solution agitation (mechanical or inert gas) in the presence of either a Class I or Class II stabilizer causes inhibition of the EN deposition reaction, as exhibited by a decrease in plating rate. Similar results were observed by Feldstein and Amodio (41). Each adsorbed stabilizer molecule or ion reduces the number of catalytic sites available for the dehydrogenation of hypophosphite. Solution agitation increases the rate of diffusion of stabilizer to the surface and, hence, increases its concentration at the surface. This condition leads to the shifting of the critical concentration of stabilizer to a lower value; for example, if the critical concentration of a particular stabilizer is 10 ppm in a quiescent solution, it may be lowered to 2 ppm when the solution is agitated vigorously. The actual stabilizer concentration can be safely maintained at concentrations below 10 ppm, say 5 to 8 ppm, in the unstirred case, whereas in the strongly agitated case, the actual concentration must be kept below 2 ppm. Skip-plating (edge-pullback) at sharp edges or corners is a visual example of the effect agitation has on the effective stabilizer concentration. In many cases, the so-called "edge

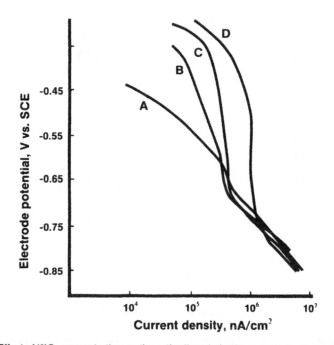

Fig. 1.14—Effect of KIO₃ concentration on the cathodic polarization curves. A = 0 ppm KIO₃; B = 50 ppm; C = 100 ppm; D = 400 ppm.

effect" can be eliminated by either decreasing or discontinuing agitation until the actual stabilizer concentration in the bath is reduced below the effective critical concentration for the agitated case.

Class III (e.g., Pb^{++}, Sn^{++}) stabilizers have a minimal effect on the steady-state potential of an EN plating bath, even at stabilizer concentrations that greatly reduce the plating rate; for example, when lead concentration in an EN bath was increased from 1 ppm to 10 ppm, the mixed potential of the working (plating) electrode changed from -625 mV to -609 mV, respectively, a E_{Mix} of only 16 mV. However, for the same change in Pb^{++} concentration, the rate of deposition decreased from 15 μm/hr to less than 2 μm/hr. Hence, it appears that lead is only loosely adsorbed onto the catalyst, but its diffusion in the double layer adjacent to the catalytic surface has a profound effect on bath stability.

In the Cavalotti and Salvago mechanism for electroless nickel deposition with $H_2PO_2^-$, Eqs. 23 and 26 were seen to represent competing reactions. When reaction 26 occurs:

$$Ni(OH)_{ads} + H_2O \rightarrow [Ni(OH)_2]_{aq} + H_{ads} \qquad [26]$$

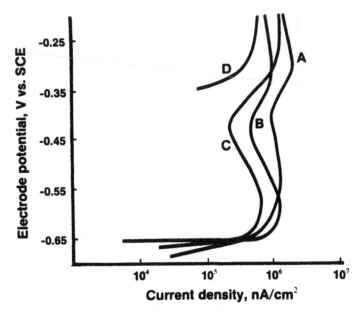

Fig. 1.15—Effect of KIO₃ concentration on the anodic polarization curves. A = 0 ppm KIO₃ B = 50 ppm; C = 100 ppm; D = 400 ppm.

the hydrolyzed Ni^{+2} species can desorb from the catalytic surface into the double layer, forming colloidal particles. The heavy metal stabilizers are adsorbed on these particles, importing a net positive charge to each particle, which causes them to repel each other. The action of the heavy metal thus prevents the aggregation of the particles to form micelles. In the absence of these stabilizers, the formation of micelles can result in the decomposition of the plating bath.

In a further attempt to characterize the observed phenomena, it is suggested that Class I and Class II stabilizers, which are adsorbed on the catalytic surface, inhibit the reaction expressed by Eq. 26. This allows Eq. 23 to proceed at the expense of Eq. 25. Since adsorption is reversible, the net effect of the addition of Class I or II stabilizers is to lower the frequency at which reactions 25 and 26 occur. This would result in a lower phosphorus content in the deposit. It is well known that EN plating baths utilizing Class I stabilizers yield deposits with lower phosphorus content than similar baths containing a Class III or no stabilizer.

In order for an unsaturated carboxylic acid to become hydrogenated during the operation of an EN plating bath, the acid must be adsorbed onto the catalytic surface. If the acid is hydrogenated, it can no longer function as a stabilizer; however, if the acid is isomerized at the surface, the isomer retains the stabilizing property; for example, both maleic acid and its isomer fumaric acid function as EN stabilizers.

At this point, it is necessary to digress and briefly discuss adsorption of substances on metallic nickel surfaces, including EN deposits. The nickel atoms at the surface of an EN deposit have unsatisfied coordinate valences in a direction perpendicular to, and away from the interior of the deposit, as seen in Fig. 1.16. These surface nickel atoms will combine with a variety of substances known as *electron donors*. When these electron donors are adsorbed on the surface, they share their unpaired electrons with the nickel atoms, thereby satisfying the coordinate valency of the nickel atoms. Examples of Class I, II, and III stabilizers that are electron donors are given below:

Sodium arsenite

Thiourea

Potassium antimony tartrate

The hydrogenation and isomerism of unsaturated carboxylic acids on the the depositing nickel surface are examples of heterogeneous or surface catalysis. The exact mechanisms of these reactions are not fully understood, however, information obtained for the catalytic hydrogenation of ethylene at nickel surfaces should be helpful in constructing a model for the reactions of interest.

It has been shown experimentally that ethylene combines exothermically (-60 kcal/mole) and reversibly with nickel surfaces (42). The bonding of the adsorbed ethylene to the surface nickel atoms most likely involves the electrons in the double bond, which satisfies the coordinate valency of nickel (33,42). Figure 1.17 is a schematic representation of ethylene adsorbed at the nickel surface. This model can be used as a basis for constructing a model of the heterogeneous catalysis of the unsaturated carboxylic acids on the EN film as it is being deposited.

The hydrogenation reaction takes place when a hydrogen molecule and an anion of the unsaturated acid, e.g., maleic acid, occupy sites on the nickel surface close enough to react, which is shown in Fig. 1.18. After hydrogenation, the saturated acid anion desorbs and diffuses into the bulk of the solution.

Because of the geometry of the carbon-carbon double bond, rotation about this bond is restricted. When the double bond of the anion is adsorbed on the

Unsatisfied valences

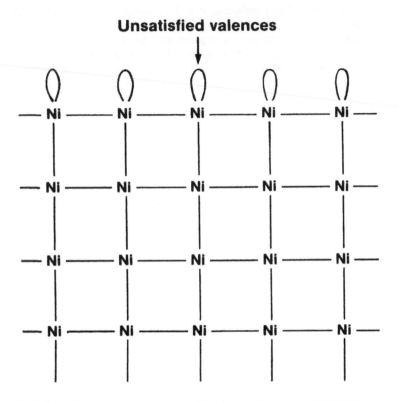

Fig. 1.16—Schematic representation of unsatisfied valences at surface of nickel deposit.

metal surface, one-half of the bond is broken in order to donate electrons to the surface nickel atoms. Now the two parts of the anion are free to rotate about the residual carbon-carbon single bond. Hence, if no hydrogen is available to react with adsorbed anions, isomerism can occur, as in Fig. 1.19, which is a representation of the conversion of maleic acid to the fumaric structure.

The stabilizing action exhibited by unsaturated compounds can be assumed to be the result of their adsorption on the nickel surface, where they inhibit one or more of the elementary EN deposition reactions in a manner analogous to that of the Class I or II stabilizers. Another method of stabilizing the EN bath might consist of the double bond combining with any colloids that are present, and preventing the formation of micelles. The two proposed methods, singularly or together, explain the stabilizing activity of the Class IV compounds.

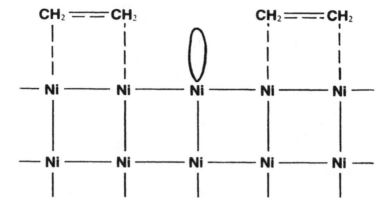

Fig. 1.17—Schematic representation of ethylene adsorbed at EN deposit surface. Dashed lines indicate sharing of electrons in double bond of ethylene with nickel.

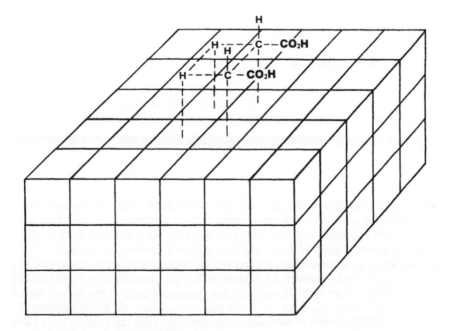

Fig. 1.18—Schematic representation of hydrogenation of maleic acid on EN deposit surface.

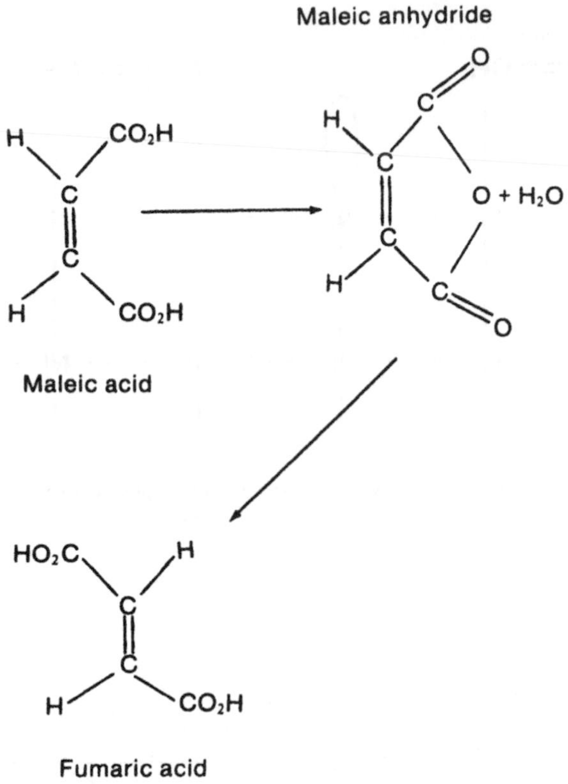

Fig. 1.19—Isomeric conversion of maleic acid to fumaric acid.

Although the above discussion on stabilizers was concerned with hypophosphite-reduced electroless nickel plating solutions, the principles also apply to amine borane-reduced baths. Stabilizers such as lead, and divalent sulfur compounds such as thiodiglycolic acid are commonly found in amine borane plating solutions. When a borohydride reducing agent is employed, the situation is different, however. As a result of the combination of high pH ($>$12.0) and high temperature (90° C), some of the stabilizers used in the former cases are not applicable because of their decomposition or precipitation. In borohydride baths, thallium salts at concentrations in the range of 40 to 50 ppm have proven to be extremely effective (16).

Energy

Catalytic reactions, such as electroless nickel plating, require energy in order to proceed. The energy is supplied in the form of heat. Temperature is a measure of the energy (heat) content of the plating solution. In the context of this chapter, energy, which is added to the plating bath, is considered a bath variable like the other reactants. The quantity of energy required by the system or added to it is one of the most important factors affecting the kinetics and rate of the deposition reaction. The dependence of reaction rate on temperature is illustrated in Fig. 1.20. The rate-energy relation is shown schematically because the exponential nature of the relation is common to all electroless nickel plating systems, even though their absolute plating rates will differ.

The rate of a chemical reaction may be studied by measuring the rate of decrease in the concentration of the reactants or the rate of increase in the concentration of the products. In electroless nickel deposition, we measure the rate of formation of the deposit. These relationships are expressed mathematically by:

$$\text{Rate} = \pm \frac{dC_i}{dt} = K \left(\prod_{i,j} C_i^j \right) \tag{96}$$

where (+) is used if C_i is a product and (-) if C_i is a reactant; K is a numerical proportionality constant called the rate constant; C_i^j is the concentration of the *i*th reactant (product) present in the system at time t; and j is the reaction order of the *i*th species.

To reiterate, the dominant variable affecting reaction rate is the temperature at which the process occurs. Many reactions that take place close to room temperature double their rates of reaction for each 10° C rise in temperature. Note that 100° C is considered close to room temperature.

The variation of rate constant K (in Eq. 96) with temperature T is given by the Arrhenius equation:

$$\frac{d}{dT} \ln K = \frac{E_a}{RT^2} \tag{97}$$

The quantity E_a is the activation energy of the reaction, and R is the gas constant. Since it is assumed that E_a is sometimes independent of temperature, T, integration of Eq. 97 yields:

$$\ln K = - \frac{E_2}{RT} + 1_{nA} \tag{98}$$

where l_{nA} is the constant of integration.

Equation 98 is of the form y = mx + b, which is an equation for a straight line. Hence, a plot of l_{nk} vs. the reciprocal of the absolute temperature, 1/T, should be a straight line whose slope is $-E_a/R$.

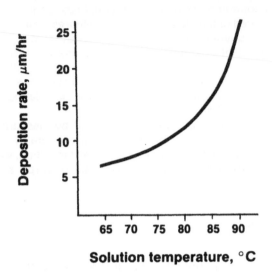

Fig. 1.20—Effect of temperature on plating rate.

Equation 98 yields an expression for the rate constant K:

$$K = A \exp\left(-\frac{E_c}{RT}\right)$$ [99]

Here A is called the *pre-exponential* or *frequency* factor. According to this equation, the reacting molecules or ions must acquire a certain critical energy E_a before they can react. The term $\exp(-E_a/RT)$ is the Boltzmann factor, which denotes the fraction of the reacting species that have managed to attain activation energy E_a. The activation energy can be considered as a potential energy barrier that the reactants must climb before they can react.

Substituting the expression for K (Eq. 98), Eq. 96 can now be written as:

$$\text{Rate} = \pm\frac{dC_i}{dt} = A\left(\pi_{i,j} \, C_j^i\right) \exp\left(-\frac{E_a}{RT}\right)$$ [100]

Equation 100 shows that the reaction rate is dependent on the addition of energy to the system as measured by temperature.

When operating an electroless nickel plating bath, it is necessary to know which conditions yield deposits with the desired properties. The deposits should be produced with minimum cost and difficulty and optimum efficiency. In some cases, it is more efficacious to operate the plating bath at a temperature that does not give the maximum plating rate. The composition of Ni-P deposits can be altered slightly by merely raising (decrease P) or lowering (increase P) the temperature of hypophosphite-reduced nickel plating solutions. The amine borane reducing agents are more temperature sensitive and will hydrolyze excessively at high temperatures, causing wasteful side reactions.

In this chapter we have discussed, in some detail, the chemistry of the principal components of electroless nickel plating solutions. Each component was shown to have a unique effect on the course of the deposition reaction. The discourse was mainly concerned with the theoretical, or better, ideal aspects of the nickel reduction reaction.

However, in real time, EN solutions are used to plate a myriad of commercial and industrial substrates. In subsequent chapters, the real time operation of EN plating solutions will be discussed.

REFERENCES

1. A. Brenner and G. Riddell, *J. Res. Nat. Bur. Std.*, **37,** 31 (1946); ibid., **39,** 385 (1947).
2. A. Wurtz, *Ann. Chim. et Phys.*, **3,** 11 (1844).
3. P. Breteau, *Bull. Soc. Chim.*, p. 9 (1911).
4. F.A. Roux, U.S. patent 1,207,218 (1916).
5. G. Gutzeit, *Plating*, **46,** 1158 (1959); ibid., **47,** 63 (1960).
6. P. Hersch, *Trans. Inst. Metal Finishing*, **33,** 417 (1955-56).
7. R.M. Lukes, *Plating*, **51,** 969 (1964).
8. P. Cavallotti and G. Salvago, *Electrochim. Metall.*, **3,** 239 (1968).
9. J.P. Randin and H.E. Hintermann, *J. Electrochem. Soc.*, p. 117, 160 (1970).
10. G. Salvago and P. Cavallotti, *Plating*, **59,** 665 (1972).
11. W. Franke and J. Moench, *Liebigs Ann. Chim.*, **64,** 29 (1941).
12. A.A. Sutyagina, K.M. Gorbunova and P.M. Glasunov, *Russian J. Phys. Chem.*, p. 37 (1963).
13. K. Lang, **56**(6), 347 (1965).
14. L.E. Tsupak, *Kandidatskaya Dissertatsiya*, Moscow (1969).
15. A. Prokopchik, I. Valsyunene, P. Butkyavichyus and D. Kimtene, *Zaschita Metal.*, **6,** 517 (1970).
16. K. Gorbunova, M. Ivanov and V. Moiseev, *J. Electrochem. Soc.*, **120,** 613 (1973).

17. G. Mallory, *Plating*, **58**, 319 (1971).
18. M. Lelental, *J. Electrochem. Soc.*, **122**, 436 (1975).
19. L. Pessel, U.S. patent 2,430,581 (1947).
20. D.J. Levy, *J. Electrochem. Technol.*, (1), 38 (1863).
21. J. Dini and P. Coronado, *Plating*, **54**, 385 (1967).
22. J.E.A.M. Van den Meerakker, *J. Appl. Electrochem.*, 395 (1981).
23. F. Pearlstein and R.F. Weightman, *Electrochem. Technol.*, **6**, 427 (1968).
24. M. Paunovic, *Plating*, **55**, 1161 (1968).
25. C.H. de Minjer, *Electrodep. and Sur. Treatment*, **3**, 261 (1975).
26. S.M. El-Raghy and A.A. Aba-Salama, *J. Electrochem. Soc.*, **126**, 171 (1979).
27. F. Pearlstein, *Plat. and Surf. Finish.*, **70**(10), 42 (1983).
28. I. Ohno, O. Wakabayashi and S. Haruyama, *J. Electrochem. Soc.*, **132**, 2323 (1985).
29. M. Paunovic, *AES 1st Electroless Plating Symp.*, Mar. 1982.
30. M. Paunovic, *Plat. and Surf. Finish.*, **70**(2), 62 (1983).
31. C.K. Mital, P.B. Shrivastava and R.G. Dhaneshwar, *Metal Finishing*, **85**(6), 87 (1987).
32. S. Chaberak and A. Martell, *Organic Sequestering Agents*, John Wiley & Sons, New York, NY, 1959; p. 125.
33. F.A. Cotton and G. Wilkinson, *Advanced Inorganic Chemistry*, Interscience Publishers, New York, NY, 1972; p. 652.
34. N. Feldstein and T.S. Lancsek, *J. Electrochem. Soc.*, **118**, 869 (1971).
35. C. Gabrielli and F. Raulin, *J. Applied Electrochemistry*, **1**, 167 (1971).
36. D. Altura, *Electrochemical Evaluation of Electroless Nickelplating*, Proc. 11th World Congress on Metal Finishing, Interfinish 84, Jerusalem, Israel (1984).
37. K. Fontenals, *Pint y Acabados*, **6**(36), 61 (1964).
38. N. Feldstein and P.R. Amadio, *J. Electrochem. Soc.*, **117**, 1110 (1970).
39. V.I. Velemitzina and A.V. Riabchenkov, *3rd Tr. Mezhclunar Kongr. Korroz. Metal.* (1966).
40. G. Linka and W. Riedel, *Galvanotechnik*, **77**(3), 568 (1987).
41. N. Feldstein and P.R. Amadio, *Plating*, **56**, 1246 (1969).
42. J.D. Roberts and M.C. Caserio, *Basic Principles of Organic Chemistry*, W.A. Benjamin, Inc., New York, NY, 1965; p. 171.

Chapter 2
The Electroless Nickel Plating Bath:
Effect of Variables on the Process

G.O. Mallory

In the previous chapter, the components of electroless nickel plating baths were discussed from the viewpoint of the function each performs in the bath, with little attention paid to their effect on the plating process. The metal and the electron source (the reducing agent) are consumed in the electroless plating reaction and so their concentrations in the bath are continuously decreasing. There are no anodes available to maintain a near-constant metal concentration, and no external electron source (rectifier) to keep a constant flow of electrons moving into the system, as in an electrolytic plating process. In order to have a continuous and consistent electroless plating process, the reactants must be replenished. The frequency at which additions of the reactants are made to the bath depends on how far the concentrations of the reacting species can be allowed to vary from their optimum concentrations without adversely affecting the plating process, or concurrently the deposit.

The electroless plating reaction not only yields a nickel alloy deposit; it also generates by-products, which accumulate in the solution. As the concentrations of the by-products increase, their influence on the plating reaction also increases. In the following sections, the effects on the plating reaction of the reactants and by-products, as variables, will be discussed.

ELECTROLESS NICKEL PLATING
WITH HYPOPHOSPHITE

Acid Plating Baths
A chemical equation that represents the aggregate electroless Ni-P plating reaction can be written as:

$$[NiL_m]^{-(n-2)}$$

$$\Updownarrow$$

$$[Ni^{+2} + mL^{-n}] + 4H_2PO_2^- + H_2O \xrightarrow{\text{cat}} Ni^0 + P + 2HPO_3^{-2} + H_2PO_3^- + 3H^+ + mL^{-n} + 3/2H_2$$

where $[NiL_m^{-(n-2)}]$ denotes the nickel complex and mL^{-n} the "free" complexing agent.

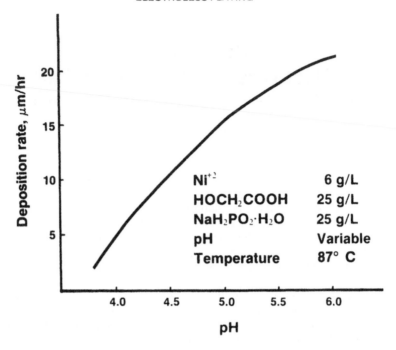

Fig. 2.1—Effect of pH on rate of deposition.

Equation 1 shows that there are at least five variables (components) that influence the deposition reaction: the reactants Ni^{+2} and $H_2PO_3^-$; the products H^+, $H_2PO_3^-$, and L^{-n}. To these five variables, at least three more are added: temperature, the anion of the nickel salt used (e.g., sulfate from $NiSO_4 \cdot 6H_2O$), and one or more stabilizers. Hence, there are at least eight variables that must be monitored during electroless nickel plating.

The number of variables illustrates the complexity of operating and maintaining these solutions. As an aside, it should be pointed out that the suppliers of proprietary electroless nickel plating solutions have reduced the number of variables the plater must be concerned with through judicious formulations and clever packaging of replenishing solutions.

The Influence of H^+

In laboratory tests, it is observed that for every mole of Ni^{+2} deposited, three moles of H^+ are generated, which is in agreement with Eq. 1. The accumulation of hydrogen ions (H^+) in the plating bath lowers the pH of the solution. When the pH decreases, the most noticeable change in the plating process is a concurrent decrease in the rate of deposition. If the pH is allowed to drop too far in acid solutions (pH <4.0 for commercial plating baths), a very low plating rate is observed. Figure 2.1 illustrates the effect of solution pH on deposition rate. The

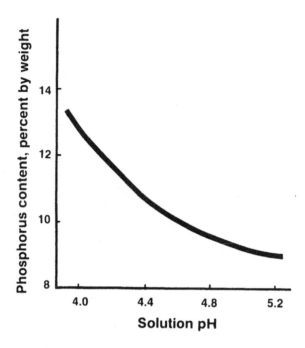

Fig. 2.2—Effect of pH on phosphorus content.

most dramatic effect of lowering the pH is on the composition, and concomitantly, the properties of the Ni-P deposit. The effect of solution pH on alloy composition is shown in Fig. 2.2.

The physical and chemical properties of electroless Ni-P deposits are dependent on the composition of the alloy in question. (A detailed discussion of the properties of electroless nickel-phosphorus deposits is covered elsewhere in this book.) As a rule of thumb, when the phosphorus content of the deposit is in excess of 10 percent by weight, the Ni-P alloy has the following characteristics:

- Low internal intrinsic stress, usually near zero or slightly compressive.
- Good corrosion resistance; low porosity.
- Non-magnetic in as-plated state.

If one or more of these properties is required, then a decrease in plating bath pH is not harmful, since the phosphorus content will be increased, allowing the deposit to achieve and maintain the desired property. On the other hand, if a particular application requires that the phosphorus content remain invariant or within a narrow range, say between 5.5 and 6.0 percent by weight, then a moderate decrease in pH can result in increasing the phosphorus content beyond the specified limit.

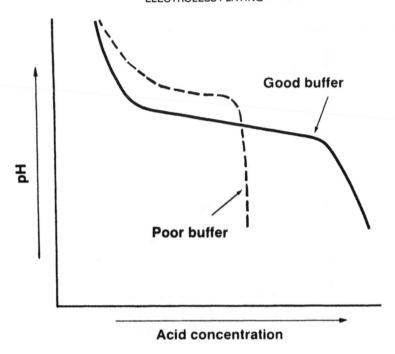

Fig. 2.3—Effect of buffering agents.

The amount that the pH changes as a result of the formation of H^+ is related to the buffer capacity of the complexing agents and certain other materials called *buffers* that are present in the plating bath. In simple terms, a buffer is a substance or mixture of substances that, added to a solution, is capable of neutralizing both acids and bases without appreciably changing the original pH of the solution. A measure of buffer capacity is the amount of acid (H^+) required to change the pH by a given amount. The greater the amount of acid required, the better the buffer. The practical buffer capacity, P_B, of an electroless nickel plating bath can be determined by titrating the bath with a standard acid solution and plotting pH vs. acid concentration. Figure 2.3 is a schematic illustration of data obtained by this technique. The slope of the linear portion of the curve can be expressed mathematically by:

$$|\alpha| = \frac{\Delta pH}{\Delta[H^+]}$$

Table 2.1
Organic Acids Used as Buffers
In Electroless Nickel Plating Solutions

Acid	Structure	Dissociation constants at 25° C	
		K_1	K_2
Acetic	CH_3CO_2H	1.75×10^{-5}	—
Propionic	$CH_3CH_2CO_2H$	1.4×10^{-5}	—
Succinic	$HO_2C(CH_2)_2CO_2H$	6.6×10^{-5}	2.5×10^{-6}
Glutaric	$HO_2C(CH_2)_3CO_2H$	4.7×10^{-5}	2.9×10^{-6}
Adipic	$HO_2C(CH_2)_4CO_2H$	3.7×10^{-5}	2.4×10^{-6}

where $\triangle pH = pH_{initial} - pH_{final}$ and $\triangle[H^+] = [H^+]_{initial} - [H^+]_{final}$. P_B can then be defined as $1/\alpha$. The larger P_B is, the greater the buffer capacity of the plating bath. This method can be used to compare the buffer capacity of various combinations of complexing agents and buffers.

The most efficient buffers will not prevent the pH from eventually decreasing, therefore it is necessary to monitor the pH and neutralize the excess H^+ being generated in the solution by the addition of ammonium or alkali metal hydroxides or carbonates. The time interval between additions of base solutions to an electroless nickel bath with a large P_B is less critical than for a bath with a much smaller P_B. A list of the commonly used buffers is given in Table 2.1. The aliphatic carboxylic acids also function as monodentate ligands (complexing agents); however, their principal action is that of buffer. Additionally, the anions of propionic, succinic, and glutaric acids are said to *exalt* the plating reaction. Exaltants are defined as those compounds that activate the hypophosphite anion and enhance the EN plating rate.

With few exceptions, it would appear that the ideal operating pH range for an acid Ni-P plating bath would be about 5.0 to 7.0. However, Gutzeit (1) points out that there are two important factors to be considered in the choice of the optimum pH range:

- The solubility of the orthophosphite ($H_2PO_3^-$) produced by the reduction reaction.
- The adhesion of the coating, particularly on ferrous substrates.

The first factor will be discussed in a subsequent section of this chapter. Data obtained from comparative adhesion tests on steel (1) reveal that the adhesion of a specimen plated at pH 4.4 was 60,000 psi, compared to values of less than 30,000 psi for a test specimen plated at pH 6.0. The optimum pH range for acid hypophosphite EN plating baths is usually 4.5 to 5.2. In the majority of EN applications, satisfactory deposits will be obtained if the solution pH is

Table 2.2
Effect of pH Change on Electroless Nickel Process

Change	Effect on solution	Effect on deposit
Raise pH	Increased deposition rate; lower phosphite solubility.	Decreased P content; shift in stress to tensile direction.
	Decreased stability with resultant plateout.	Poorer adhesion on steel.
Lower pH	Decreased deposition rate; improved phosphite solubility.	Increased P content; shift in stress to compressive direction.
		Improved adhesion on steel.

maintained within the stated range. The effects of raising or lowering the pH on the process itself as well as the resultant deposits are discussed in Table 2.2.

The importance of maintaining an essentially constant solution pH during the plating sequence should now be readily apparent; therefore, in the discussions on the remaining bath variables, a constant pH should be inferred unless otherwise stated.

The Influence of Nickel and Hypophosphite Ion Concentration

Brenner and Riddell (2) discovered that autocatalytic nickel plating would proceed on an immersed catalytic substrate at temperatures near 90° C in the pH range of 4 to 6. Plating was shown to occur over a wide range of nickel and hypophosphite concentrations—3 to 100 g/L (0.05 to 1.7M) for nickel and 10 to 100 g/L (0.09 to 0.94M) for hypophosphite. Under these conditions, the authors found that nickel deposition would occur on a suitable substrate without the simultaneous random reduction of nickel throughout the plating solution.

Gutzeit and Krieg (3) observed that to achieve optimum plating conditions, a narrow, more limited concentration range for hypophosphite is required. They also found that the molar ratio of $Ni^{+2}/H_2PO_2^-$ should be maintained within a limited range of 0.25 to 0.60, with the preferred range being 0.30 to 0.45.

The nickel concentration of commercial acid-type (pH 4 to 6) electroless nickel solutions lies within the range of 4.5 to 11 g/L (0.08 to 0.19M). Simple calculations based on the nickel concentration range and the preferred molar ratio of $Ni^{+2}/H_2PO_2^-$ yields a molar concentration range of 0.18 to 0.27M for sodium hypophosphite.

The nickel concentration of EN solutions used for most industrial applications is usually 6.5 ±1.0 g/L (0.09 to 0.13M). When the nickel concentration is equal to or greater than approximately 5 g/L (0.085M), it has little or no effect on the plating rate (4); the plating reaction is said to be zero order with respect to the

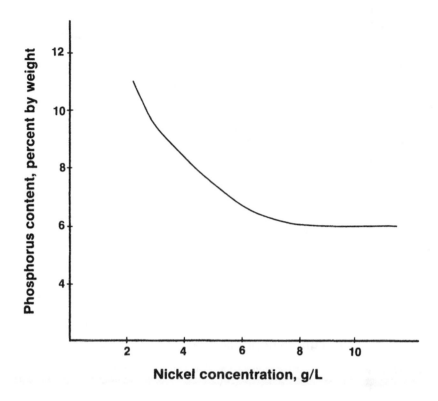

Fig. 2.4—Effect of nickel concentration on Ni-P alloy composition (4).

nickel ion concentration. The phosphorus content of Ni-P deposits is influenced by the nickel concentration in the plating baths only when the nickel concentration is less than about 0.1M. If the nickel concentration is increased beyond 0.1M (5.8 g/L), the phosphorus content will remain invariant, provided that the hypophosphite concentration is held constant. Figure 2.4 shows the effect of the nickel concentration on Ni-P alloy composition.

Lee (5) reported the influence of the reactants on the phosphorus content of the Ni-P coatings. Generally, increasing the H_2PO_2 concentration in the plating bath results in an increase in the phosphorus content of the deposit, as shown in Fig. 2.5.

Whereas nickel deposition is first order with respect to the hypophosphite concentration, Lee found a second order dependence on hypophosphite for phosphorus deposition. He derived the following empirical kinetic equation from experimental data:

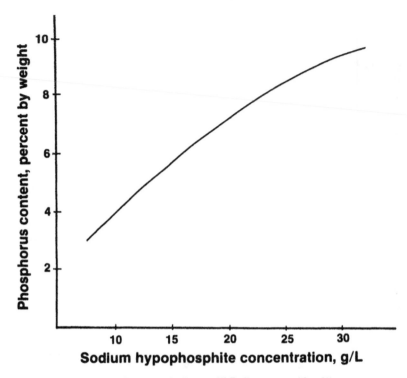

Fig. 2.5—Effect of hypophosphite concentration on Ni-P alloy composition (4).

$$\frac{dp}{dt} = K[H_2PO_2^-]^{1.91}\ [H^+]^{0.25} \qquad\qquad [2]$$

Equation 2 indicates that phosphorus deposition is acid-catalyzed (increase phosphorus content by increasing H^+). On the other hand, experimental data for nickel deposition leads to an empirical rate equation of the form:

$$\frac{dNi^0}{dt} = K\ \frac{[H_2PO_2^-]}{[H^+]^\beta} \qquad\qquad [3]$$

where β is the order of the reaction with respect to H^+ concentration. Hence Eq. 3 shows that nickel reduction is base-catalyzed (increase nickel deposition by decreasing H^+).

In addition to determining the effects of individual reactants on the electroless nickel plating process, it is equally instructive to ascertain the role their interdependence plays on the plating reaction. This is usually done by observing how the relationship expressed by the molar ratio $H^+/H_2PO_2^-$ affects the plating

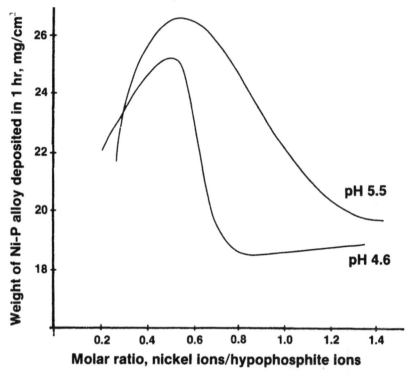

Fig. 2.6—Rate of deposition from acetate baths as a function of $Ni^{+2}/H_2PO_2^-$ molar ratio at two different pH values (1).

rate. Figure 2.6, taken from Gutzeit (1), illustrates the effect of the molar ratio on the plating rate (constant $H_2PO_2^-$, varying Ni^{+2}). The results given in Fig. 2.6 were obtained from an uncomplexed acetate bath, and it in no way infers that all electroless nickel solutions, especially those containing chelating agents, will yield equivalent results.

The Phosphite Anion
The phosphite anion is generated by the oxidation of hypophosphite during the plating reaction (see Eq. 1). For each nickel ion reduced to metal, approximately three phosphite ions are produced; or, for each gram of Ni^{+2} reduced, approximately four grams of phosphite are formed. For continuous operation of the plating bath, the depleted hypophosphite (as well as nickel) must be replenished. Hence, phosphite steadily accumulates in the plating bath. With time, phosphite becomes an important solution component. As the phosphite concentration increases, it will begin to compete with the complexing agent for nickel ions. This usually occurs when the phosphite concentration reaches 30 g/L. Nickel phosphite is relatively insoluble and exhibits the inverse solubility

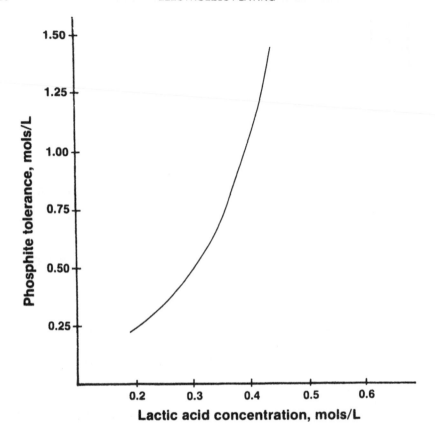

Fig. 2.7—Phosphite tolerance of plating bath as a function of lactic acid concentration (1).

phenomenon. That is, nickel phosphite is more soluble at low temperatures than at higher temperatures.

When the formation of nickel phosphite is first observed, the addition of excess complexing agent will solubilize the compound and prevent its precipitation. The hydroxy carboxylic acids (lactic and hydroxyacetic) are very effective in preventing the precipitation of nickel phosphite. Figure 2.7 illustrates the variation in phosphite tolerance with lactic acid.

Ultimately, the phosphite ion concentration becomes so large that further additions of complexing agent cannot prevent the precipitation of nickel phosphite. At this point the plating bath is discarded. Since nickel phosphite is more soluble at lower pH values, it may be more efficacious to sometimes operate the plating bath at lower pH values, say, 4.6 rather than 5.0. The solubility of phosphite vs. bath pH for two lactic acid concentrations is shown in Fig. 2.8.

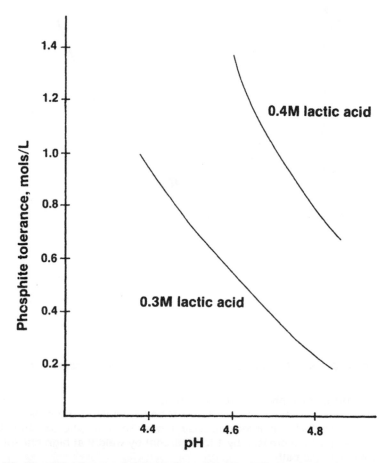

Fig. 2.8—Phosphite tolerance of plating bath as a function of pH at two lactic acid concentrations.

Nickel phosphite precipitation, which is visually observable, was considered the only harmful result of phosphite accumulation in the plating solution; however, there is a more serious consequence of phosphite presence in the solution. A definite correlation between the phosphite concentration and the internal stress of Ni-P deposits was observed by Baldwin and Such (4) and later verified by others (6). The effect of phosphite concentration on the internal intrinsic stress of Ni-P deposits is shown in Fig. 2.9. As the phosphite concentration increases, the stress becomes more tensile. In many applications (e.g., corrosion, memory disks), tensile stress is to be avoided. The data reported in Fig. 2.9 indicates that solutions used for engineering purposes have a limited useful life expectancy, up to three metal turnovers.

Some plating baths are able to tolerate moderate concentrations of phosphite better than others. This depends on the formulation of the bath and the nature of

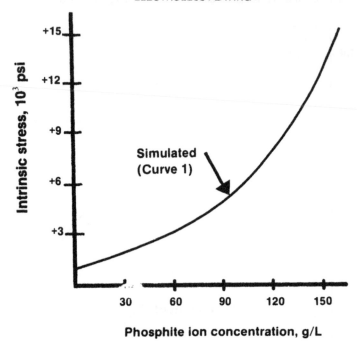

Fig. 2.9—Variation of intrinsic stress with phosphite ion concentration.

the complexing agent. A phosphite concentration of about 60 g/L will cause only a slight decrease in plating rate from some baths, while in other baths this concentration results in a marked decrease in rate. The phosphorus content in Ni-P deposits may be increased by 1 to 2 percent by weight at high phosphite concentrations in the bath.

To extend the life of an electroless nickel plating bath beyond a few metal turnovers, it is essential that the phosphite be removed from the solution, or at least reduced to a tolerable level. Methods of phosphite removal from EN plating baths have been investigated for many years.

The use of ferric chloride ($FeCl_3$) to precipitate the accumulated phosphite was proposed by Gorbunova and Nikiforova (7). Ferric chloride reacts with phosphite to form an insoluble complex compound according to the following equation:

$$2NaH_2PO_3 + FeCl_3 \xrightarrow{H_2O} Na_2[Fe(OH)(HPO_3)_2] \cdot xH_2O + 3HCl \qquad [4]$$

To avoid contamination of the plating bath with an excess of ferric chloride, the amount of $FeCl_3$ added must be less than the stoichiometric quantity required to precipitate all the $H_2PO_3^-$ in the bath. A considerable quantity of hypophosphite is separated from the solution along with the phosphite that is precipitated; however, very little nickel is removed when the solution is treated with $FeCl_3$.

The formation of hydrochloric acid (HCl), as seen in Eq. 4, lowers the pH of the plating bath to inoperative levels. After removal of the insoluble iron complex by filtration, the pH of the bath must be raised to its optimum value. It is recommended that an acid neutralizing reagent be added simultaneously with the addition of ferric chloride.

Parker (8) discusses several procedures for precipitating the phosphite from spent EN plating baths. Of particular interest is his discussion of the ion-exchange treatment for the removal of phosphite. His data show that weak-base anionic resins are the most effective in adsorbing phosphite, as follows:

$$ROH + H_2PO_3^- \rightarrow RH_2PO_3 + OH^- \tag{5}$$

Parker claims that the ion exchange procedure is technically feasible and cost-effective, while on the other hand, the precipitation methods require further development to be practical.

The Role of the Complexing Agent

In the previous chapter, complexing agents were found to perform three principal functions in electroless nickel plating baths:

- Reduce the concentration of free nickel ions.
- Prevent the precipitation of basic nickel salts and nickel phosphite.
- Exert a buffering action.

de Minjer and Brenner (9) studied the relationship between ligand concentration and the plating rate. The data given in Fig. 2.10 are the results of their investigations.

Initially, a small addition of complexing agent is accompanied by an increase in plating rate. As the concentration of ligand is increased, the plating rate passes through a maximum. When the concentration of ligand is increased beyond the value where the rate maximum occurs, the plating rate decreases.

Several theories have been proposed to explain the maximum in the rate vs. ligand concentration curves. First, de Minjer and Brenner (9) argue that the maximum is the result of the low adsorption of ligand on the catalytic surface at low concentrations, which accelerates the reaction. At higher concentrations, there is a high adsorption of ligand on the surface, which poisons the reaction.

A more plausible explanation suggests that the rate increasing portion of the curve is due to the buffering action of the complexing agents. Maximum rates occur while there are uncoordinated (solvated) sites remaining on the nickel ions, that is, the nickel ions are only partially complexed or chelated. It must be remembered that partially complexed nickel ions retain some of the properties of free, solvated nickel ions. Hence, for ligand concentrations up to values where the maximum plating rates occur, buffering is the dominant function of the various ligands.

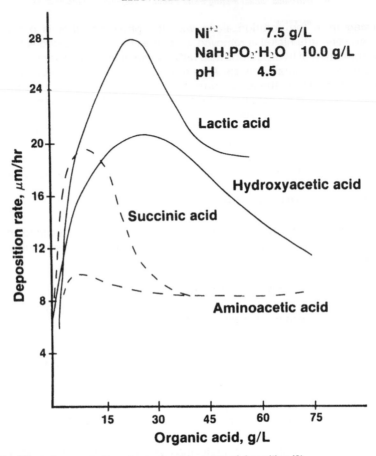

Fig. 2.10—Effect of concentration of organic acids on rate of deposition (9).

An explanation based on pH is inadequate to explain the decrease in rate for further increases in ligand concentration beyond the maximum point. The pH of the plating bath changes very little with continued plating. Moreover, the buffer capacity of the complexing agent does not parallel the plating rates obtained with them.

The decrease in plating rate, to the right of the maximum, is most likely due to the coordination of the remaining solvated sites with ligand atoms. The concentration of free nickel ions decreases and the metal ions take on the characteristics of complexed or chelated nickel ions.

The large differences in the rate maxima between the various complexing agents, as seen in Fig. 2.10, points to a specific property of ligands, vis-a-vis the nickel complex, that affects the rate of the plating reaction. The property in question is the stability constant of each nickel complex. Variations in the maxima relative to the ordinate and absicca of each curve in Fig. 2.10 can be explained on the basis of the strength of the stability constant for each complex.

The difference in the rate of deposition between two plating baths containing different complexing agents is attributed to the differences in magnitude of the stability constants of the two nickel complexes. The nickel complex controls the number of free ions that are available to take part in the deposition reaction. Nickel complexes with low-value stability constants yield a larger quantity of free nickel ions when compared to complexes with numerically larger stability constants. It follows, then, that the larger the number of free nickel ions adsorbed on the catalytic surface, the greater the nickel plating rate.

The relationship between the nature of the nickel complex and the composition of the Ni-P deposit can be explained on the basis of the strength of the nickel complex. When the fraction of catalytic sites onto which nickel ions have been adsorbed is large (small stability constant), the number of sites available for phosphorus reduction is small and the phosphorus content of the deposit is lowered. On the other hand, when the stability constant is relatively large, the free nickel ion concentration is low, and the fraction of sites covered by adsorbed nickel is small; hence, in the latter case, the number of sites available for phosphorus reduction is large and the phosphorus content of the coating is increased.

In summary, electroless nickel complexes with low-value (weak) stability constants will show higher deposition rates and yield deposits with lower phosphorus content than baths with numerically larger (strong) stability constants. Electroless nickel plating baths based on lactic acid and citric acid are examples of complexing agents that form weak and strong nickel complexes, respectively.

The Selection and Replenishment Of EN Stabilizers

Certain special compounds, called *stabilizers*, prevent or retard the spontaneous decomposition of electroless nickel plating solutions. This is especially true for acid EN solutions. Not only do these compounds stabilize plating solutions, but some of them also accelerate the plating rate. Other stabilizers can affect the phosphorus content and the internal stress of the electroless nickel deposits. If the stabilizer increases the reducing efficiency of the reducing agent, increases in plating rate and nickel content are observed. For example, thiourea increases the reducing efficiency of hypophosphite in acid EN plating solutions, with concomitant increases in plating rate and nickel content (lower phosphorus content) of the resultant deposits. Before incorporating a stabilizer in a production plating solution, it may be necessary to determine if it will affect the composition of the deposit.

Methods for determining the optimum stabilizer concentration were discussed in the previous chapter. The optimum stabilizer concentration is usually determined for some fixed set of plating conditions, e.g., temperature, pH, rate of agitation, and the surface area to solution volume ratio. However, in actual practice, these conditions vary and additionally, the stabilizer concentration decreases as a result of its consumption during the deposition reaction. It is obvious that the control and replenishment of the stabilizer presents a difficult task.

The initial stabilizer concentration in the production bath is usually a compromise between the optimum concentration determined for the fixed set of conditions mentioned above, and the midpoint of a rate-vs.-concentration or potential-vs.-concentration curve in an unstirred solution.

The rate of consumption of the stabilizer is determined relative to the rate of nickel deposition, or even better, by the amount of nickel consumed. The stabilizer concentration can be monitored by several techniques, e.g., polarography, voltammetry, or atomic adsorption. Thus, when the nickel concentration is replenished to its initial value, a proportional quantity of stabilizer is also replenished. If more than one stabilizer is used, the above techniques are used to determine the working concentration of each stabilizer separately. However, when the stabilizers are combined in the working solution, synergistic effects must be compensated for by lowering the concentration of stabilizer component.

The working concentration of stabilizer(s) must be sufficient to prevent plate-out of nickel on the walls of the plating tank as well as solution decomposition. The concentration must be below the value where edge pull-back occurs on sharp edges of the parts being plated. Finally, replenishment additions must be able to maintain bath stability without poisoning the deposition reaction.

The Effect of Temperature

The temperature at which the plating reaction occurs is the principle variable that determines the rate of the reaction. Readers with intimate knowledge or experience with electroless nickel plating know that very little, if any, plating occurs at temperatures below about 60° C (140° F). As the temperature is increased, the plating rate increases exponentially.

Acid hypophosphite plating solutions are operated between 85 and 90° C. If the temperature is allowed to increase much beyond 90° C, the possibility of solution plate-out or even solution decomposition increases. Baldwin and Such (4) reported that the phosphorus content of deposits is decreased when the temperature of the plating bath is increased beyond its normal operating range, with all other variables held constant.

MISCELLANEOUS IONS

The term *miscellaneous* is used to describe certain cations and anions that accumulate in the plating bath by replenishment of consumed reactants, and the neutralization of H^+ ions that are generated during the plating reaction. The cations Na^+, K^+, and NH_4^+ are introduced into the solution by additions of sodium hypophosphite ($NaH_2PO_2 \cdot H_2O$), sodium or potassium hydroxide (NaOH, KOH), and ammonium hydroxide or ammonium carbonate [NH_4OH or $(NH_4)_2CO_3$]. Nonreactive anions, such as SO_4^{-2} and Cl^- build up in the plating bath as the result of nickel ion replenishment in the form of nickel salts such as nickel sulfate or nickel chloride ($NiSO_4 \cdot 7H_2O$ or $NiCl_2 \cdot 6H_2O$).

The actual effect that these so-called miscellaneous ions have on the plating process has not been clearly established in all cases. However, Haydu and Yarkowsky (10) show that sulfate anions ($SO_4^=$) in sufficient quantities will increase the phosphite tolerance of the plating bath. They found that the amount of phosphite required to precipitate metal ions from a new plating bath was increased from 28 g/L (0.25M) to 70 g/L (0.7M) by the addition of 152 g/L sodium sulfate at pH 5.0. This suggests that additional sodium sulfate be added along with maintenance additions of reactants to enhance phosphite solubility.

One can speculate as to the effect that the alkali metals ions (Na^+ and K^+) have on nickel deposition. From a previous section, it is known that Ni^{++} and H^+ ions compete for association with complexing and buffering agents. Sodium and potassium ions can be included in that competition. Once the concentration of Na^+ or K^+ exceeds some statistically significant quantity, the probability of an alkali metal reacting with the complexing agent or buffer is much greater than the probability of either the nickel or hydrogen ion reacting. Hence, as alkali metal ions accumulate in the bath, the equilibrium between "free" nickel ions and nickel complex is more likely to be affected, as shown in the following scheme:

$$M^{+2} = Ni^{+2} \longrightarrow NiL_m^{2-n} \qquad \text{[6a]}$$

$$L_m^{-n} + M^+ \underset{M^{+2} = H^+}{\rightleftharpoons} HL_m^{1-n} \qquad \text{[6b]}$$

$$M^{+2} = Na^+ \longrightarrow NaL^{1-n} \qquad \text{[6c]}$$

The decrease in plating rate observed in aged (replenished) plating baths may be attributed to the predominance of Eq. 6c over 6b.

Several authors have ascribed changes in solution and deposit properties to phosphite buildup in the plating solution as a result of prolonged replenishment of the solution. It is imputed, for example, that the decrease in plating rate in the bath and the increase in phosphorus content of the deposits from said bath, are due to phosphite buildup. Other researchers have added quantities of phosphite to new plating solutions (equivalent to the amount of phosphite formed corresponding to the number of turnovers and replenishments reported in prior studies) and found no correlation between phosphite and deposition rate or phosphorus content. However, the addition of both sodium sulfate (Na_2SO_4) along with phosphite ($NaHPO_3$) to new plating solutions does change the deposition rate and alter the phosphorus content. It must be noted that the addition of Na_2SO_4 does increase the alkali metal ion content of the bath. These

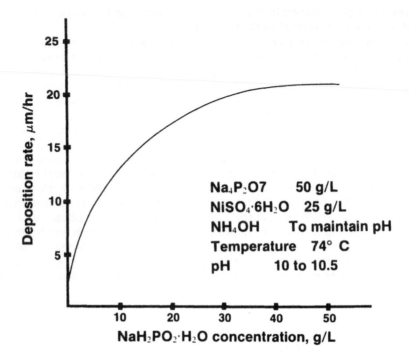

Fig. 2.11—Effect of sodium hypophosphite concentration on rate of deposition.

additions (Na_2SO_4 and Na_2HPO_3) do not account for all of the alkali metal ions that could be added over an equivalent number of replenishments.

Neutralization of H^+ by NaOH or K_2CO_3 increases the concentration of Na^+ or K^+. Hence, the changes in rate and percentage of phosphorus can just as well be justified by the alkali cation buildup; i.e., if the buffer associates with the excess sodium or potassium cations in preference to H^+ ions, then there are more H^+ ions available to lower the pH of the reaction layer adjacent to the substrate. This surplus of H^+ ions is also available to increase the phosphorus reduction reaction:

$$2H_2PO_2^- + H^+ \rightarrow P + H_2PO_3^- + H_2O + \tfrac{1}{2}H_2 \qquad [7]$$

An increase in phosphorus reduction is concurrently accompanied by a decrease in nickel deposition.

ALKALINE HYPOPHOSPHITE
PLATING SOLUTIONS

The industrial uses of alkaline plating solutions have been confined to a very few special applications, usually those that require plating at low temperatures, such as plating of various polymeric materials. In general, the phosphorus content of deposits obtained from alkaline solutions is lower than those produced in acid solutions. Alkaline solutions were used exclusively to produce low-phosphorus deposits. (Recently developed solutions that are slightly acidic [pH 6.0 to 6.6] yield deposits containing 1 to 2 percent phosphorus).

The plating reaction in alkaline solutions proceeds in a manner analogous to that in acid solutions; i.e., nickel deposition is accompanied by the production of hydrogen ions (H^+), hydrogen gas (H_2) evolution, and the oxidation of hypophosphite to phosphite. Further, the hypophosphite anion is a much more powerful reducing agent in basic solutions than in acid solutions, as shown by the following equations:

Acidic
$$H_2PO_2^- + H_2O \rightarrow H_2PO_3^- + 2H^+ + 2e, \ E^0 = 0.499 \text{ V} \tag{8}$$

Basic
$$H_2PO_2^- + 3OH^- \rightarrow HPO_3^{2-} + 2H_2O + 2e, \ E^0 = 1.57 \text{ V} \tag{9}$$

The early investigations of Brenner and Riddell (2) were performed with alkaline solutions. As far as alkaline plating solutions were concerned, no significant advances were made until Schwartz (11) reported on the use of pyrophosphate solutions. The unique aspect of the alkaline pyrophosphate system is that it is composed of inorganic materials.

The Effects of Reactants on Alkaline
Electroless Nickel Plating Baths

The rate of deposition in alkaline electroless nickel solutions is proportional to the hypophosphite concentration; the plating rate increases with increasing reductant concentration. Although raising the hypophosphite concentration improves the plating rate, the stability of the plating bath is lowered, necessitating the use of stabilizers. Figure 2.11 is a graphical presentation of data for the pyrophosphate bath. It is evident that the hypophosphite ion concentration has a pronounced effect upon the plating rate.

The influence of the nickel concentration on plating rate in the citrate and pyrophosphate baths was found by Schwartz to be at variance with the results reported by Brenner and Riddell (2); i.e., in the citrate bath, the hypophosphite

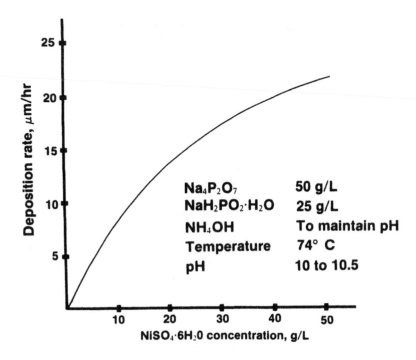

Fig. 2.12—Effect of nickel concentration on deposition rate (4).

ion has an appreciable effect on the rate, whereas the rate is only slightly increased by increasing the nickel ion concentration. The rate is increased significantly by increasing the nickel ion concentration in the pyrophosphate bath. The curve in Fig. 2.12 was plotted from data obtained by Schwartz for the pyrophosphate bath.

The Effect of Temperature on Rate
The rate of the deposition reaction is dependent on temperature in an exponential manner, and the relationship is independent of the acidity or alkalinity of the solution. A direct comparison between the two types of alkaline baths (citrate and pyrophosphate) is presented in Fig. 2.13. The deposition rate is seen to be considerably higher in the pyrophosphate bath than in citrate baths at equivalent temperatures; e.g., at 80° C, the rate in the pyrophosphate solution is nearly four times the rate in a citrate bath. It is also important to note that the initiation temperature is substantially lower in the alkaline pyrophosphate bath than in the alkaline citrate bath. In either case, the initiation temperatures are considerably lower for alkaline electroless nickel plating solutions than for

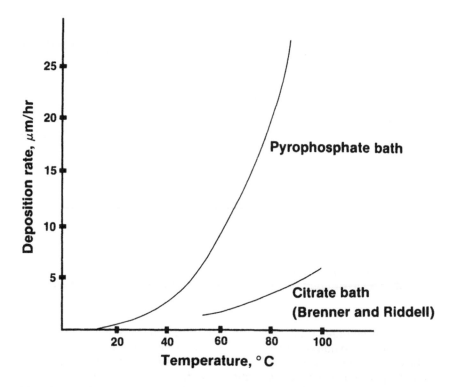

Fig. 2.13—Effect of temperature on deposition rate (11).

acidic plating systems. The fact that the pyrophosphate bath yields significant plating rates at "low" temperatures has led to its use in the so-called "low-temperature" applications.

The Effects of pH
As the plating reaction proceeds, the pH of an alkaline solution decreases. Figure 2.14 (taken from Schwartz) shows the effect of pH on plating rate in the pyrophosphate solution. Schwartz (11) points out that there is a lower limit for pH (about 8.5), below which the pH of the pyrophosphate solution must not be allowed to drop. Between pH 8.0 and 8.5, the operating solution turns turbid as a result of the precipitation of a basic nickel salt or complex.

The plating rates of non-proprietary alkaline electroless nickel solutions formulated with citrate are, with few exceptions, almost constant over the pH range of 8.0 to 11.0. The plating rate at 90° C lies in the range of 8 to 10 μm/hr. The maintenance of solution pH within its operating range is made easy by the intense blue color of the nickel ammine complex (a result of the addition of ammonium hydroxide) in the citrate baths. When the pH decreases, the color of

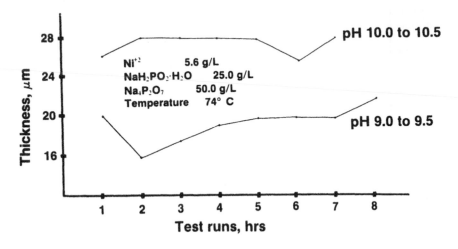

Fig. 2.14—Effect of pH on deposition rate (11).

the solution changes from dark blue to blue-green, to green, at which point ammonium hydroxide is added to restore the pH. Hence, the color, and intensity of the color, of the nickel-ammonia complex can serve as a qualitative indicator for both pH and nickel concentration. On the other hand, the color of the pyrophosphate nickel plating solution turns from intense dark blue to green when heated from ambient to operating temperature, even at optimum concentrations of the bath components.

The Effect of Complexing Agents
Based on the high reduction potential given in Eq. 9, it might be assumed that the rate of deposition would be correspondingly higher in alkaline plating solutions than in acid plating solutions. However, the strong complexing agents used to prevent the precipitation of nickel hydroxide or basic nickel salts have a negative effect on the plating rate.

The complexing agents used to sequester nickel ions alkaline plating baths are generally combinations of citrate and ammonium ions or pyrophosphate and ammonium ions. EDTA and its substituted homologues are excellent chelators in basic solutions, but are seldom used as primary complexing agents because of the extremely low plating rate obtained in solutions containing these materials.

In 1975, Booze (12) reported a "nitrogen-free" alkaline plating solution, containing the so-called "borogluconate" addition compound. He observed that the complexing capacity of gluconate was greatly enhanced when sodium gluconate is reacted with boric acid to form the borogluconate complex. The mannitol/boric acid complex (from analytical chemistry) is the best known homologue of this type of complex. The addition of boric acid to polyhydric materials is illustrated as:

Mannitol/boric acid complex

and:

Sodium gluconate   Boric acid   Borogluconate complex

The pH of alkaline electroless plating baths based on borogluconate can be adjusted with alkali metal hydroxides or carbonates. The rate of deposition of these baths is a function of the borogluconate concentration, as shown in Fig. 2.15. Deposition rates in borogluconate solutions are comparable to those of pyrophosphate plating baths. Typical alkaline electroless nickel plating bath formulations are provided in Table 2.3.

Referring to Table 2.3, Baths 3, 4, and 5 are of particular interest. Iwasa et al. (13) observed that when electroless nickel plating silicon wafers with p-n junctions, and using conventional Brenner-type alkaline plating solutions, there was a pronounced difference in plating rate between the p- and n-type surfaces. They found that additions of EDTA to alkaline citrate solutions (Bath 3) resulted in an increase in deposition rate on p-type surfaces until it was essentially equal to the rate at n-type surfaces. The effect of EDTA concentration on plating rate of p-and n-type silicon as found by Iwasa et al. is shown in Fig. 2.16.

Electroless nickel plating of zinc die castings has generated considerable interest in alkaline pyrophosphate solutions (Bath 4) for use in this application. The pyrophosphate solution is claimed to have excellent tolerance to the zinc ions that build up in the solution. The pyrophosphate bath is also used as a

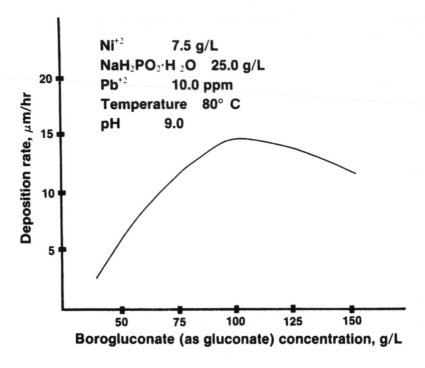

Fig. 2.15—Effect of borogluconate concentration on deposition rate (12).

Table 2.3
Typical Alkaline Electroless Nickel Plating Baths

Concentration, g/L

	Bath				
Component	1	2	3	4	5
Ni^{+2}	7.0	7.5	7.5	6.0	7.5
$NaH_2PO_2 \cdot H_2O$	20.0	10.0	10.0	30.0	25.0
$C_6H_5ONa^1 \cdot 2H_2O$	36.0	100.0	—	—	—
$C_6H_6O_7(NH_3)_2$	—	—	65.0	—	—
$K_4P_2O_7$	—	—	—	60.0	—
$C_6H_{11}O_7Na$	—	—	—	—	110.0
H_3BO_3	—	—	—	—	30.0
$C_{10}H_{10}N_2O_8$	—	—	5.0	—	—
NH_4C_1	—	50.0	50.0	—	—
pH (NH_4OH)	8 to 9	8 to 9	8 to 10	10.0	9.0
Temp., °C	90	90	90	75	80

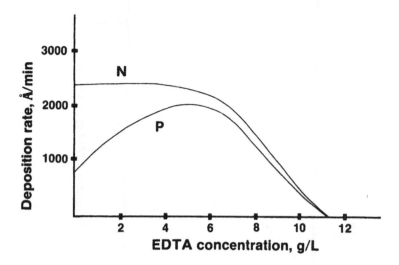

Fig. 2.16—Effect of EDTA concentration on plating rate on p- and n-type silicon (13).

strike-plate over zincated aluminum prior to plating with acid-type electroless nickel solutions. It was pointed out earlier that the pyrophosphate plating solution will initiate deposition at much lower temperatures than other electroless nickel solutions, either acidic or basic.

The fact that Bath 5 can be operated without ammonium hydroxide should encourage further development on the efforts of Booze, and lead to an increase in the use of alkaline plating solutions.

PLATING WITH SODIUM BOROHYDRIDE

The borohydride anion hydrolyzes almost spontaneously in neutral or acidic aqueous solutions:

Neutral
$$BH_4 + 4H_2O \rightarrow B(OH)_4 + 4H_2 \tag{10}$$

Acidic
$$BH_4 + H^+ \rightarrow BH_3 + H_2 \tag{11}$$

When nickel ions are present in the solution, the homogeneous reaction represented by the following equation occurs, precipitating nickel boride:

$$8BH_4 + 4Ni^{+2} + 18H_2O \rightarrow 2Ni_2B + 6H_3BO_3 + 25H_2 \tag{12}$$

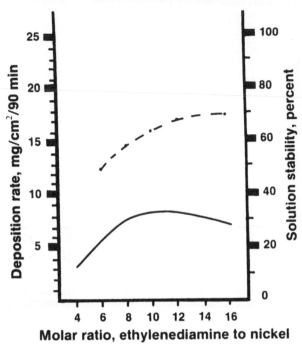

Fig. 2.17—Effect of the molar ratio of ethylenediamine to nickel on plating rate (solid line) and solution stability (dashed line). No stabilizer present (14).

The formation of the finely divided and extremely reactive Ni_2B is the basis for a method of waste-treating spent hypophosphite-reduced electroless nickel plating baths.

When the pH of an aqueous solution containing both nickel and borohydride ions is adjusted to values between 12 and 14, the formation of nickel boride is suppressed and the principal products of the reduction reaction are given by:

$$2Ni^{+2} + 2BH_4^- + 4H_2O \xrightarrow{n[OH^-]} 2Ni^0 + B + B(OH)_4 + 3H^4 + 9/2\ H_2 \qquad [13]$$

To prevent precipitation of basic nickel salts or nickel hydroxide, ligands that are effective in sequestering nickel ions within the operating pH range of 12 to 14 must be used. Ethylenediamine tetraacetic acid (EDTA), its analogues, and ethylenediamine, have been found to be the most effective complexing agents for use in these plating baths. On the basis of the results of experiments comparing several complexors, ethylenediamine is favored. The formation of completely coordinated nickel complexes with either zero or net positive charge imparts conditions that are favorable for the reduction process. Access to the nickel ions by water molecules or the borohydride ion is hindered by the

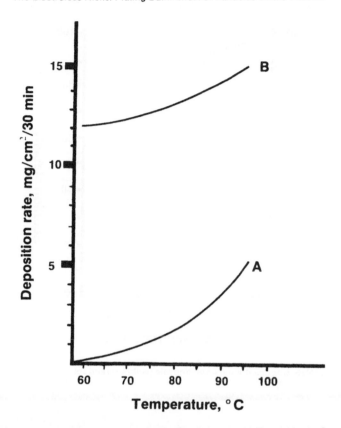

Fig. 2.18—Effect of temperature on plating rate. Curve A—no stabilizer in bath. Curve B—bath contains 100 ppm TINO₁ (14).

structure of the complex. Under these conditions, the probability is quite small for the homogeneous reaction to occur in the bulk solution. Further, the zero or net positive charge on the complex favors the absorption of the complex at the negatively charged cathodic sites on the surface of the metal substrate catalyst. Figure 2.17 shows the influence of the molar ratio of ethylenediamine to nickel on the plating rate and solution stability (no stabilizer present).

As in all catalytic reactions, the temperature is the most important factor that influences the rate of the plating reaction. Borohydride-reduced plating solutions, which are operated at temperatures between 85 and 95° C, are no exceptions. The deposition rate will increase exponentially with an increase in the temperature of the plating bath, as illustrated in Fig. 2.18.

The plating rate is also affected by the concentration of borohydride. Figure 2.19 shows the relationships between BH_4^- concentration vs. plating rate, and BH_4^- concentration vs. bath stability. It is evident that as the concentration of BH_4^-

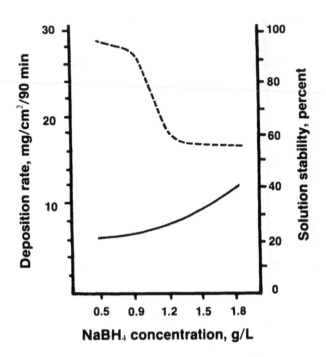

Fig. 2.19—Effect NaBH₄ concentration on plating rate (solid line) and solution stability (dashed line). No stabilizer present (14).

increases, the stability of the plating bath decreases. Lowering the operating temperature to accommodate high concentrations of borohydride ion enhances bath stability, but the lower temperature results in a sharp reduction in plating rate. It has been recommended that small quantities of the reducing media be added at very close intervals, along with additions of sodium or potassium hydroxide to maintain the concentration and activity of the borohydride ion, as well as the pH and the plating rate.

Studies on the effects of some stabilizers on the process indicates that bath stability can be sustained even at high BH₄ concentrations and high operating temperatures in the presence of certain stabilizing additives. The data obtained by Gorbunova, Ivanov and Moisseev (14) from their investigation of thallium nitrate, lead chloride, and 2-mercaptobenzothiazole (2-MBT) in these baths is very informative. The composition and operating conditions of several borohydride plating baths are given in Table 2.4. The effects of NaBH₄ concentration on plating rate and bath stability in the presence of stabilizers are given in Fig. 2.20. The results shown in Fig. 2.20 reveal a marked improvement in plating rate and bath stability when thallium is present, as compared to the data given in Fig. 2.19. The influence of the molar ratio of ethylenediamine to nickel

Table 2.4
Composition of Sodium Borohydride Electroless Nickel Plating Solutions

Concentration, g/L

Component	Bath					
	1	2	3	4	5	6
Ni^{+2}	7.4	7.4	2.5	7.6	5.0	6.0
Ethylenediamine (98%)	45	40	—	—	—	—
EDTA disodium salt	—	—	35	—	—	—
Triethylene-tetramine	—	—	—	87	—	—
Sodium potassium tartrate	—	—	—	—	65	—
Ammonium hydroxide (28%)	—	—	—	—	—	120 mL/L
Sodium hydroxide	40	40	40	40	40	—
Sodium borohydride	1.5	1.2	0.5	1.0	0.75	0.4
$TINO_2$, mg/L	100	—	50	50	—	—
$Pb(NO_3)_2$, mg/L	—	40	—	—	10	—
2-MBT, mg/L	—	5	—	—	—	20
Temp., °C	95	95	95	97	92	60
pH	14	13	14	14	13	12

on plating rate and stability in the presence of stabilizers is presented in Fig. 2.21. The data of Fig. 2.17 is included in Fig. 2.21 to show by comparison the specific influence of thallium nitrate on the deposition rate. The use of $TINO_3$ allows the process to be carried out at much lower temperatures, but with a considerably higher plating rate. The *overriding* aspect of the data in Figs. 2.19, 2.20 and 2.21 is that it draws attention to the fact that one stabilizer can so dramatically influence the process. It is not surprising that thallium codeposits into the alloy in significant quantities. Some nickel-boron alloys contain as much as 6 percent thallium.

The physical and chemical properties of the nickel-boron alloys, in particular those containing thallium, have generated considerable interest for their use in several applications. On the other hand, the baths used to produce these unique deposits present some operating limitations and disadvantages. The solution pH must be maintained in excess of 12 (preferably at 14) in order to suppress the homogeneous reaction expressed in Eq. 12. As a result of the high alkalinity of the plating bath, only substrate materials that are alkali-resistant can be plated in such solutions. Only substrates that are thermally stable at the operating temperature of borohydride-reduced baths can be coated in these plating

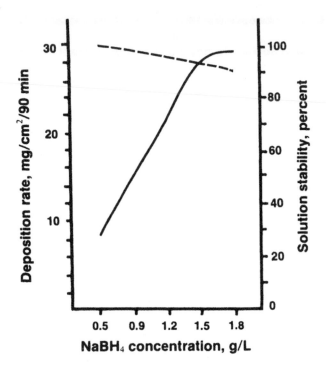

Fig. 2.20—Effect of NaBH₄ concentration on plating rate (solid line) and solution stability (dashed line). Bath contains 100 ppm TlNO₃ (14).

solutions. Finally, the use of thallium compounds is considerably more hazardous than, say, the use of lead salts and should be avoided whenever possible or whenever an alternative process is available. To circumvent the disadvantages and limitations proffered by BH₄⁻ plating baths and still retain most of the physical and chemical properties of the Ni-B coatings, the amine borane-reduced nickel baths are proposed as alternatives.

PLATING WITH AMINE BORANES

In general, the techniques of electroless nickel plating with hypophosphite as the reducing medium are applicable to plating with amine boranes (15); i.e., the substitution of water-soluble dimethylamine borane (DMAB) in a standard hypophosphite electroless nickel plating formulation will yield a workable plating bath. Typical acid and alkaline plating bath formulations are illustrated in Table 2.5.

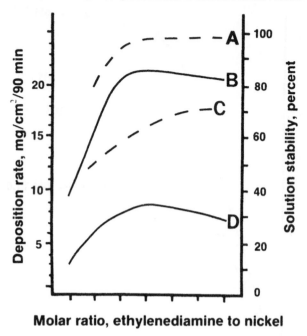

Fig. 2.21—Effect of molar ratio of ethylenediamine to nickel on plating rate (solid line) and solution stability (dashed line). Curves A and B contain 100 ppm TINO₃; curves C and D contain no stabilizer (14).

Table 2.5
Typical DMAB Electroless Nickel Plating Solutions

	Concentration, g/L			
	Bath			
Component	**1**	**2**	**3**	**4**
Ni²⁺	6.0	11.0	7.5	7.5
Lactic acid (88%)	30.0	25.0	—	—
Citric acid	—	25.0	—	—
Sodium succinate	—	—	20	—
Sodium acetate	15.0	—	—	—
Sodium glycolate	—	—	40	—
Sodium pyrophosphate	—	—	—	60
DMAB	2.5	2.5	2.5	2.5
Thiourea, mg	1	—	2	—
Pb(NO₃)₂, mg	—	2	—	—
Thiodiglycolic acid, mg	—	70	—	50
pH (with NH₄OH)	6.1	6.3	7.0	9.0
Temp., °C	60	50	65	40

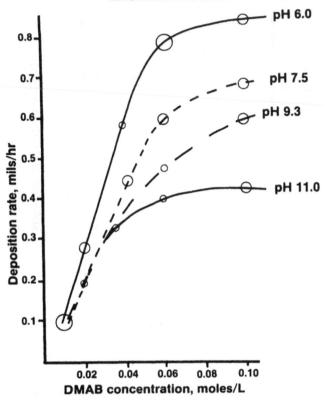

Fig. 2.22—Rate vs. DMAB concentration at various pH values. Bath operating temperature was 71° C (160° F) (15).

INFLUENCE OF VARIABLES ON PLATING RATE

Effect of DMAB Concentration

The concentration of the reducing medium is an important rate-determining factor, as shown in Fig. 2.22. The rate of deposition increases linearly with increasing DMAB concentration until the concentration is about 0.06M; above this point, increasing the DMAB concentration produces only small increases in the plating rate. The magnitude of the rate at the inflection point on the four curves in Fig. 2.22 is related to pH, since the lower the pH, the greater the magnitude of the rate at the inflection point.

Effect of Nickel Concentration

The rate of deposition is independent of nickel concentration when the nickel concentration is >0.06M (about 3.5 g/L). When the nickel concentration is less

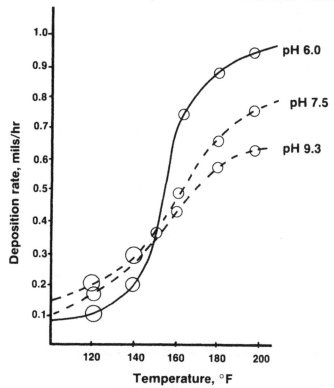

Fig. 2.23—Effect of temperature on deposition rate. DMAB concentration 0.06M (15).

than 0.06M, there is a strong dependence of rate on nickel concentration. However, plating baths are not operated at these low concentrations of Ni^{-2} ions. Detailed studies on the effect of the molar ratio of nickel ions to DMAB are not available in the literature.

Effect of Temperature

DMAB-reduced nickel plating solutions show the same exponential dependence of plating rate on temperature that is typical of all catalytic nickel reduction reactions. Rate vs. temperature data are displayed as curves in Fig. 2.23. The exponential character of the rate vs. temperature relationship is maintained until the temperature exceeds some critical value that corresponds to the first inflection points on the curves. At this critical temperature, the rate of deposition becomes less dependent on temperature. The phenomena can be explained by the temperature dependence of DMAB hydrolysis reactions:

Acidic
$$(CH_3)_2NHBH_3 + 3H_2O \xrightarrow{H^+} (CH_3)_2NH_2 + B(OH)_3 + 3H_2 \qquad [14]$$

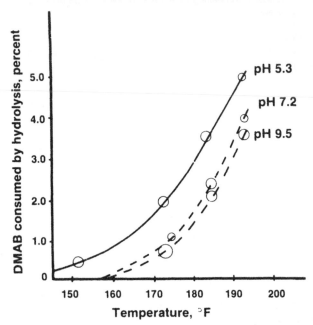

Fig. 2.24—Effect of temperature on hydrolysis of DMAB. Initial DMAB concentration 0.06M, operating time 90 minutes (15).

Neutral

$$(CH_3)_2NHBH_3 + 3H_2O \rightarrow (CH_3)_2NH_2 + B(OH)_3 + 3H_2 \qquad [15]$$

The effect of temperature on the hydrolysis of DMAB is shown in Fig. 2.24. When the temperature of the plating solution reaches a temperature that is equivalent to the first inflection point on the rate vs. temperature curve, the hydrolysis of DMAB becomes important; i.e., the hydrolysis of DMAB begins to compete with nickel reduction. At the second inflection point on each curve, hydrolysis of DMAB becomes the dominant reaction.

Effect of pH

The hydrolysis reaction expressed in Eq. 14 suggests that there is a practical lower limit for the operating pH of a DMAB plating solution. Figure 2.25 shows that above pH 5.0, the quantity of DMAB consumed by hydrolysis approaches a minimum almost asymptotically. The pH of a DMAB electroless plating bath, formulated to be operated in the acid pH range, should be maintained above pH 5.0 to minimize DMAB hydrolysis. The preferred operating pH range is the so-called "near-neutral" pH range (6.0 to 7.0).

At pH values below 5.0, the hydrolysis reaction is very temperature-dependent. Above pH 5.0, the hydrolysis of DMAB does not become a major consideration until the temperature exceeds 70° C. It has been shown experimentally that in the presence of reducible ions, such as Ni^{+2}, the hydrolysis of DMAB becomes

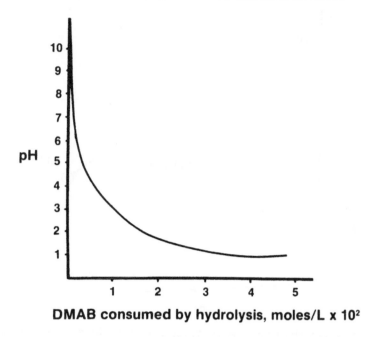

DMAB consumed by hydrolysis, moles/L x 10²

Fig. 2.25—Effect of pH on hydrolysis of DMAB. Temperature 65° C (150° F), operating time 1 hour (15).

more temperature-dependent, regardless of the pH.

The effect of pH on the plating rate in DMAB-reduced electroless nickel plating baths was found to be opposite to that observed in hypophosphite-reduced plating solutions. This result was alluded to in an earlier section. It was previously noted that the inflection point corresponding to the highest plating rate occurred on the curve (Fig. 2.22) representing data obtained from the plating bath with the lowest pH. Each successively lower plating rate (inflection point) occurred on a curve representing a successively higher pH value. In other words, the rate of deposition increases when the pH of the plating bath is decreased. The relationship of rate vs. pH is illustrated in Fig. 2.26.

There are probably several mechanisms that can be used to explain the effect pH has on plating rate in DMAB-nickel plating solutions. Two such reactions schemes are offered here:

• During the dissociation of the nickel complex to "free" solvated nickel ions and ligand molecules, a nickel-dimethylamine complex is formed as an intermediate; i.e., water molecules and dimethylamine molecules compete for nickel coordination sites that are available as the result of ligand groups leaving:

$$NiY_6 \underset{Y_n}{\overset{nDA}{\rightleftharpoons}} NiY_{6-n}(DA)_n + Y_n \underset{nDA}{\overset{nH_2O}{\rightleftharpoons}} NiY_{6-n}(H_2O)_n + nDA \qquad [16]$$

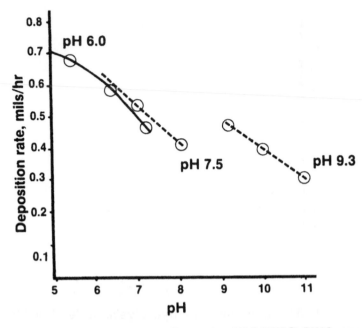

Fig. 2.26—Influence of pH on deposition rate. Temperature 71° C (160° F), DMAB concentration 0.06M (15).

where Y_6 represents the number of ligand atoms required to completely coordinate nickel, and DA denotes the dimethylamine molecule.

• After the dissociation of the nickel complex, dimethylamine (DA) molecules compete with hydroxyl ions (from dissociated water molecules) for the "free" solvated nickel ions:

$$Ni(H_2O)_6^{+2} \underset{-2DA}{\overset{+2DA}{\rightleftharpoons}} \left[Ni_{aq} \begin{matrix} DA \\ DA \end{matrix} \right]^{+2} + 2H_2O \qquad [17a]$$

$$Ni(H_2O)_6^{+2} \underset{-2OH^-}{\overset{+2OH^-}{\rightleftharpoons}} \left[Ni_{aq} \begin{matrix} OH \\ OH \end{matrix} \right]^{+2} + 2H_2O \qquad [17b]$$

Either mechanism will affect the kinetics of the reduction reaction, and because intermediates are formed, the overall reaction rate will decrease. In a manner analogous to the formation of the blue nickel-ammonium complex, as the pH of the solution is raised, the formation of the nickel-dimethylamine complex increases.

The practical operating limitations recommended for pH and temperature are unequivocal in their intent; at the present cost of DMAB, the consumption of the reducing media through nonproductive side-reactions can be hazardous to the bottom line.

Effects of Stabilizers

The compositions of Ni-B alloys produced with DMAB are 95 to 99.8 percent by weight nickel, the remainder is essentially boron. If these deposits are formed from solutions that do not contain stress reducers, their internal stress is extremely high and tensile in nature. Divalent sulfur compounds, such as thiourea and thiodiglycolic acid, are among the most effective stress reducing materials available. Not only do the organic divalent sulfur compounds reduce stress, they also act as stabilizers, preventing the homogeneous decomposition reaction from occurring. However, a sometimes-unwanted side effect occurs as a result of the use of the sulfur-containing materials: the boron content of the deposit is dramatically reduced. When the plating solution contains a divalent sulfur compound, its deposits usually contain less than 1 percent by weight boron. The high boron deposits ($>.3$ percent) produced with DMAB are, in general, so high in tensile stress that they cannot be used in the same applications as the borohydride-generated deposits. However, Mallory and Hohrn (18) were recently issued a patent detailing the deposition with DMAB of Ni-B alloys with low stress and whose boron content ranges from 3 to 8 percent by weight.

Effect of Reaction Products

Nickel plating with DMAB is accompanied by the formation of hydrogen ions, the evolution of hydrogen gas, and the formation of boric acid and dimethylamine. Boric acid and its salts, along with the amine, act as buffering agents in the plating bath and tend to decrease the effect of the H^+ ions. In certain plating baths, the pH actually increased slightly after using the baths for about one hour.

Electroless nickel plating baths utilizing DMAB are usually operated for a number of metal turnovers that exceeds, by a wide margin, the number of turnovers experienced with hypophosphite baths. One group of investigators (15) claim to have operated a DMAB-reduced plating bath for over 65 metal turnovers. One of the factors contributing to the exceptionally long operating life of these plating baths is that the reaction by-products are soluble, at least up to 65 metal turnovers.

Based on experimental data, it has been suggested that dimethylamine and boric acid form a polar, complexed amine borate. This proposition is a plausible explanation for two phenomena peculiar to these baths:

• The greatly enhanced solubility of boric acid in these aqueous plating solutions.

• The accumulation of reaction by-products does not cause a decrease in plating rate.

KINETICS OF ELECTROLESS NICKEL DEPOSITION

Empirical Rate Laws

The aim of chemical kinetics is to predict the rates of chemical reactions and to delineate the path or paths by which reactants proceed to products, i.e., the reaction mechanisms. Usually, reaction rates are determined under conditions of constant temperature, preferably at several temperatures. In a typical kinetic experiment, the data are records of concentrations of reactant and product species taken at various times and at constant temperatures.

Theoretical expressions for reaction rates as functions of concentrations of reactants and products are differential equations of the form:

$$\text{Rate} = \pm \frac{dC_i}{dt} = f(C_A, C_B, \ldots C_N) \tag{18}$$

where t denotes the time, and C_i is the time-dependent concentration of the ith species (+ for product, - for reactant) that is followed to determine the reaction rate.

Generally, the concentrations of the reactants are known at the moment of mixing. The progress of the reaction is then studied as a function of time by any of a number of available techniques. Similar experiments may then be performed by using different initial concentrations of the reactants at different temperatures.

The rate of a reaction may be determined by measuring the rate of decrease in the concentration of a particular reactant, or, conversely, the rate of increase of by-products. In the case of electroless nickel plating, the rate of deposition is most commonly determined by measuring the rate of weight gain of the deposit in $mg/cm^2/hr$, or the rate of deposit thickness increase in $\mu m/hr$.

The next step is to express the resultant data from the rate measurements in some simple mathematical form; i.e., an expression (represented by the function f in Eq. 18) is needed for reactants and products. Usually, the relationship is expressed as:

$$\text{Rate} = \frac{dC_A}{dt} = k' C_A{}^\alpha C_B{}^\beta C_C{}^\gamma \ldots \tag{19}$$

where K' is a numerical proportionality constant called the *rate constant*; C_A, C_B, C_C, etc. refer to the concentrations of the chemical species A, B, C and so forth (reactants and products) present in the reacting system at time t; and α, β, γ, etc. are called the *reaction orders* of the respective chemical species.

In terms of this empirical rate equation, the particular reaction is said to α order with respect to A, β order with respect to B, etc.; and $\alpha + \beta + \gamma \ldots$ is the overall reaction order. The individual time derivatives dC_A/dt, dC_B/dt, dC_C/dt are referred to as the rate of the reaction with respect to species A, B, and C, respectively.

The dominant variable in most reactions is the temperature. Many reactions occurring close to room temperature double their rates for each 10° C rise in temperature. Note that 100° C is considered close to room temperature. The expression for the rate constant K' in Eq. 19 was given in Chapter 1 in the form:

$$K' = K_1 \exp \left(-\frac{E_a}{RT} \right) \qquad [20]$$

where K_1 is the frequency factor, E_a is the activation energy, T absolute temperature ° K, and R the universal gas constant. The empirical rate law, Eq. 19, can now be written as:

$$\text{Rate} = \frac{dC_A}{dt} = K_1[C_A]^\alpha [C_B]^\beta \ldots [C_N]^\gamma \exp \left(-\frac{E_a}{RT} \right) \qquad [21]$$

Empirical Rate Equation for Ni-P Deposition

An empirical rate equation for Ni-P plating, based on those variables that affect the plating rate, can be derived from Eqs. 1 and 21, and written as follows:

$$\text{Rate} = \frac{dNi^0}{dt} = K_1[H_2PO_2^-]^\alpha [Ni^{+2}]^\beta [L]^\delta [H^+]^\gamma [H_2PO_3^-]^\epsilon \exp \left(-\frac{E_a}{RT} \right) \qquad [22]$$

Putting Eq. 22 into logarithmic form yields:

$$\log \text{rate} = K_2 + \alpha \log[H_2PO_2^-] + \beta \log[Ni^{+2}] + \delta \log[L^{-n}] - \gamma pH + \epsilon \log[H_2PO_3^-] - \frac{E_a}{2.3\,RT} \qquad [23]$$

where $K_2 = \log K_1$, and $\log (H^+) = -pH$.

From Eq. 23, the reaction orders α, β, δ, γ, ϵ, and the activation energy E_a can be evaluated:

$$\left(\frac{\partial \log \text{rate}}{\partial \log[H_2PO_2^-]} \right)_q = \alpha \qquad [24a]$$

$$\left(\frac{\partial \log \text{ rate}}{\partial \log [\text{Ni}^{+2}]} \right)_{u} = \beta \qquad [24b]$$

$$\left(\frac{\partial \log \text{ rate}}{\partial \log [\text{L}^{-n}]} \right)_{w} = \delta \qquad [24c]$$

$$\left(\frac{\partial \log \text{ rate}}{\partial \text{pH}} \right)_{x} = -\gamma \qquad [24d]$$

$$\left(\frac{\partial \log \text{ rate}}{\partial \log [\text{H}_2\text{PO}_3]} \right)_{y} = \epsilon \qquad [24e]$$

$$\left(\frac{\partial \log \text{ rate}}{\partial (1/t)} \right)_{z} = -\frac{E_a}{2.3R} \qquad [24f]$$

The subscripts q, u, w, x, y, and z denote each set of variables held constant for the particular partial derivative.

The values of certain of the reaction orders can be ascertained, though not rigorously, from the discussions of the effect of the respective variables on the plating rate. For example, it was observed that the plating rate is independent of the nickel concentration in the concentration range where most hypophosphite-reduced baths are operated. Hence, in Eq. 24b $\beta = 0$, so that the term $[\text{Ni}^{+2}]$ in Eq. 22 reduces to $[\text{Ni}^{+2}] = 1$. Gutzeit (1) arrived at the same conclusion in a more elegant and rigorous treatment.

It was noted earlier that the buildup of phosphite in electroless nickel baths causes changes in the intrinsic stress of the Ni-P deposits, even at moderate concentrations. As the phosphite continues to accumulate in the bath, the relatively insoluble nickel phosphite will commence to precipitate. However, while these phosphite-induced phenomena are occurring, the soluble phosphite has little or no effect on the plating rate. In other words, the partial derivative, Eq. 24e, vanishes, i.e., $\epsilon = 0$ and $[\text{H}_2\text{PO}_3] = 1$. The remaining reaction orders, α, β, and γ can be determined by plotting log rate vs. log (concentration of species). The slopes of the straight lines obtained in these plots are the reaction orders.

Equation 22 can be simplified to:

$$\text{Rate} = K_1 [\text{L}^{-n}]^{\delta} \frac{[\text{H}_2\text{PO}_2]^{\alpha}}{[\text{H}^+]^{\gamma}} \exp \left(-\frac{E_a}{RT} \right) \qquad [25]$$

Gutzeit (1) and later Lee (8) (see Eq. 3) have shown that the rate of nickel reduction is first order with respect to the hypophosphite ion concentration; i.e.,

$\alpha = 1$. Mallory and Lloyd (16) have evaluated γ for several electroless nickel solutions containing different complexing agents and found its value, in all but one case, to be $\gamma = -0.4$.

The activation energy, E_a, is computed from the slope of the straight lines obtained in log rate vs. $1/T$ plots, i.e.:

$$\left(\frac{\partial \log \text{rate}}{\partial 1/T}\right) = -\frac{E_a}{2.3R} = -\text{slope}$$

$$E_a = 2.3(1.98 \text{ cal deg}^{-1} \text{ mole}^{-1})(\text{slope})$$

Table 2.6 gives the activation energies for several electroless nickel baths with different complexing agents.

In order to simplify computations, Eq. 25 can be put in the following form:

$$\text{Rate} = K_1 [L^{-n}]^\delta \frac{[H_2PO_2^-]}{[H^+]^{0.4}} \exp\left(-\frac{E_a}{RT}\right) \times \left[\exp\left(\frac{E_a}{360R}\right) \exp\left(-\frac{E_a}{360R}\right)\right]$$

since,

$$\left[\exp\left(\frac{E_a}{360R}\right) \exp\left(-\frac{E_a}{360R}\right)\right] = 1$$

so that after rearranging:

$$\text{Rate} = K_2 \frac{[H_2PO_2^-]}{[H^+]^{0.4}} \exp E \left(\frac{T-360}{T}\right) \qquad [26]$$

where:

$$K_2 = K_1 [L^{-n}]^\delta \exp\left(-\frac{E_a}{360R}\right)$$

and:

$$E = \frac{E_a}{360R}$$

The rate constant K_2 can be computed directly from Eq. 26. The straight line plots for the reaction orders are extrapolated to the point where the concentrations of the appropriate components are unity. At this point, the respective deposition rates can be read from the ordinate. The following identities are derived from Eq. 26, when T = 87° C (360° K):

$$\text{Rate} = K_2 \, \frac{1}{[H^+]^{0.4}} \quad [H_2PO_2] = 1 \tag{27}$$

$$\text{Rate} = K_2 \, [H_2PO_2] \, [H^1] = 1 \tag{28}$$

The rate constant K_2 is computed by solving Eqs. 27 and 28.

The specific rate constant, K_1, can be determined, if necessary, by solving the expression:

$$K_1 = \frac{K_2}{[L^{-n}]^\delta} \, \exp\left(\frac{E_a}{360R}\right)$$

It has been shown (17) that the agreement between observed plating rates and those computed using Eq. 26 are in excellent agreement. The deviation from perfect correlation was found to be less than 5 percent.

Using arguments similar to those above, an empirical rate law for DMAB-reduced nickel baths can be derived. Since the development will not be given here, the interested reader is referred to Ref. 18.

Table 2.6
Activation Energies for Ni-P Baths
Containing Different Ligands

Ligand	Chelate ring size	Coordinating atoms	Activation energy, Kcal/g-mde
Hydroxyacetate	5	(O,O)	23
Lactate	5	(O,O)	16
Aminoacetate	5	(O,N)	17
β-alanine	6	(O,N)	13
Malate	5,6	(O,O,O)	17
Citrate	5,6	(O,O,O)	16

REFERENCES

1. G. Gutzeit, *Plating,* **46,** 1275, 1377 (1959).
2. A. Brenner and G. Riddell, *J. Res. Nat. Bur. Std.,* **37,** 31 (1946).
3. G. Gutzeit and A. Kreig, U.S. patent 2,658,841 (1953).
4. C. Baldwin and T. Such, *Trans. Inst. Met. Fin.,* **46,** 73 (1968).
5. W.G. Lee, *2nd Int. Congress on Met. Corrosion,* NACE, Houston, TX (Mar. 1963).
6. G. Mallory and D. Altura, *S.A.E. Technical Paper 830693* (1983).
7. K.M. Gorbunova and A.H. Nikiforova, "Physicochemical Principles of Nickel Plating," *Acad. Sci. USSR* (1960).
8. K. Parker, *Plat. and Surf. Fin.,* **67**(3), 48 (1980).
9. C. de Minjer and A. Brenner, *Plating,* **44,** 63 (1960).
10. J. Haydu and E. Yarkosky, *Proc. AESF 68th Ann. Tech. Conf.,* paper 62 (1981).
11. M. Schwartz, *Proc. AES,* **47,** 176 (1960).
12. C.A. Booze, *Trans. Inst. Met. Fin.,* **53,** 49 (1975).
13. H. Iwasa, M. Yokozawa and I. Teramoto, *J. Electrochem. Soc.,* **115**(5) (1968).
14. K. Gorbunova, M. Ivanov and V. Moiseev, ibid., **120,** 613 (1973).
15. G. Mallory, *Plating,* **58**(4) (1971).
16. W. Cooper, J. Renforth, R. Vargo and A. Weber, "Electroless Plating with Amine Boranes," Tech. Bulletin CCC-AB1, Callery Chemical Co., Calery, PA (Feb. 1968).
17. G. Mallory and V. Lloyd, *Plat. and Surf. Fin.,* **72**(9), 65 (1985).
18. G. Mallory and V. Lloyd, ibid., **72**(8), 65 (1985).

Chapter 3
Troubleshooting Electroless Nickel Plating Solutions

Michael J. Aleksinas

The chemical reduction potential of the electroless nickel plating reaction can be affected by many factors; hence, troubleshooting these solutions can be difficult and time-consuming. A working knowledge of the chemistry of the electroless process, coupled with the selection of the proper pretreatment techniques, is necessary to insure reliable, consistent results.

In order quickly and effectively to solve the technical problems that may arise, it is first necessary to outline the typical problems that can arise. These problems can be categorized into four basic groups:

- Bath chemistry imbalance
- Improper substrate preparation/activation
- Equipment/mechanical problems
- Solution contamination

By taking a logical approach to these four categories, troubleshooting can be accomplished effectively and efficiently.

BATH CHEMISTRY

Components of the electroless bath include an aqueous solution of nickel ions, reducing agent(s) (NaH_2PO_2, DMAB, etc.), complexing agent(s), buffer(s), and stabilizer(s). These components work in concert with each other and operate in specific concentration, temperature and pH ranges. Optimizing these parameters can be different from solution to solution. Maintaining the optimum metal content, reducer concentration, pH and temperature of the solution can minimize many of the technical problems that can arise. Analytical techniques to determine these parameters can be obtained from text books or from suppliers.

One of the first and easiest bath parameters to be checked is pH. The pH is determined by pH meter or pH paper. Caution should be used here because some pH papers may differ from the electrometric readings by as much as 0.5 pH units. In some electroless processes, pH papers can give erroneous values because of the "salt-ion" effect of the solution. As an electroless bath ages, there is a buildup of ions such as NH_4^+, Na^+, K^+, $SO_4^=$, Cl^-, etc., and all can contribute to the change in ionic strength of the solution. Consequently, pH paper may act differently in new solutions than in old solutions. Electrometric pH readings are preferred.

Calibration of the pH meter is mandatory for reliable information. Electrometric pH meters require standardization between two units of the pH scale (e.g., 4 and 7). One buffer solution should be above the expected pH of the sample to be measured, while the second buffer should be below. Once the meter is standardized to both of these standards, the solution pH can be measured. A common error in reading pH accurately is temperature variations between the bath sample and the two buffers. The proper procedure is to have identical temperatures for both the buffers and the sample. Thus, a hot bath sample should be cooled to room temperature or the buffer solutions should be heated to the same temperature as the sample. A high pH value of the plating solution can cause abnormally high plating rates that may lead to roughness, pitting, and/or cloudy deposits. Too low of a pH will cause slow deposition rates and matte/dull finishes.

The temperature of the solution should also be closely monitored for consistent high quality deposits. Accurate temperature controllers are necessary. Checking temperature controllers with certified thermometers provides further assurance that temperatures are maintained at proper levels. Unchecked temperatures can lead to decomposition of the plating bath (temperature too high) or no plating at all (temperature too low).

Finally, solution imbalance as a result of poor maintenance of the nickel and/or reducing agent concentrations can give slow plating rates, poor coverage, and dull deposits. Maintaining the nickel and reducing agent concentrations within 10 to 15 percent will usually provide consistency in terms of rate, color, and stability. Periodic analysis of these constituents is required, with the frequency of analysis proportional to the workload being processed. Excessively large replenishments can lead to overstabilization or suppressed rates, as many of the components used for replenishment often contain high concentrations of stabilizers and/or brighteners, which may act as catalytic poisons.

In "home-brew" situations, where commodities such as pH adjusters, metal salts, and reducing agents are added to the plating solutions, only high-purity chemicals from an approved source should be used. Low cost chemicals have often been found to contain impurities that may lead to pitting, dullness, or suppressed plating rates. Insoluble materials can also be present, which will cause roughness or act as nuclei for the spontaneous decomposition of the bath.

Keeping records of replenishment additions, pH adjustments, and temperature readings provides valuable information for maintaining the chemistry balance of the solution. Automatic controllers have proven useful if they themselves are closely monitored. The use of controllers avoids large additions to the plating bath and eliminates the possibility of bath over-stabilization. When bath parameters are found to fall within the given ranges and problems still exist, then such circumstances as inadequate surface preparation and/or contamination of the bath should be examined.

SUBSTRATE ACTIVATION

Proper preparation of the substrate to be plated is vital for quality results. Poor surface preparation can cause lack of adhesion, deposit porosity, roughness, non-uniform coatings and/or dark deposits.

A properly prepared substrate is one whereby surface contamination is removed, which leaves a clean, nominally oxide-free surface. Typical surface contaminants that must be removed (or replaced) prior to plating usually include one or more of the following:

- Oils, lubricants
- Buffing compounds
- Oxide films (replaced on Al)
- Weld scale
- Fluxes

Depending on the type of soils present, different pretreatments are needed. Pretreatment choice should be the best available for the specific substrate and should be closely monitored. Cleaners and pickling solutions should be changed at predetermined intervals to eliminate the possibility of ineffective cleaners and descalers, which will cause poor adhesion, streaky deposits, or blistering.

The quality of the substrate itself also must be checked carefully as a potential problem source. Often plating problems resulting from inferior substrates are wrongfully diagnosed as pretreatment or bath chemistry problems; for example, intermetallic compounds at aluminum substrate surfaces can manifest themselves as nodules or pits in the final plate. In other cases, porous castings or powdered metal substrates can entrap solutions, which result in bleedout or voids.

The way in which a part is stamped, cast, drilled, or machined can have a great impact on the final plated product. Improperly stamped parts can imbed difficult-to-remove oils or compounds into the surface of the part. This will lead to dull and often non-adherent coatings. Improper temperature control while casting aluminum or zinc die cast parts can cause lamination of the substrate, which will lead to uneven appearance in the end plate. Obviously then, much attention must be focused upon the raw part itself. By noticing imperfections early, problems can be avoided later in the plating process itself. Pretreatment of various substrates is covered extensively in another section of this book.

EQUIPMENT/MECHANICAL NEEDS

When bath parameters appear to be satisfactory and pretreatment processes are in order, the next area of concern is the type of equipment and the mechanical techniques that are being used. Electroless nickel solutions should have constant filtration to eliminate any particulate bath impurities such as dust, sand, or loose maskant material. With proper filtration ($\geqslant$10 turnovers/hr at 5 μm or less), roughness of the final deposit can be greatly reduced or even eliminated. Replacement of these filters on a regular basis is recommended so that retention of pore size of these filters is maintained. Since electroless plating is an autocatalytic process, it is vital to remove all foreign particulates so that plate-out will be virtually eliminated. Plate-out, if allowed to occur, can lead to roughness of the deposit, bath decomposition, and excessive plating costs.

The type of heating that is used in the plating process is also of great concern. Electroless nickel plating tanks can be heated internally or externally. Excessive localized overheating can cause plate-out, roughness, or even bath decomposition.

Lack of agitation of the plating solution can also cause problems. Solution stratification can occur, resulting in gas pitting, patterns, and/or streaking of the deposit. Proper agitation allows uniform distribution of plating chemicals and helps to eliminate localized overheating. Work-rod agitation or clean, filtered air is suitable for most electroless nickel solutions. Agitation of the solution or parts is necessary to provide a fresh supply of solution to the parts, and to enhance the removal of hydrogen gas produced during deposition.

In most cases, high-temperature, stress-relieved polypropylene is the material of tank construction. Etched tanks can become more active towards electroless processes and should be replaced to minimize plate-out on the tank walls. Stainless steel may also be used as an alternative for tank construction. A small anodic charge will minimize plate-out on these tank walls. Caution must be used when plating in stainless steel tanks (even anodically charged tanks), because the walls may become catalyzed and plate, accompanied by an excessive usage of metal. Plastic drop-in liners are convenient to use, provided all sizing materials are leached out prior to use. This is usually accomplished by soaking the liners in hot deionized water or dilute sulfuric acid and then neutralizing them with acid or alkali. Retained sizing materials in the liner can cause pitting and roughness, besides creating a foam on the surface of the solution.

CONTAMINATION OF THE SOLUTION

When all other parameters appear to be in order, extraneous solution contamination can be a reality. The electroless nickel plating reaction is affected by many impurities. Trace impurities can be organic in nature, such as oils or solvents, or inorganic, such as silicates or nitrates. Metallic contaminants such as lead, copper, cadmium, bismuth, etc., can cause severe problems if they are introduced into the bath in excessive quantities. Most metallic ions will plate out and have little effect on the electroless nickel process if they are introduced into the plating solution in small quantities.

The organic contaminants can come from degreasing solvents, oil residues, mold releases, drag-in of cleaners or acid inhibitors, and unleached equipment or filters. These contaminants will manifest themselves in cloudy, streaked deposits, along with poor adhesion.

Inorganic ions, such as nitrates, can be introduced from improperly neutralizing tanks after stripping with nitric acid. High levels of nitrates can reduce the plating rate or even stop deposition entirely. Silicates are equally detrimental. Drag-in of these ions, usually from preplate cleaners, can form gelatinous films on the work, which is manifested by cloudy deposits or pitting. Improved rinsing will decrease this occurrence.

Metallic impurities can have a profound effect on electroless nickel plating and can be introduced by drag-in from previous tanks, dissolution of base metal of the substrate (i.e., leaded alloys), poorly cleaned or exposed plating racks, or the water itself. Heavy metals that are of special concern include, but are not limited to, lead, cadmium, copper, bismuth, arsenic, and palladium.

Lead can be built up in an electroless bath by dissolution of leaded alloy substrates and by the improper use of lead as a masking material. Lead concentrations of greater than 5 ppm can cause dark deposits, skip plating, pitting, short bath life, and even cessation of plating.

Copper contamination of electroless nickel baths can be equally detrimental. Copper concentrations of ⩾100 ppm will cause immersion-deposit on ferrous alloy parts, which in turn causes adhesion problems of the electroless nickel plate. Poor pretreatment, which leads to poor initiation on copper, may allow excessive amounts of copper to dissolve in electroless nickel baths. Checking preplate cycles, where acid pickles leave an immersion copper deposit on ferrous substrates, may also be the source of copper ions.

Cadmium is usually introduced by the use of plating racks that had been previously used in cadmium plating. Contaminated cleaners may also contain cadmium from stripping operations. Cadmium and lead can also build up in the bath when either or both are used as brightener and stabilizer, respectively. Cadmium concentrations of >3 ppm can cause dark deposits, feathering around the holes, and skip plating.

Calcium and magnesium are introduced from the water supply system. If allowed to build up through evaporation, these contaminants will cause rough, hazy, pitted deposits. Precipitation of insoluble compounds may even lead to spontaneous decomposition of the plating solutions. The proper corrective action is to check the deionized water system, or if using tap water, to change to deionized water.

Palladium is used in the activation of non-catalytic substrates. If not rinsed properly, palladium ions or particles may be introduced into the electroless bath, which form nuclei and cause spontaneous decomposition of the plating bath.

Most of the kinds of contamination described above can be reduced by carbon treatment (for organic contaminants) or dummy plating (for heavy metals). The effectiveness of these techniques depends on the type and quantity of the contaminant and the age of the plating solution. Carbon treatment can leave residual carbon, which could lead to considerable roughness or dullness. Some contamination can even be leached from the carbon itself. Carbon treatment can lead to destabilization of some electroless nickel baths because of its ability to remove the stabilizers and brighteners necessary to the bath chemistry itself. Excessive dummying of the solution may lead to severe reduction of other bath constituents, such as nickel, hypophosphite, and stabilizers. Maintaining optimum ratios of these constituents is imperative to successful operation.

CONCLUSION

In addition to the aforementioned observations, a list of the most common problems found in electroless nickel plating is included at the end of this section. Sources of these problems and their remedies are also listed.

When electroless nickel-coated parts are rejected because of roughness, lack of adhesion, poor coverage or lack of uniformity, stripping and replating is feasible. The stripping should be done with as much care and planning as required in the original plating process.

To choose wisely from the many strippers available, it is important to take into consideration the nature of the substrate the deposit is to be stripped from, time, thickness of the final plate, and overall cost. Stripping from steel usually can be accomplished with nitric acid solutions, cyanide/nitroaromatic/caustic solutions or amine-based strippers. In situations where concentrated nitric acid solutions are used, parts to be stripped should be dry prior to stripping, and rinsed immediately after stripping. Close attention to this process is required to minimize water drag-in. Severe etching can occur if the stripping solution does become diluted with water. Copper and its alloys can also be stripped with inhibited strippers, which tend to contain some type of sulfur inhibitor. These strippers may also contain amines or other complexors along with oxidizing nitroaromatic compounds. Aluminum alloys are easily stripped in concentrated nitric acid solutions. Minimum immersion times are preferred here to minimize any possible attack of the aluminum surface itself.

What is evident throughout this chapter is that troubleshooting is not an easy task. First, classifying the problem into one of the four categories, then identifying the cause of the problem are the two most important steps to a viable troubleshooting strategy. When the problem is identified, the economics of remedying the solution should be addressed. Depending on the nature of the problem, the difficulty or cost to remedy the solution, and the downtime required to cure the problem will determine whether the bath should be discarded or if it should be "troubleshooted".

Troubleshooting Guide
Electroless Nickel

Problem	Probable Causes	Suggested Remedy
Poor adhesion and/or blistering	1) Improper surface conditioning	1) Improve cleaning and pickling cycle
	2) Poor rinsing	2) Improve rinse and transfer time
	3) On aluminum, poor zincating	3) Analyze and correct zincating solution
	4) Metallic contamination	4) Dilute or dummy plate solution
	5) Organic contamination	5) Carbon treat solution
	6) Reoxidation	6) Reduce transfer times
	7) Improper heat treatment	7) Correct time and/or temperature of heat treatment
Roughness	1) Suspended solids	1) Filter solution and locate source of solids
	2) Improper cleaning	2) Improve cleaning and rinsing
	3) Too high pH	3) Lower pH
	4) Drag-in of solids	4) Improve rinsing, clean rinse tanks
	5) Contaminated liner or filter cartridges	5) Leach tanks and filters prior to use
	6) Inadequate nickel chelation	6) Reduce drag-out, check replenishment cycles
	7) Contaminated water supply	7) Use deionized or distilled water
Pitting	1) Suspended solids	1) Improve filtration
	2) Excess loading	2) Reduce workload, lower pH
	3) Organic contamination	3) Carbon treat solution
	4) Metallic contamination	4) Dummy plate
	5) Poor agitation	5) Improve agitation, work rod preferred
Dullness	1) Too low temperature	1) Raise temperature
	2) Too low pH	2) Raise pH
	3) Low nickel or hypo concentration	3) Check and correct
	4) Metallic contamination	4) Dummy plate
	5) Organic contamination	5) Carbon treat
	6) Aged bath	6) Replace with new bath
Patterns/streaking	1) Poor agitation	1) Improve agitation
	2) Poor surface preparation	2) Improve and correct cleaning cycle
	3) Metallic contamination	3) Dummy plate

	4) Surface residue	4) Improve rinsing, minimize silicate drag-in
	5) Gas patterns	5) Reposition work, increase agitation
Step/skip plating	1) Metallic contamination	1) Dummy plate
	2) Substrate effect (i.e., leaded alloy	2) Copper or nickel strike prior to plating
	3) Bath overstabilized	3) Dummy plate or dilute bath
Low deposition rate	1) Low temperature	1) Increase operating bath temperature
	2) Low pH	2) Raise pH
	3) Low nickel or hypo content	3) Analyze and correct
	4) High orthophosphite content	4) Discard all or part of bath
	5) Too small a workload	5) Increase workload or reduce agitation
	6) Overstabilization	6) Dilute or dummy bath
Instability	1) Bath temperature too high	1) Lower temperature
	2) Too high pH	2) Lower pH
	3) Localized overheating	3) Locate and correct
	4) Improper passivation of tank	4) Improve passivation solutions and times
	5) Airborne contamination	5) Clean area of dust and loose dirt
	6) Drag-in of catalytic metals	6) Improve rinsing
	7) Large additions made of replenishers	7) Use more frequent additions to maintain consistent stabilizer concentration
No deposition	1) Overstabilization	1) Dilute bath, avoid large additions of replenishers
	2) Improper substrate surface	2) Substrate may not be autocatalytic and require a nickel or copper strike, i.e., stainless steel, copper
	3) Too low temperature or pH	3) Analyze and correct
	4) Metallic contamination	4) Electrolytically dummy solution
Dark deposits	1) Contaminated rinse after EN	1) Improve rinsing
	2) Improper surface preparation	2) Improve pretreatment
	3) Too low pH and/or temperature	3) Check and correct
	4) Low bath activity	4) Analyze bath constituents and correct
	5) Organic contamination	5) Carbon treat
Rapid pH changes	1) Drag-in of pretreatment	1) Improve rinsing
	2) Excessive workload	2) Reduce workload and check plate-out on tank and heaters

	3) Bath not at proper pH range	3) Check pH and adjust to optimum buffered range
High nickel usage	1) High drag-out	1) Reduce drag-out with replacement of stagnant rinse after plating tank
	2) Bath decomposition	2) Cool and filter solution
	3) Plating on tank and equipment	3) Filter solution and strip tank and heaters
	4) High surface area	4) Reduce workload size
Cloudy plating solution	1) pH too high	1) Lower pH with dilute sulfuric acid or hydroxyacetic acid
	2) Drag-out losses excessive	2) Reduce sources of drag-out losses. Add more make-up additive
	3) Under-complexed solution	3) Add more make-up additive

Chapter 4
The Properties of Electroless Nickel

Rolf Weil and Konrad Parker

Electroless nickel deposits are used by many industries as functional coatings because they possess a unique combination of both corrosion and wear resistance. The ability to coat parts, regardless of their size, with a deposit of uniform thickness is another desirable property of electroless nickel. As is the case for electrodeposits, most of the properties of electroless nickel are structure-dependent. However, as its structure and composition are quite different from those of electrodeposited nickel, its properties differ also.

The structures and properties of two groups of electroless nickel (EN) deposits will be considered in this chapter. In the first group are nickel-phosphorus alloys, which are deposited from solutions in which sodium hypophosphite is the reducing agent. These deposits typically have phosphorus contents of 7 to 11 percent (all compositions in this chapter are in weight percent). Nickel-boron alloys constitute the second group. They are obtained from solutions in which either an organic aminoborane or sodium borohydride is the reducing agent. Typical boron contents of such deposits are 0.2 to 4 percent or 4 to 7 percent, respectively. The deposit plated in sodium borohydride solutions usually also contain 4 to 5 percent thallium. EN deposits containing particles of materials such as silicon carbide, diamonds, alumina, and PTFE are discussed elsewhere in this book.

STRUCTURE OF ELECTROLESS NICKEL

As-deposited electroless nickel is a metastable, supersaturated alloy. The equilibrium phase diagrams (1) of the Ni-P and Ni-B systems shown in Figs. 4.1 and 4.2 exhibit essentially no solid solubility of phosphorus or boron in nickel at ambient temperatures. Under equilibrium conditions, therefore, the alloys consist of essentially pure nickel and the intermetallic compound Ni_3P or Ni_3B. The conditions existing during plating, however, do not permit the formation of the intermetallic compounds. Even the growth of very tiny crystals would involve the movement of large numbers of atoms by surface diffusion in order to achieve the correct stoichiometry of three nickel atoms to one phosphorus or boron atom. This movement cannot occur before the next layer of atoms has been laid down; thus, the phosphorus or boron atoms are trapped between nickel atoms, resulting in supersaturation. There is experimental evidence (2,3) that

ELECTROLESS PLATING

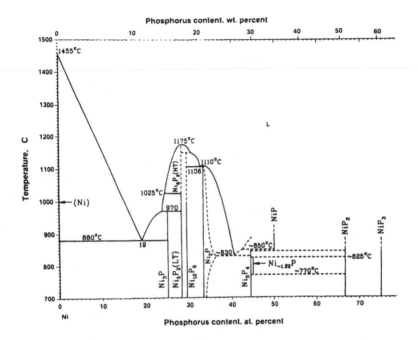

Fig. 4.1—Nickel-phosphorus phase diagram (1).

phosphorus atoms so trapped are not uniformly distributed. There are very small regions where one-third of the atoms are phosphorus, and others where there is essentially pure nickel.

The crystal structure of nickel is face-centered-cubic (fcc), in which each atom has 12 near neighbors. The entrapment of phosphorus or boron makes it impossible for this atom arrangement to extend over large surfaces. The volume of material within which the fcc atom arrangement can be maintained is called a *grain*. Grain sizes are very small in electroless nickel. If the fcc structure cannot be maintained at all, the structure is equivalent to that of a liquid and is considered to be *amorphous*.

Low-alloy electroless nickel deposits are *microcrystalline*, that is, they consist of many very small grains. Values of about 2 and 6 nm (1 nm = 10 angstroms), as determined from the broadening of diffraction peaks, have been reported (4) for Ni-B and Ni-P, respectively. Other investigators (5), have also reported a grain size of about 2 nm for Ni-P. Phosphorus contents of <7 percent result in microcrystalline deposits (6). Deposits having higher alloy contents can be considered amorphous, with their diffraction patterns consisting of only one broad peak or ring. An electron diffraction pattern showing one such broad ring is illustrated in Fig. 4.3.

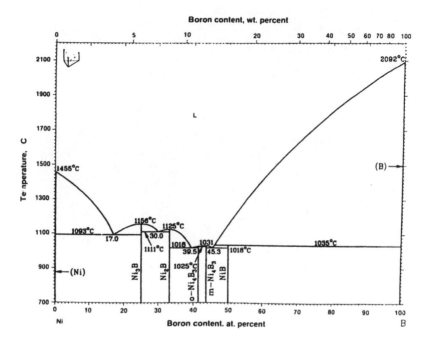

Fig. 4.2—Nickel-boron phase diagram (1).

Careful analysis (7) of X-ray diffraction patterns of electrodeposits containing 12 to 13 percent P has shown, however, that even high-alloy deposits may contain small regions that are microcrystalline. Nevertheless, since the diffraction patterns of the high-alloy electroless nickel deposits are very similar to those of materials that are rapidly cooled from the liquid state (6) and that are considered to be glasses, electroless nickel can justifiably be considered amorphous. A typical structure of a high-phosphorus electroless nickel deposit, as seen in a thin foil by transmission electron microscopy (TEM), is shown in Fig. 4.3. No grains or other prominent structural features are seen.

The structure, as seen by optical microscopy, of metallographically polished and properly etched cross sections of electroless nickel show the so-called *banded* or *lamellar* structure depicted in Fig. 4.4. This structure has been attributed to compositional variations with deposit thickness (8,9). The cause of the compositional variation has been explained (10) in terms of periodic fluctuations in the pH of the plating solution adjacent to the deposit surface. These fluctuations, which result in variations of the phosphorus content, were postulated as being caused by hydrogen evolution, which raised the pH. The resulting stirring action then lowered the pH by mixing the plating solution adjacent to the deposit with the bulk of the bath. Markings perpendicular to the

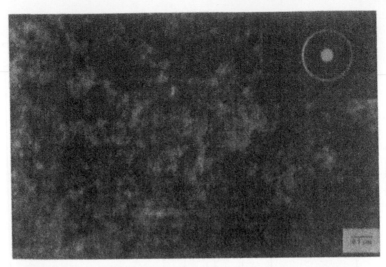

Fig. 4.3—TEM photo of a high-phosphorus electroless nickel deposit. Inset shows electron diffraction pattern.

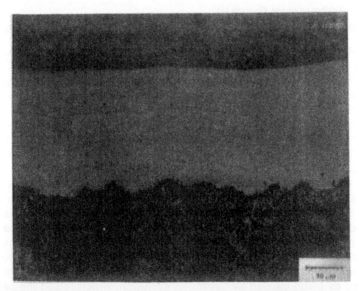

Fig. 4.4—Optical micrograph of metallographically polished and etched cross-section of an amorphous, unannealed electroless nickel deposit. Banded structure is evident.

banded structure can also be seen in Fig. 4.4. Sometimes these markings can appear similar to those of a columnar structure. However, as the grain size is much too small to be resolved with an optical microscope, the structure is not the columnar one exhibited by electrodeposits.

The structural changes that occur when amorphous electroless nickel deposits are annealed above 300° C are illustrated in Figs. 4.5 and 4.6. First, the amorphous structure crystallizes, as seen in Fig. 4.5. This structural change is accompanied by the evolution of the heat of crystallization. Figure 4.6 shows the completion of crystallization. The grains have become somewhat larger but are still very small.

Intermediate processes during the formation of intermetallic compounds can result in precipitation or age hardening. Particles of a precipitate form when sufficient atoms diffuse to a particular location to form a volume of material that has the correct stoichiometric composition and that is large enough so that a boundary can form around it. Before the volume of material reaches this critical size, its atom arrangement and that of the surrounding matrix must remain continuous. This coherency between regions having different interatomic spacings results in severe straining, which hardens the alloy. The initial stage of the precipitation of the intermetallic phase, which is still coherent with the matrix, can be seen in Fig. 4.6. The small dark particles are the coherent precipitate.

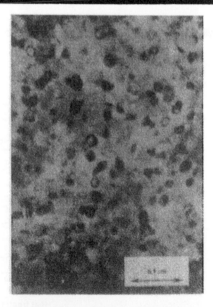

Fig. 4.5—TEM photo of originally amorphous Ni-P deposit annealed at 310° C, resulting in crystallization.

Fig. 4.6—TEM of originally amorphous Ni-P deposit annealed at 375° C. Crystallization is complete and a coherent precipitate is evident.

Graphs of hardness vs. aging time at several temperatures for an electroless Ni-P alloy (11) are shown in Fig. 4.7. These graphs are very similar to those of age-hardenable aluminum alloys. Above 450° C the times are too short to be practical for age hardening. An aging temperature of 400° C appears to be the most practical for deposits on steel, as only a short time is needed. A hardness of 900 to 1000 HV can also be attained by heating at lower temperatures for longer periods, e.g., >260° C for more than 6 hr. Age-hardening curves for Ni-B alloys (13) are shown in Fig. 4.8. Again, it is seen that in a relatively short annealing time, the maximum hardness can be attained at 400° C.

Deposit overaging occurs when hardness values peak and then decline with increasing heat treatment, due to the formation of Ni_3P crystals, as shown in Fig. 4.9. The increase in the sizes of grains of the nickel matrix and the Ni_3P phase further reduces the hardness, as does diffusion of phosphorus to the surface. Ni-B alloys tend to maintain hardness better than Ni-P when they are overaged. The banded structure that was depicted in Fig. 4.4 is seen to have disappeared in Fig. 4.9. A layer of an intermetallic phase, which formed as a result of interdiffusion of nickel and the steel substrate is also shown in Fig. 4.9. This layer develops upon heating above 600° C and thickens with increasing annealing time. While in this case adhesion was improved by interdiffusion, other substrates may form brittle intermetallic phases with nickel. Such layers could spall off and result in loss of adhesion.

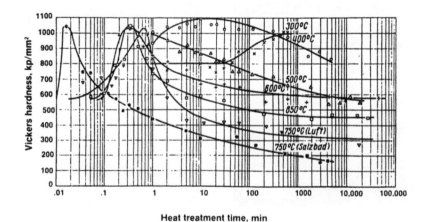

Heat treatment time, min

Fig. 4.7—Effects of aging temperature and time on hardness for a Ni-P deposit (11).

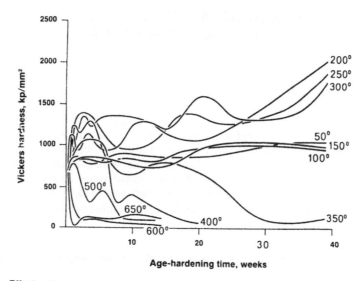

Fig. 4.8—Effects of aging temperature and time on hardness for a Ni-B deposit (13).

Fig. 4.9—Optical micrograph of cross-section of overaged Ni-P deposit on steel. Ni₃P grains and an intermetallic compound layer are evident. The banded structure that was seen in Fig. 4.4 has disappeared.

MECHANICAL PROPERTIES
OF ELECTROLESS NICKEL

The mechanical properties of electroless nickel plated in acidic solutions as a function of phosphorus content are listed in Table 4.1 (14). These properties were determined with a tensile-testing device described by Kim and Weil (15). It is evident that electroless nickel is a relatively strong but brittle material. The low ductility is due to the microcrystalline and amorphous structures that essentially preclude plastic deformation (14). The deformation is therefore mostly elastic until fracture occurs. The modulus of elasticity appears to have a maximum in the intermediate composition range. A much lower value of 120 MPa of the modulus of elasticity was reported by other investigators (4,11) for an alloy of 5 percent P. However, a 9-percent P deposit had a modulus of 50 GPa (16), which is in the same range as the values in Table 4.1.

Table 4.1
Mechanical Properties of Nickel-Phosphorus Deposits

Composition, percent P	Modulus of elasticity		Tensile strength		Elongation, percent	HK$_{100}$
	GPa	Msi	MPa	Ksi		
1 to 3	50-60	7.1-7.4	150-200	21-29	<1	650
5 to 7	62-66	8.9-9.4	420-700	60-100	<1	580
7 to 9	50-60	7.1-8.6	800-1100	114-157	1	550
10 to 12	50-70	7.1	650-900	93-129	1	500

It is evident from Table 4.1 that the tensile strength of the low-phosphorus, microcrystalline alloys is lower than that of the amorphous ones. Graham, Lindsay and Read (17) measured the mechanical properties of Ni-P deposits from alkaline solutions by bulge testing and found a sharp increase in the significant stress at fracture (which is essentially the same as the tensile strength) at about 7 percent P, where the transition from microcrystalline to amorphous occurs (6). The reported significant fracture stress values (17) of about 650 to 850 MPa for >7-percent P alloys agree fairly well with those listed in Table 4.1. Significant fracture stresses of 300 to 500 MPa were reported (17) for 5-percent-P deposits. Table 4.1 shows the ductility of electroless nickel as manifested by the elongation to be 1 percent or less over the whole range of compositions. All the deposits are therefore brittle in the as-plated condition.

An increase in the tensile strength of Ni-P specimens plated on thin copper tubes, which were subsequently dissolved, from 450 MPa to 550 MPa has been observed (11) after heating to 200° C for 15 minutes. Heating to 750° C for 2 hours was found (17) to considerably increase the significant fracture stress for low-phosphorus deposits while that of the higher-phosphorus alloys decreased. Thus, the fracture stress of a 5-percent P deposit increased to about 800 MPa, while that of a 9-percent P alloy decreased to about 250 MPa.

Hardness, the resistance of a material to permanent deformation by indentation, is an easily measured property, and therefore is most frequently determined by investigators of electroless nickel. It is useful in predicting abrasive-wear properties but bears no direct relationship to the strength of a material (18). The hardness of as-plated Ni-P deposits falls in the range of 500 to 600 kg/mm^2 as measured with a Knoop or Vickers indenter using a 100 g load. The Vickers numbers are generally greater than the Knoop hardnesses (18). There are a few special plating solutions that yield deposits of higher hardness. A gradual decrease in hardness with increasing phosphorus content has also been observed (19). As-plated Ni-B deposits have a Knoop hardness number of about 700 (19). The composition appears not to have a great effect on the hardness of Ni-B deposits, which tend to be harder than Ni-P, especially after a long heat treatment. (See Fig. 4.8)

Comparison of Figs. 4.1 and 4.2 shows that the eutectic temperature (where melting starts) is 1140° C for Ni-B alloys, but only 880° C for Ni-P. As softening occurs at these temperatures, Ni-B deposits possess a higher so-called *hot hardness*.

The ability of electroless nickel to maintain its hardness at elevated temperatures is illustrated in Fig. 4.10 (20), which shows the hardness of Ni-P deposits measured at various temperatures. It is seen that hardness decreases linearly with increasing heat-treatment temperature up to 385° C. Above 385° C, the hardness decreases rapidly with increasing temperature and holding time. High-phosphorus deposits harden more slowly in the 260 to 300° C range, but also take longer to soften at 400 to 600° C than those with lower alloy contents. The slower decrease in the deposits with higher alloy contents is due to their containing a larger volume of the Ni$_3$P phase. For example, under equilibrium conditions, which are essentially attained after prolonged heat treatment, the volume percentage of Ni$_3$P in an 8-percent P alloy is about 50 percent, and 100 percent in a 15-percent P deposit.

A third element can also affect hardness. For instance, the addition of molybdenum increases the hardness of nickel-boron deposits (21). The hardness of Ni-Mo-P alloys heat treated at 400° C, where they also attain their maximum value, increases with increasing molybdenum content. The temperature at which Ni-Mo-B alloys reach their maximum hardness is a function of composition. For example, an alloy containing 17 percent Mo and 0.3 percent B attains its maximum hardness of about 1000 HV when heat treated at 600° C.

The adhesion between electroless nickel and non-metallic surfaces depends on an essentially mechanical bond. Deposition into micro-crevices results in a keying action. The preparation of the substrate surface therefore affects adhesion. The adhesive strength increases with increasing surface roughness

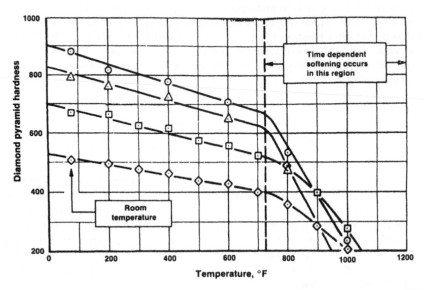

Fig. 4.10—Hot hardness of electroless nickel (20). ○ = 10.5 percent P, prior heat treatment at 600° C for 1 hr; △ = 8.0 percent P, 1 hr at 600° C; □ = 6.8 percent P, 1 hr at 600° C; ◇ = 6.8 percent P, 1 hr at 600° C plus 1 hr at 750° C.

(22). Adhesion to some metals can thus be enhanced by roughening their surfaces. Such was found to be the case when electroless nickel was plated on aluminum alloys (23). Loss of adhesion has been attributed to a difference in the coefficients of thermal expansion of the deposit and substrate (24). It has already been pointed out that annealing above 600° C so as to form a diffusion layer can improve the bond between electroless nickel and steel. Annealing deposits on aluminum substrates at 150° C and those on steel at 200° C also improves adhesion. However, deposits on age-hardened aluminum should not be annealed above 130° C, in order to maintain the hardness of the substrate.

The mechanical properties of parts can be adversely affected by coating them with electroless nickel. Metals such as high-strength steels, which are sensitive to hydrogen embrittlement, belong to this group. It is possible for sufficient hydrogen to be absorbed during electroless plating to cause failure in high-strength steels (25). Baking at about 200° C for several hours often relieves the hydrogen embrittlement. However, sometimes baking does not help, indicating that hydrogen is not the sole cause of poor mechanical properties in the substrate. Cracks that develop in the brittle electroless nickel coating during fatigue loading can propagate into the substrate and lower its endurance limit, i.e., the largest stress at which it did not fail in fatigue loading after 10^7 cycles. For example, a low-carbon, high-strength steel originally had an endurance limit of 342 MPa, which was reduced to 260 MPa when the steel was coated with 25 μm of electroless nickel. Heat treatment at 400° C further reduced the endurance limit to 176 MPa. The reversal of the sign of the internal stress from compressive to tensile upon heat treating may have been the cause of the reduction in the endurance limit after annealing (19). Shot peening of parts prior to plating can improve the fatigue properties by putting the surface in a state of compression.

INTERNAL STRESSES IN ELECTROLESS NICKEL DEPOSITS

Internal stresses develop within deposits and are not due to externally applied forces. They can be classified as *extrinsic* and *intrinsic*. Extrinsic stresses are primarily due to differences in the thermal expansion coefficients of electroless nickel and the parts to be coated. As electroless nickel plating takes place at elevated solution temperatures, differences in shrinkage result when the coated parts are cooled to room temperature. If there is no loss of adhesion, tensile stresses develop in the component with the larger coefficient of thermal expansion, while the other one is in compression. Electroless nickel shrinks about 0.1 percent on cooling from 90° C (19). Deposits on brass and aluminum, which have larger expansion coefficients than electroless nickel, are in a state of tensile stress. Deposits on beryllium and titanium, which have smaller coefficients, are in compression. The substrate surface is then in the opposite state of stress.

Intrinsic stresses develop due to the deposition processes. In general, electroless nickel is not deposited initially as extended atom layers, but as discrete particles having the appearance of islands. The formation and subsequent joining of particles rather than lateral spreading of layers probably continues to be the growth mode as the deposit thickens. When these particles are pulled together by surface tension before the spaces between them are filled by plated metal, tensile stresses develop. Atom rearrangement can occur in a surface layer as it becomes covered during continuing deposition or during annealing, so as to change the interatomic distances. If the interatomic spacing decreases, the stresses are tensile. Compressive stresses can result when codeposited gases, such as hydrogen, diffuse into microvoids and expand them. If gases diffuse out of the deposit or into the substrate, tensile stresses develop in the deposit.

The variations of the internal stress with phosphorus content of the electroless nickel plated on steel and aluminum are shown in Fig. 4.11 (26). A linear relationship between internal stress and the pH of the plating solution has also been found (27). At a pH of 4, where high-phosphorus deposits are produced, the stress was about -50 MPa, became tensile at a pH of 4.6, and reached a value of about 110 MPa when the pH was 6. On aluminum substrates, the stresses are generally compressive. A higher phosphorus content results in a lower tensile or a higher compressive stress. The values of the intrinsic plus extrinsic internal stresses shown in Fig. 4.11 are typical for electroless nickel. However, they can vary with the condition of the substrate and the instrumentation used for stress measurement. Chelating agents and additives to the plating solution can also change the intrinsic stresses significantly.

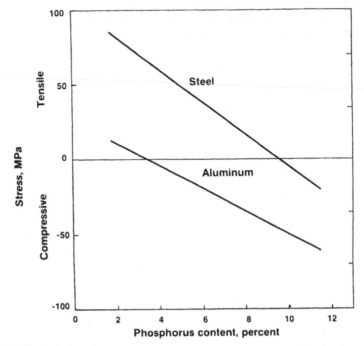

Fig. 4.11—Effects of phosphorus content on internal stress of Ni-P deposits plated on steel and aluminum substrates.

Annealing increases tensile stresses and decreases compressive ones. Nickel-boron deposits generally exhibit higher tensile stresses than Ni-P alloys (28). Values of about 480 MPa have been measured for deposits containing 0.6 percent B. The tensile stresses decrease with increasing boron content to 310 MPa and 120 MPa for alloys of 1.3 percent B and 4.7 percent B, respectively (28).

PHYSICAL PROPERTIES OF ELECTROLESS NICKEL

The density of electroless nickel depends on the interatomic spacing and on the amount of porosity. The variation of the density with alloy contents is depicted in Fig. 4.12 (29). Data obtained at the National Bureau of Standards (30) are also shown. An essentially linear decrease with increasing phosphorus content is seen. As phosphorus atoms increase the spacing between those of nickel atoms, the supersaturated solid solution has the lower density. The crystalline material is denser than the amorphous type (29,31). For example, a value of 7.9 g/cm^3 has been determined for amorphous electroless nickel containing 12 percent P, as compared to 8.1 g/cm^3 if it were crystalline (29). The interatomic spacing is also

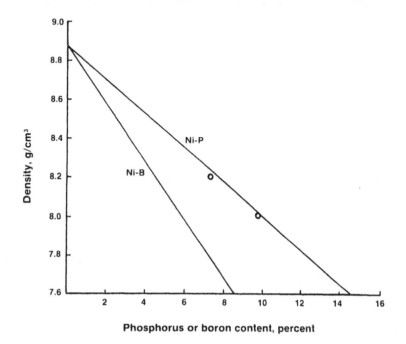

Fig. 4.12—Effects of alloy composition on density for Ni-P and Ni-B deposits (29).

larger in the grain boundaries. As microcrystalline materials contain a considerable amount of grain boundary material, their density is lower than that of larger-grained electrodeposits. The density of nickel-boron deposits also varies essentially linearly with alloy content.

Values of the electrical resistivity for electroless nickel alloys of various compositions are listed in Table 4.2. The electrical resistivity of electroless nickel is higher than that of electroplated nickel, which is about 8 μohm-cm (19), because alloying elements such as phosphorus and boron, especially when they are in solid solution, increase it. Also, as deviations from a regular crystal structure increase the electrical resistivity, amorphous deposits exhibit larger values than crystalline ones. Heat treatment, which results in the formation of intermetallic compounds and the crystallization of amorphous structures, substantially reduces the electrical resistivity. A sharp decrease in the resistivity in the temperature range where the amorphous structure crystallizes has been observed (33).

The thermal conductivity is generally proportional to the electrical conductivity. It is thus possible to calculate the values of the thermal conductivity from the reciprocals of the electrical resistivity. For high-phosphorus alloys, the

Table 4.2
Physical Properties of Electroless Nickel

Alloy	Density, g/cm³	Electrical resistivity, μohm-cm	Coefficient of thermal expansion, μm/m/°K
1-3% P	8.6	30	
0.1-2% B	—	5-13	
5-7% P	8.3	50-70	
4-5% B	8.5	—	12
8-9% P	8.1	70-90	13
5% B	8.3	90	12
>10% P	<8	<110 (19)	11 (32)
7% B	7.8	190 (26)	11 (13)

thermal conductivity ranges from 0.01 to 0.02 cal/cm/sec/°K (34,35), compared to 0.2 cal/cm/sec/°K for electroplated nickel.

Values of the coefficient of thermal expansion are also listed in Table 4.2. It should be noted that on heating, when crystallization takes place and intermetallic phases form, volume changes occur that are reflected in the thermal expansion coefficient. A deposit containing 11 to 12 percent P had shrunk by 0.11 percent after heating to 300° C and cooling back to room temperature (36). Most of this shrinkage was due to structural changes. After a second heating and cooling cycle, the shrinkage was only 0.013 percent, indicating a more stable structure. The reported value of the thermal expansion coefficients for a 5 percent B deposit of 12 μm/m/°K decreased after heat treatment to 10.8 μm/m/°K (13).

The magnetic properties of electroless NiP deposits depend on whether they are crystalline or amorphous. Crystalline deposits are ferromagnetic, while those having an amorphous structure are essentially non-magnetic (37). The amorphous, high-phosphorus alloys are often used under cobalt deposits on memory disks. The low-phosphorus, microcrystalline deposits have a relatively low coercivity, which is the external magnetic field that has to be applied for demagnetization. Values varying from 0 to 80 oersteds have been reported (19). The coercivity of electroplated nickel varies from 40 to 120 oersteds (19). The remnant flux density, which is proportional to the magnetization that remains when the outside field is removed, has reported values of about 400 to 600 gauss, as compared to 2500 to 3300 gauss for nickel electroplated in a sulfamate solution. Most nickel-boron deposits are weakly ferromagnetic.

After non-magnetic, electroless nickel deposits have been heat treated at 300 to 400° C, they are weakly ferromagnetic because of crystallization. The coercivities of the as-plated, microcrystalline deposits also increase when they are heat treated. A reason for the higher coercivities after heat treatment is that the paramagnetic intermetallic compound, Ni_3P, impedes the movement of domain walls (6). The remnant flux density also increases after heat treatment to values ranging from 1000 to 3000 gauss (37). The Curie temperature, at which magnetism is lost on heating, decreases with increasing phosphorus content from 360° C for pure nickel, to 77° C when the alloy contains 7 percent P (38).

Corrosion Resistance of Electroless Nickel

Electroless nickel coatings are primarily used in engineering applications to provide protection for common metal surfaces exposed to corrosion as well as to wear. The corrosion protection is due to the low porosity of the coatings and the excellent resistance of nickel to many liquids and most atmospheric conditions. Electroless nickel prevents the contamination and discoloration of plasticizers, solvents, oils, glycols, monomers, and gases by metals such as iron, copper and aluminum. Significant savings in equipment costs are possible by replacing stainless steel with ordinary steel plated with electroless nickel. Stainless steel parts have been coated with electroless nickel to prevent pitting in the presence of chloride ions.

The most important factors that determine the corrosion resistance of electroless nickel are:

- Substrate composition, structure and surface finish.
- Pretreatment of the substrate to achieve a clean, uniform surface.
- Adequate deposit thickness for the severity and time of exposure to corrosive conditions.
- The properties of the deposit (composition, porosity, internal stress) as determined by the pH, formulation and use (turnover) of the plating solution.
- Postplating treatments of the coating, such as passivation and annealing.
- The aggressiveness of the corrosive conditions.

Generally, electroless nickel coatings have lower porosity and a more uniform thickness than the equivalent electroplated nickel and therefore provide better corrosion protection. In order for electroless nickel to be corrosion resistant, the pretreatment and plating conditions must achieve good adhesion and continuity of the coating. The integrity and corrosion resistance of thin coatings are generally evaluated by a neutral 5 percent salt spray test (ASTM B117) and outdoor exposures. Electroless nickel coatings of about 9 percent P were found to provide longer protection than electroplated nickel. However, electroplated and electroless coatings containing 9 percent P behaved similarly (39).

The difference in the electrochemical potential between electroless nickel and the substrate greatly affects what happens if there is a void or pore in the coating. The potential difference varies in different corrosive media, as seen in Table 4.3 (40). In an electrolytic cell, electroless nickel is the cathode against aluminum and steel in most media, as indicated by the unbracketed numbers in Table 4.3. If a small area of the substrate, which is the anode, is exposed, the current density, which is proportional to the corrosion rate, is very high there. A 300 mV difference will initiate and sustain corrosion of steel or aluminum substrates if the electroless nickel coating is not continuous. If the electroless nickel is the anode, as is the case for the bracketed numbers in Table 4.3, the coating corrodes sacrificially. However, as the anode area in this case is large, the current density and the corrosion rate are relatively low.

Using the salt spray test, it was found (41) that periodic reverse alkaline cleaning prior to plating produced the most corrosion-resistant coating. If pickling is necessary, 10 percent sulfuric acid is preferable to hydrochloric acid. The pretreatment should be adequate to achieve a "clean", film-free surface (42).

Table 4.3
Potential Differences Between Electroless Nickel and Other Metals (40)

Environment	Potential difference, mV		
	Aluminum	Mild steel	304 stainless steel
Seawater	450	450	[90]*
Aerated tap water	400	650	[120]
Deareated tap water	710	450	[350]
0.1N HCl	320	200	[520]
0.1N HNO$_3$	170	440	[340]
0.1N NaOH	1160	[20]	[60]
3.5% CO$_2$ brine	530	430	[440]
H$_2$S brine	260	380	40

Outdoor exposure and salt spray tests have shown that electroless nickel coatings containing more than 9 percent phosphorus possess superior corrosion resistance to those with lower alloy contents. Compressively stressed deposits with higher phosphorus contents can be achieved by a pH less than 4.5 and the presence of strong chelating agents in the plating solution (43,44). One reason for the better corrosion resistance of coatings having higher phosphorus contents is that their more noble and passive polarization potential, which makes them less prone to pitting in such media as 3 percent NaCl and 10 percent HCl solutions (45). Another reason is that the amorphous microstructure is more homogeneous and there are no grain boundaries (46).

The presence of a fibrous-appearing microstructure seen in the cross section of some electroless nickel deposits renders them more susceptible to corrosion, as evidenced in potentiodynamic and salt spray tests (47). The fibrous structure can result from sulfur-containing stabilizers, such as 2-mercaptobenzothiazole.

The banded structure shown in Fig. 4.4, which is a result of variations in the phosphorus content with thickness, can be beneficial (46) in that it can behave similarly to duplex nickel coatings. Coatings consisting of a sacrificial layer of a lower phosphorus deposit over one of a higher alloy content can also behave similarly to duplex nickel (48).

Tensile internal stresses render electroless nickel coatings more susceptible to corrosive attack. Internal stresses result in increased potential differences and, therefore, in lower breakdown potentials and higher corrosion currents (49). Such stresses can result from the deterioration of the plating solution due to the accumulation of orthophosphite and sulfate ions after several turnovers (50). It can be readily seen that differences in the composition, structure and properties can result in the large variations of from 24 to 1000 hours resistance observed in salt spray tests of thin (10 to 12 μm) electroless nickel coatings. The coating thickness also affects the corrosion resistance. On rough steel or aluminum surfaces, thicknesses of 50 to 75 μm are required for extended corrosion protection.

Heat treatment in the temperature range of 300 to 400° C to achieve maximum hardness lowers the corrosion resistance of electroless nickel, probably as a result of microcracking. However, heat treatments at lower temperatures (in the 150 to 200° C range) have little effect on corrosion resistance. Improved corrosion resistance has been observed (51) after annealing at 600 to 700° C, because of improved bonding to steel substrates and improved integrity of the coating. Care must also be taken that heat treatment does not cause cracking of the coating as a result of shrinkage on metals with low thermal expansion coefficients, such as titanium and certain stainless steels.

The salt spray resistance of most electroless nickel deposits can be improved significantly by passivation of the surface in a warm 1 percent chromic acid solution for 15 minutes. Considerable improvement in the results from the salt spray test after such a treatment was observed (52).

Because of the many variables that affect the properties of electroless nickel coatings, it is difficult to predict their corrosion resistance. Electroless nickel should not be used in environments that attack unalloyed nickel. These environments include most organic and inorganic acids having a pH less than 4, as well as ammonium and halide salt solutions. Included also are solutions of chemical compounds that complex nickel. Electroless nickel has good resistance to alkaline solutions except ammonia and very concentrated (70 percent) hot caustic, which dissolves the phosphorus. Other liquids that attack nickel are carbonated and alcoholic beverages, coffee, fruit juices, vinegar, dextrose, bleach and fertilizers. Table 4.4 (53) shows laboratory corrosion rates of electroless nickel containing 8 to 9 percent P in various liquids. Aeration of the liquids increased the corrosion rates, while heat treatment of the deposits at 750° C for 1 hour can improve resistance, probably because of the formation of passive surface layers (51).

Field performance does not always correspond to observed laboratory test results. Therefore, Table 4.4 should be used cautiously. Service tests in the actual environment should be conducted if possible. There are many cases in

ELECTROLESS PLATING

Table 4.4
Corrosion Resistance Tests of Electroless Nickel (9% P)

Corrodant	Temp., °F	Aeration	Heat Treated**	Corrosion rate* Mil/yr	Corrosion rate* μm/yr
Acetaldehyde, waterfree	150	Yes	No	0.02	0.5
Acetone	Room	No	No	0.003	0.076
Acrylonitrile	150	Yes	No	0.016	0.4
Ammonium sulfide	120	Yes	No	0.15	3.8***
Amyl acetate	Room	No	No	0.0019	0.048
Amyl chloride	Room	No	No	0.013	0.33
Beer	45	No	Yes	0.0078	0.2
Benzene	Room	No	No	0.0016	0.04
Benzyl acetate	Room	No	No	Zero	Zero
Benzyl alcohol	Room	No	No	0.0036	0.092
Calcium chloride, 48.5%	Room	No	No	0.02	0.5
Caprolactam monomer	185	Yes	No	0.03	0.76
Caprylic acid	150	Yes	No	0.06	1.52
Carbon disulfide	Room	No	No	Zero	Zero
Carbon tetrachloride	Room	No	No	Zero	Zero
Cresylic acid	120	No	No	0.01	0.25
Detergent, 5%	Room	No	No	0.037	0.94
Ethyl alcohol	Room	No	No	0.0065	0.16
Ethylene glycol	Room	Yes	Yes	0.025	0.64
Formaldehyde, 37%	Room	No	Yes	0.012	0.30
Gasoline	Room	No	No	0.022	0.56
Lactic acid, 80%	Room	Yes	Yes	0.145	3.7***
Methyl alcohol	Room	No	No	Zero	Zero
Naptha	Room	No	No	Zero	Zero
Naphthenic acid	150 max.	Yes	No	0.02	0.5
Oleic acid	Room	No	No	0.012	0.3
Orange juice	Room	No	No	0.013	0.33
Perchloroethylene	150 max.	Yes	No	0.15	3.8***
Petroleum (sour crude)	Room	No	No	Zero	Zero
Polyvinyl acetate	150 max.	Yes	No	0.05	1.27
Rosin size	210	Yes	Yes	Zero	Zero
Sodium carbonate, 10%	Room	No	No	Zero	Zero
Sodium chloride, 3%	Room	No	Yes	0.04	1.0
Sodium hydroxide, 50%	250	No	No	Zero	Zero
Sorbitol	150 max.	Yes	No	0.1	2.5***
Stearic acid	158	No	No	0.02	0.5
Sugar (liquid)	Room	No	No	Zero	Zero
Tall oil (crude)	150	Yes	No	0.04	1.0
Toluol (nitration)	150 max.	Yes	No	0.023	0.58
Trichloroethylene	150	Yes	No	0.02	0.5
Urea, 25 % solution	Room	Yes	No	Zero	Zero
Water, deionized	180	Yes	No	Zero	Zero
Water, distilled	Room	No	No	0.029	0.74

*Corrosion rates are based on laboratory tests. Commodities listed in the table may vary from batch to batch in trace constituents that could affect corrosion rates. Where large-scale use is contemplated, specific corrosion data should be obtained for the commodity in question.

**Heat treated at 750° C for 1 hour.

***Not recommended.

which electroless nickel has provided excellent protection from corrosion under many service conditions. Coatings containing 10 to 11 percent P have performed well on tools used in oil and gas wells when exposed to sour (carbon dioxide and hydrogen sulfide) gases (54,55). Meat-cutting equipment has been protected by electroless nickel for many years. Electroless nickel has been substituted for stainless steel in some food packaging and processing applications, such as dough-mixing troughs, baking pans, electric grills, and wire baskets for deep frying.

The Ni-B alloys generally have poor salt spray and corrosion resistance, probably because they are porous and possess a fibrous structure.

WEAR PROPERTIES OF ELECTROLESS NICKEL

As was already pointed out, a primary reason for using electroless nickel is to protect substrates not only from corrosion, but also from wear. Wear is the gradual mechanical deterioration of contacting surfaces. One type, adhesive wear, results from the welding of the mating surfaces. As no actual surface is atomically smooth, it consists of asperities and depressions, i.e., hills and valleys. When two surfaces are brought together, the area of contact consists primarily of the tops of the asperities, and is therefore only a small fraction of the total area. A relatively light load applied to the two mating parts thus results in a large stress, i.e., the load divided by the contact area. The lateral movement of one surface relative to the other can remove any oxide or other soils on the tops of the asperities. Thus, the conditions for welding, namely intimate contact between two clean surfaces exist. Welding may be increased when the metals of the contacting surfaces are similar and mutually soluble in each other. The lateral movement can shear the welds again. If the shear fracture does not occur at the original weld, material from one surface adheres to the other. The resulting weight loss is adhesive wear. It is also possible that the shear fracture takes place on both sides of the original weld producing a separate particle. Then both surfaces undergo adhesive wear. Such particles, as well as ones from other sources can abrade the surfaces, causing a second type, namely abrasive wear. Abrasive wear can be minimized by making the surface harder and smoother. Hardening of a surface and lubrication reduce adhesive wear. If strong bonds develop under stress over a large portion of the contacting area, the mating surfaces may gall or seize, resulting in gross damage. Selecting dissimilar mating materials helps to avoid welding.

Adhesive and abrasive wear are related, though not directly, to the hardness of a surface, which is an indication of how much the tops of the asperities deform plastically. The greater the hardness, the less deformation, and consequently, there is less intimate contact and less welding and friction. Hard surfaces are also less likely to have the shear fracture occur at the original weld, resulting in less wear. Polishing the contacting surfaces also reduces wear because the contact area is larger and the stress that produces the intimate contact leading to welding is reduced.

Lubrication also reduces friction and wear by inhibiting the intimate contact. When hard particles, such as oxides, carbides, and diamonds are included in electroless nickel, they constitute the principal areas of contact. As these particles are less likely to weld, they reduce adhesive wear (18). However, if they are pulled out of the matrix they can cause abrasive wear. The inclusion of PTFE, which would tend to be smeared over the contacting surfaces, reduces friction by preventing the intimate contact that leads to welding and abrasion (56).

The abrasive wear properties of electroless nickel can be improved by heat treatment. However, such heat treatment may lower its corrosion resistance. In general, thin electroless nickel coatings are only effective under relatively mild wear conditions. Severe or sudden loading should be avoided. Good adhesion to the substrate is essential for satisfactory wear performance. A hard coating on a soft substrate such as aluminum is easily disrupted and penetrated by a hard, rough contacting surface. Therefore, a hard substrate provides better support for electroless nickel coatings.

Wear under working conditions is a very complex phenomenon. Among the many parameters that affect wear and make it difficult to control are surface hardness and finish; the microstructures and bulk properties of the mating materials; the contact area and shape; the type of motion, its velocity and duration; the temperature; the environment; the type of lubrication; and the coefficient of friction. The various laboratory tests can only provide indications of how parts will behave in service. Values of hardness and coefficients of friction can serve as a guide in selection of materials for a specific application. Trials under actual service conditions are still the only reliable indicators of how a part will wear.

The coefficients of friction of electroless nickel in the as-plated condition are generally higher than those of electroplated chromium (19). Some values of the friction coefficient of a 9 percent P electroless nickel alloy against three different mating materials, with and without lubrication, are shown in Table 4.5 (19). The coefficients of friction of pins coated with Ni-P or Ni-B deposits are compared in Table 4.6 against various plates of different metals. Friction coefficients for lubricated Ni-B coated pins are also presented. It is evident that under dry conditions the Ni-B deposits have higher coefficients of friction, especially against copper and bronze. It can also be seen from Table 4.6 that the coefficients of friction do not bear a direct relationship to the hardness of the plates. The coefficient of friction for electroless nickel against steel appears to be relatively independent of phosphorus content. The least wear of electroless nickel against quenched steel is obtained when the deposit contains between 8 and 12 percent P after heat treatment at 400 to 600° C (57).

Other wear test results are presented in Tables 4.7 and 4.8 (18). The wear data in Table 4.7 were obtained using a LFW-1 tester according to ASTM D2714-68. It compares the wear of a block against a rotating ring. The electroless nickel data in Table 4.8 were obtained with a Falex tester according to ASTM D2670-67. In this tester, a pin (journal) is rotated between two V-shaped blocks. The data in Tables 4.7 and 4.8 show a good correlation of hardness of the electroless nickel deposit to wear. The least wear resulted when the electroless nickel had been

Table 4.5
Friction Coefficients of Several Mating Surfaces (19)

Mating surfaces	Coefficients of friction	
	Unlubricated	Lubricated
Electroless nickel (EN) vs. steel	0.38	0.19-0.21
Cr vs. steel	0.19-0.23	0.12-0.13
Steel vs. steel	galling	0.2
EN vs. EN	0.45	0.25
Cr vs. Cr	0.43	0.26
EN vs. Cr	0.43	0.30

Table 4.6
Friction Coefficients of Electroless Nickel
Against Different Metals (19)

Metal	Plate properties		Coefficients of plated pin vs. plate		
	Hardness, kg/mm²	Finish, μm	Ni-P (11) dry	Ni-B (13) dry	lubricated
Steel*	960	2.6	0.32	0.44	0.12
Steel**	310	3.0	0.37	0.43	0.13
Gray iron	210	2.9	0.30	0.38	0.12
Hard Cr	960	1.2	0.38	0.36	0.10
Cu (plated)	105	2.7	0.33	0.70***	0.10
Bronze	200	1.0	0.26	0.65	0.09

*Hardened 105WCr6 steel.
**Normalized 105WCr6 steel.
***After surface was abraded.

Table 4.7
LFW-1 Wear Test Results

Heat treatment		EN-plated ring vs. block			EN-plated block vs. ring		
°C	hr	HVN$_{100}$	Coefficient of friction	Weight loss, mg	HVN$_{100}$	Coefficient of friction	Weight loss, mg
As plated		523	0.09	102	585	0.13	9.0
260	1	743	0.09	50	724	0.13	8.8
260	10	1010	0.11	13	988	0.13	2.8
400	1	1060	0.11	1	1064	0.10	2.3
540	1	not tested			892	0.10	1.7

*The hardness of the unplated steel block and ring were Rc 60 and 65, respectively. In the tests of the plated ring vs. block, the load was 284 kg and the ring rotated for 25,000 revolutions. In the plated block vs. ring test, the load was 68 kg and the ring rotated for 5000 revolutions. The lubricant was USP white oil. The rotational velocity of the ring was 72 rpm.

Table 4.8
Falex Test Data of Rotating Pins vs. V Blocks (18)*

EN deposit Heat treatment			Unplated pin	Plated block	Plated pin	Unplated block
°C	hr	HV	W.L.*	W.L.	W.L.	W.L.
As plated		590	0.2	6.6	—	—
288	2	880	0.1	1.2	3.9	0.3
288	16	1050	0.1	1.2	3.7	0.7
400	1	1100	0.2	0.5	3.0	1.5
540	1	750	0.1	0.4	5.6	0.8
Chromium		1050	1.9	0.5	6.2	15.0

*The unplated V blocks and pins had hardnesses of Rc 20 to 24 and 60, respectively. The loads were 94 kg applied for 1 hr and 182 kg applied for 40 min. The pin rotated at a speed of 290 rpm. The lubricant was white oil.
**Weight loss, mg.

heat treated to HV 850 to 1100, as obtained at 260 to 290° C in 2 to 10 hours, or at 400° C in less than 1 hour. A minimum hardness of HV 850 is necessary to obtain wear resistance comparable to that of hard chromium.

The unlubricated wear rates of pins plated with electroless nickel containing 8.5 percent P against blocks of different metals determined in a testing device like the one used to obtain the data given in Table 4.8 are shown in Table 4.9 (58). Wear rates exceeding 10 mg/1000 cycles are considered to be unsatisfactory. It is thus seen that all the block materials, except chromium, which were tested against as-plated electroless nickel failed. Heat-treated, electroless nickel plated pins behave unsatisfactorily against chromium, stainless steel, and as-plated electroless nickel. The wear rates of blocks of steel and those coated with electroless nickel heat treated at 400 and 600° C against electroless nickel pins heat treated at 400 and 600° C were satisfactory. It is apparent that electroless nickel should be used cautiously under unlubricated conditions.

Table 4.9
Falex Unlubricated Wear Rates
Of Electroless Nickel Coated Pins (58)

Block material	Wear rates (wt. loss, mg/1000 cycles)		
	As-plated	Heat treated, 400° C	Heat treated, 600° C
Steel	70	3	4
Cr	10	20	20
Stainless steel	F*	F	20
EN, as-plated	F	F	10
EN, heated treated, 400° C	75	5	5
EN, heat treated, 600° C	F	4	2

*Failure load—22.5 kg; rotation—290 rpm.

In unlubricated Falex wear tests on blocks with a hardness of Rc 20 to 24, Ni-B plated pins from an acidic (DMAB) solution performed poorly as compared to those from a basic (sodium borohydride) solution, as well as Ni-P deposits from either acidic or alkaline sodium hypophosphite solutions. The data are shown in Table 4.10 (57). All deposits were heat treated for 1 hr at 400° C. The maximum load was 182 kg.

Some Taber abrasion test data (16) are presented in Table 4.11 for chromium and electroless nickel alloys containing 10 percent P or 5 percent B. The Taber Wear Index is the weight loss in mg per 1000 revolutions with a 1 kg load using resilient, abrasive (CS-10) wheels.

Table 4.10
Unlubricated Falex Wear and Hardness Data (59)

Solution	Knoop hardness, KN$_{50}$	Plated journal weight loss, mg	Time to failure, min
Acid Ni-P	1088	64	6*
Basic Ni-P	990	16	3*
Acid Ni-B	1284	419	—
Basic Ni-B	1195	68	7*

*Maximum load 182 g.

Table 4.11
Taber Abrasion Data (16)

Heat treatment	Taber Wear Index*
Ni-P, as-plated	15-20
Ni-P, 1 hr, 288° C	10-15
Ni-P, 0.25-1 hr, 400° C	6
Ni-P, 0.25-1 hr, 600° C	4
Ni-B, as-plated	8-12
Ni-B, 1 hr, 400° C	4
Chromium	2

*Taber wear index = mg/weight loss/100 revs, CS-10 wheels, 1 kg load.

It is apparent from Table 4.11 that the abrasion resistance increased with increasing heat treatment temperature. This improvement may be due to the larger grain size of the deposits. It is also apparent that Ni-B alloys are more abrasion resistant than Ni-P deposits. The best resistance to scoring (equal to that of chromium) was attained in Ni-P deposits heat treated for 2 hr at 289° C (60). In fretting wear tests it was found (61) that as-plated electroless nickel coatings containing 11 to 12 percent P on low alloy steel substrates behaved better than those heat treated to the maximum hardness. A minimum coating thickness of 25 μm (1 mil) was also found to be necessary.

Duplex coatings consisting of 29-μm-thick hard Ni-P with 6 percent P deposited over a 40-μm-thick, somewhat more ductile layer of Ni-P with 12 percent P have protected hydraulic equipment subjected severe wear and corrosion conditions (62).

The wear and abrasion applications of electroless nickel coatings can be summarized as follows:

- Coatings must be heat treated to a hardness greater than HV 800.
- Coating hardness on rotating parts should be greater than that of the mating surface.
- Phosphorus content should be greater than 10 percent.
- Contacting surfaces should be smooth and lubricated.
- Electroless nickel coatings are not suitable for use under high shear and load conditions.

REFERENCES

1. *Binary Phase Diagrams*, T.B. Massalski, chief ed., American Society for Metals, Metals Park, OH, 1986.
2. J.P. Randin and H.E. Hintermann, *J. Electrochem. Soc.*, **115,** 480 (1968).
3. T. Omi, S. Kokunai and H. Yamamoto, *Trans. Jpn. Inst. Metals*, **17,** 370 (1976).
4. W. Wiegand and K. Schwitzgebel, *Metall*, **21,** 1024 (1967).
5. M. Sadeghi, P.D. Longfield and C.F. Beer, *Trans. Inst. Metal Finish.*, **61,** 141 (1983).
6. T. Yamasaki, H. Izumi and H. Sunada, *Scripta Met.*, 15, 177 (1981).
7. E. Vafaei-Makhsoos, E.L. Thomas and L.E. Toth, *Metall. Trans.*, **9A,** 1449 (1978).
8. A.W. Goldenstein, W, Rostoker, F. Schlossberger and G. Gutzeit, *J. Electrochem. Soc.*, **104,** 104 (1957).
9. F. Ogburn and C.E. Johnson, *Plating*, **60,** 1043 (1973).
10. C.C. Nee and R. Weil, *Surface Tech.*, **25,** 7 (1985).

11. H. Wiegand, G. Heinke and K. Schwitzgebel, *Metalloberflaeche*, **22**, 304 (1968).
12. W.G. Lee, *Plating*, **47** (1960).
13. H.G. Klein, H. Niederprum and E.M. Horn, *Metalloberflaeche*, **25**, 305 (1971).
14. I. Kim, R. Weil and K. Parker, *Plat. and Surf. Fin.*, **76**(2), 62 (1989).
15. J. Kim and R. Weil, *Testing of Metallic and Inorganic Coatings,* ASTM STP 947, W.B. Harding and G. DiBari, eds., Am. Soc. Testing and Materials, Philadelphia, PA; p. 11-18.
16. K. Parker, unpublished data.
17. A.H. Graham, R.W. Lindsay and H.J. Read, *J. Electrochem. Soc.*, **112**, 401 (1965).
18. K. Parker, *Plating*, **61**, 834 (1974).
19. K. Parker in *The Properties of Electrodeposited Metals and Alloys,* 2nd edition, W.H. Safranek, ed.; American Electroplaters and Surface Finishers Society, Orlando, 1986; Chapter 23.
20. K.T. Ziehlke, "Hot Hardness of Electroless Nickel Coatings," Union Carbide Nuclear Company Report K-1460 (1962).
21. G.O. Mallory, *Plat. and Surf. Fin.*, **63**(6), 34 (1976).
22. N.F. Murphy and E.F. Swansey, *Plating*, **56**, 371 (1969).
23. S. Armyanov, T. Vanegelova and R. Stoyanchev, *Surf. Tech.*, **17**, 87 (1982).
24. T.N. Khoperia, *Proc. Interfinish '80,* p. 147 (1980).
25. J.M. Hysak, "Current Solutions to Hydrogen Problems in Steels," American Soc. for Metals, Metals Park, OH (1982); p. 84.
26. K. Parker and H. Shah, *Plating*, **58**, 230 (1971).
27. C. Baldwin and T.E. Such, *Trans. Inst. Metal Finish.*, **46**, 73 (1968).
28. G.O. Mallory, *Plating*, **58**, 319 (1971).
29. T. Schmidt et al., *Nucl. Instr. and Methods*, **199**, 359 (1982).
30. F. Ogburn, R.M. Schoonover and C.E. Johnson, *Plat. and Surf. Fin.*, **68**(3), 45 (1981).
31. G.S. Cargill III, *J. Appl. Phys.*, **41**, 12 (1970).
32. G. Gutzeit, *Trans. Inst. Metal Finish.*, **33**, 383 (1956).
33. S.T. Pai, J.P. Morton and J.B. Brown, *J. Appl. Phys.*, **43**, 282 (1972).
34. G. Gutzeit and E.T. Mapp, *Corrosion Tech.*, **3**, 331 (1956).
35. A.A. Smith, Oak Ridge National Lab. Report Y2269 (1982).
36. M.W. Poore, Union Carbide Nucl. Div. Prelim. Report Y/DL-838 (1982).
37. M. Schwartz and G.O. Mallory, *J. Electrochem. Soc.*, **123**, 606 (1976).
38. A.A. Albert et al., *J. Appl. Phys.*, **38**, 1258 (1967).
39. C.H. Minjer and A. Brenner, *Plating*, **44**, 1297 (1957).
40. R.N. Duncan, *Electroless Nickel Conference IV* (1985).
41. K.B. Saubestre and J. Haydu, *Proc. Surface 66,* p. 192 (1966).
42. D.L. Snyder, *Electroless Nickel Conference III* (1983).
43. S. Yajima, Y. Togawa, S. Matsushita and T. Kanbe, *Plat. and Surf. Fin.*, **74**(8), 66 (1987).
44. G.O. Mallory, *Plating*, **61**, 1005 (1974).
45. M.L. Rothstein, *Plat. and Surf. Fin.*, **73**(11), 44 (1986).

46. M. Ratzker, D.S. Lashmore and K.W. Pratt, ibid., **73**(11), 44 (1986).
47. G. Salvago et al., *Proc. Interfinish '80*, p. 162 (1980).
48. L.L. Gruss and F. Pearlstein, *Plat. and Surf. Fin.*, **70**(2), 47 (1983).
49. G.O. Mallory, SAE Technical Paper 830693 (1983).
50. G. Linka and W. Riedel, *Galvanotechnik*, **77**, 586 (1986).
51. K. Parker, *Plat. and Surf. Fin.*, **68**(12), 71 (1981).
52. U.S. patent 3,088,846 (1986).
53. J.R. Spraul, *Plating*, **46**, 1364 (1959).
54. J.F. Colaruotulo, B.V. Tilak and R.S. Jasinski, *Electroless Nickel Conference IV* (1985).
55. R.N. Duncan, *Electroless Nickel Conference II* (1982).
56. S.S. Tulsi, *Trans. Inst. Metal Finish.*, **64**, 129 (1986).
57. J.P Randin and H.E. Hintermann, *Trans. Inst. Metal Finishing*, **72**(9), 53 (1974).
58. U. Ma and D.T. Gawne, *Trans. Inst. Metal Finish.*, **64**, 129 (1986).
59. R.F. Wrightman and F. Pearlstein, *Metal Finishing*, **72**(9), 53 (1974).
60. D. Butschkow and G. Gawrilov, *Metalloberflaeche*, **34**, 238 (1980).
61. A.J. Gould, P.J. Boden and S.J. Harris, ibid., **61**, 97 (1983).
62. E. Schmeling and G. Schmidt, *Electroless Nickel Conference III* (1983).

Chapter 5
Equipment Design
For Electroless Nickel Plating

John J. Kuczma Jr.

When designing an electroless nickel plating facility, it is necessary to look first at the chemical and operational objectives. If planning for a specific job or part to be plated, then the following factors should be considered: (a) the type of chemistry required and the shape of the part; (b) production time required per unit; (c) possible/probable expansion; and (d) space availability for current installation.

The major types of electroless nickel include nickel-phosphorus, nickel-boron, composite coatings, and others. Nickel-phosphorus solutions contain sodium hypophosphite as the reducing media and are classified according to phosphorus content—high (10 to 12 percent P), medium (6 to 9 percent P), and low (1 to 5 percent P). Nickel-boron solutions contain either sodium borohydride or DMAB as the reducing agents. Composite electroless nickel deposits commonly feature codeposited ceramic particles or codeposited Teflon. Finally, a 99-percent-Ni, hydrazine-reduced electrolyte is also in use. If non-standard solutions (e.g., composites, ammoniacal, or boron-reduced electrolytes) are contemplated, suppliers or experienced platers should be consulted before equipment is ordered.

The initial installation of an electroless nickel line in a job shop requires consideration of the following guidelines:

• The line should be designed to accommodate 90 percent of the type of work to be plated.
• A properly designed line for a single type of work will invariably be more effective than an improvised line. The occasional odd can be effectively handled in most cases.
• The selection of specific marketing target areas (e.g., large parts, bulk processing, multiple alloy, and substrate pretreatment) should precede the selection of the line.

Safety can be a distinct problem with both equipment and chemistry. Equipment safety will be seriously and thoroughly discussed throughout this chapter. Toxic chemicals are present in certain types of electroless nickel solutions (e.g., thallium in borohydride formulations). These chemicals and their mist have an impact on the total facility, as well as tank design. All plating tanks should include exhaust hoods for the safety of the operators as well as the environment.

TANK CONSTRUCTION MATERIALS

Polypropylene

Natural, stress-relieved polypropylene is well regarded for tank construction, as it is inert to plating solutions. Tanks are easily fabricated, and when using proper nitrogen welding techniques, are fairly reliable. Polypropylene lends itself to special plumbing or attached fixtures (Fig. 5.1). The relative low cost and versatility of design considerations make polypropylene the most widely used material of construction for electroless nickel tanks.

Polypropylene is not without its disadvantages, however. These tanks have a finite life span, which can be approximated by taking the inverse of the tank wall thickness (e.g., ¼-in. liners usually last 4 to 8 years, while ½-in. tank walls will need to be replaced in roughly half that time). The size of the tank also influences its useful life. Generally, larger tanks have shorter lives. Polypropylene electroless nickel tanks should be enclosed in outside support tanks for safety. As the support tank must be substantially larger, to accommodate the polypropylene expansion, the liner corners are subjected to extreme pressures both while cold, and more importantly, during heat-up and operation.

While this in no way detracts from the use of polypropylene liners for large tanks, it does necessitate the adherence to strict polypropylene fabrication techniques. Polypropylene-lined electroless nickel tanks have been fabricated

Fig. 5.1—Polypropylene tank liner with attached plumbing.

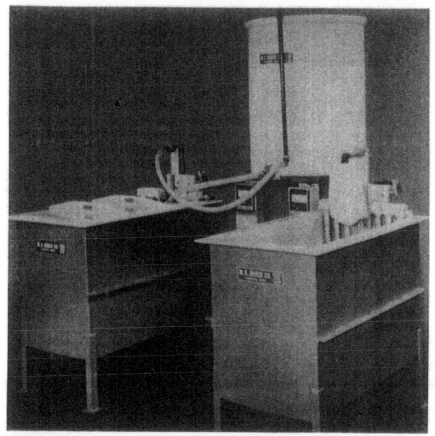

Fig. 5.2—Polypropylene-lined steel tanks for electroless nickel plating. Cylindrical tank holds nitric acid for stripping.

in sizes up to 20 feet long and 10 feet wide (Fig. 5.2). The life of the welds and of the polypropylene itself are directly related to the strength of nitric acid used in cleaning and stripping the tank, as polypropylene is oxidized by nitric acid, thereby accelerating deterioration.

Another disadvantage is that polypropylene is flammable. If exposed to electric immersion heaters with low solution level or no solution in the tank, the tank will melt and probably burn.

Chlorinated Polyvinyl Chloride (CPVC)

This material is relatively inert to electroless nickel solutions and is less flammable than polypropylene. It is not oxidized by nitric acid. It is, however, more costly and much harder to fabricate and weld than polypropylene. Glued CPVC tanks are not recommended for use with electroless nickel.

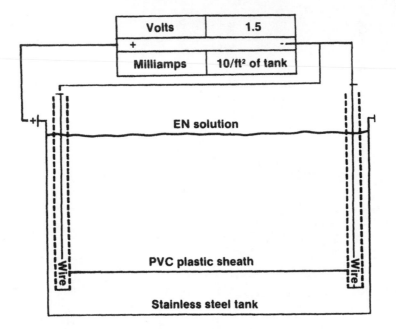

Volts	1.5
+	-
Milliamps	10/ft² of tank

Fig. 5.3—Anodic passivation of a stainless steel electroless nickel tank.

Stainless Steel

Stainless steel tanks have infinite life spans. Materials and competent fabrication techniques are readily available (Fig. 5.3). Electroless nickel, however, will plate onto the stainless steel tank walls, and these tanks may prove expensive to operate because of periodic plate out, which necessitates downtime for tank stripping and clean-up.

Satisfactory operation has been sustained using well-stabilized electroless nickel solutions with one or more of the following techniques:

Chemical Passivation

Passivation is attained by exposing the interior walls of the tank to 50 percent by vol. nitric acid for several hours at room temperature. Extreme care is necessary when using strong nitric acid (lower concentrations give less than acceptable passivation).

There has been some discussion with regard to applying anodic current to the tank surface or heating the nitric acid during passivation. This technique is not recommended because it is deleterious to the system in general, and makes areas adjacent to welds and bends more susceptible to accelerated failure at these points.

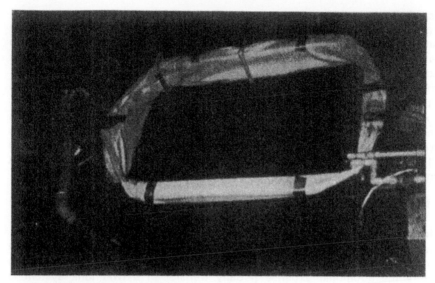

Fig. 5.4—Bag lined electroless nickel tank.

Applied Anodic Current to Tank During Operation

When stainless steel tanks are chosen, this is probably the most dependable passivation method (Fig. 5.3). The most commonly accepted applied current is 10 mA/ft^2. This process requires constant monitoring and control. Tanks, especially welded and high current density areas, will be affected after a while, and walls may start plating if the parts touch them during operation. Special care must be used with smaller and/or wired parts, since contact with the anodically charged tank can cause deplating and/or etching of the part.

Bag Liners

Liners are usually made from polyvinyl chloride (PVC) or polyethylene (PE) (Fig. 5.4). Thick (30+ mil) film bags, while the most dependable, are obviously the most expensive. Thin film (4 to 8 mil) bags, while less expensive, are susceptible to leaking in seamed areas. This leads to "double bagging" tanks, substantially increasing cost. Vibrations from plumbing or other fixtures in the tank will tear holes in the liner at contact points. This can lead to plating on the outer tank, resulting in increased costs. Hydrogen gas can also become trapped between the tank and the liner, creating a hazardous situation with the potential for a hydrogen explosion. Some thin bags act as semipermeable membranes to hydrogen, allowing gas pockets to build on tank bottoms. In some instances, these gas pockets have burst in 500 gallon tanks, sending hot (190° F) electroless nickel solution up to heights of 18 feet or more.

Coated Tanks

Many materials, ranging from Teflon to special ceramics, are used for coating tanks. While these systems may prove viable for short term usage, they are naturally fragile and susceptible to pores, scratches, and fractures. During extended use, stripping of electroless nickel particles will be necessary. Problems arise when small pockets created around these holes retain enough nitric acid to contaminate the solution.

An effective way to process parts in electroless nickel is to utilize a twin-tank operation (Fig. 5.5). Not only does this system assure that one tank will always be in operation while the other is being cleaned, but for critical applications it allows all of the solution in the tank to be transferred through a 1-μm or less filter at least once a day. Many new installations now incorporate this method of operation.

HEATING OF ELECTROLESS NICKEL SOLUTIONS

Electric Immersion Heaters

Stainless steel 304 is usually used for electric immersion heaters (Fig. 5.6). While most literature refers to SS 316, the performance difference would not seem to justify the additional expense if daily transfer in a twin tank system is contemplated. Titanium, while an excellent material, is too expensive for some installations.

The main advantages of immersion heaters is a low front-end capital cost and easy installation. The major disadvantages are potential plate out on the heaters, as well as creating a potential fire hazard when the solution level drops too low. Automatic controls should *never* be installed on electric immersion heaters. Electrical shock and stray current plating problems will be caused by improperly grounded heaters or resistance coils to stainless steel sheaths, causing electrical shorting.

Fig. 5.5—Twin tank electroless nickel system.

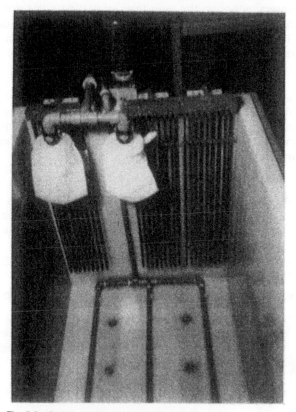

Fig. 5.6—Stainless steel immersion heaters.

Systems without adequate agitation should consider use of derated immersion heaters to lower the potential for overheating, which can cause plating solution decomposition.

Quartz immersion heaters are seldom used, except with solutions that will plate immediately on other materials (e.g., ammoniacal zinc die casting solutions). These heaters resist plate out until they become etched, which occurs after repeated exposure to nitric acid cleaning. Another disadvantage is the tendency for quartz heaters to break easily (always use guards), creating a serious safety hazard.

Steam

Steam heaters have a front end capital cost that is generally higher than electric immersion heaters, although the long term operating costs are substantially lower. The most frequently used types of steam heaters are panel coils (Fig. 5.7), Teflon coils (Fig. 5.8), heat exchangers (Figs. 5.9-11), superheated water, and double boilers (Fig. 5.12).

Panel Coils

The advantage of panel coils lies in their low cost, while the major disadvantages are the potential for plate out and the need for passivation; they take up tank space and can cause "hot spots" between coil and tank wall. Coating panel coils with a nonconductor such as Teflon cuts efficiency drastically and provides the same problems mentioned for coated tanks.

Teflon Coils

While Teflon is inert to all plating solutions, plating chemicals and cleaning chemicals, it is also very expensive and fragile, and cannot be moved after installation without the potential of leaking. Fragility increases with age. Interior buildup of hard water by-products will also cause brittleness.

The potential for porosity of Teflon coils increases with age. Hence, all installations should include valves for coil segregation. Coils should be pressure

Fig. 5.7—Stainless steel panel coil.

Fig. 5.8—Teflon-coated coil.

checked with 20 psi of air while the tank is full of water. When using spaghetti-type coils, the tank must be filled with enough nitric acid passivation solution to remove any electroless nickel particles deposited in the "tubing nest", especially at the top of the coil. If left unattended, these particles will continue to grow and eventually cause leaks. All installations of Teflon coils should include a ground wire in the tank. Until a certain quantity of water-hardness byproducts build up inside the coil, there is a potential for static electricity discharge as supersaturated steam hits the coil when it comes on. This is especially prevalent in installations using high pressure steam that has been reduced. The resulting discharge voltage has been recorded up to 1700 V. This can cause operators to drop racks, and is sometimes evidenced by pattern deplating in shielded areas on parts.

Fig. 5.9—Basic heat exchanger.

Heat Exchangers

Several different types of heat exchangers have been successfully used in electroless nickel operations. The materials most often chosen are stainless steel and titanium. Heat exchangers are mounted outside the tank, allowing maximum tank-space utilization. They heat up faster than coils; however, the pumping and filter system *must* be engineered to be compatible with the exchanger.

Flows and pressure not only affect efficiency but can cause solution deterioration. Exchangers must be designed and mounted so that they can be effectively cleaned and stripped/passivated with nitric acid, as electroless nickel can build up rapidly.

Fig. 5.10—Tube and shell type heat exchanger.

Hot Water
Superheated water can be used in place of steam in all of the aforementioned types of heaters. Many finishers like the cost factor and feel hot water is more effective than steam. In cases of very high water temperatures, heat exchangers can be too efficient and boil away the solution in the exchanger.

Double Boiler
This type of unit provides even heating for baths that are sensitive to hot spots. Double boilers have poor heat efficiency, low heat-up times, and preclude the possibility of using bottom plumbing. Glycol chambers must have large rear safety vents. Electric heater failure in glycol has caused violent reactions, resulting in operators being flooded with scalding 220° F glycol.

Fig. 5.11—Stainless steel heat exchanger.

ENERGY CONSERVATION

With energy becoming an ever-increasing portion of the overhead burden, the simplest and most seemingly insignificant conservation steps must be seriously considered. All heated tank walls must be insulated. This can be accomplished by cutting urethane or some other type of insulation sheet to size and attaching it to the outside of the tank. A protective shield or coating (e.g., plywood or fiberglass) is required to keep the insulation intact. The best method is to consider a tank liner/support tank combination with the urethane sandwiched between the liner and support tank, creating an air pocket.

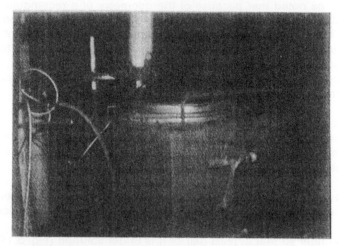

Fig. 5.12—Double-boiler electroless nickel tank.

Safety again becomes a factor when using a freestanding plastic tank. Under no circumstances should polypropylene tanks over 50 gallons be constructed without outside support and protection, either steel or fiberglass. At temperatures near 190° F, polypropylene is extremely weak, especially at weld junctions and corners. A 200-lb part striking the corner of an unsupported tank could give the operator an unexpected drenching and create an unnecessary legal problem. Metal strapping or angle around tanks *does not* solve the problem.

Approximately 85 percent of the heat loss in electroless nickel operations is from the surface of the plating solution; therefore, consideration should be given to polypropylene balls or other blanket insulation if the operation will allow it. Caution should be used when covering tanks with polyethylene sheet. Steam directed toward pumps or coated heaters will affect the life of equipment. Condensed steam on copper wires or bare steel fixtures drip large amounts of metallic contaminants into the bath.

If covers are used on operating tanks, they should be solid polypropylene for many reasons, not the least being cleanliness and safety.

FILTRATION

Filtration of electroless nickel baths should be continuous and at a rate of at least 10 times per hour. A 1 μm filter is adequate for most production facilities. Sub-μm filtration is recommended for more critical applications or for thick deposits.

It should be noted that because of the geometry of a rectangular tank, a filtration rate (tank turnover) of 100 times/hr would not expose all the solution to the filter. The only way to be assured that particles are not present when a morning run starts is to require daily bath transfers through a 1-μm or less filter. Filters should be changed daily.

There are a number of filter systems available, which can be broken down into two general groups—free flowing and closed chamber.

Bag Filters

Free flowing filters (Fig. 5.13) generally use bag-type polypropylene felt bag filters. These filters are usually the most cost effective. These factors should be considered when using bag filtration:

• Make sure the metal ring in the bag (if applicable) is stainless steel and *not* zinc- or cadmium-coated mild steel. Do not rely on the label—check with a magnet.

• Mount filter so that at least half of it is under solution (Fig. 5.14). Pump pressure impinging directly on the filter will stretch the felt bag, decreasing filter effectiveness as the solution absorbs shock. The submersion line on the bag will also display visible dirt or oil lines, giving evidence of incomplete pretreatment and sources of contamination. Any evidence of bath decomposition will also be visible, alerting the operator to problems in the bath, usually in time for correction.

• Filter chambers for bag filters (Fig. 5.15) are available. These provide better use of tank space; however, the advantage of being able to inspect the filter and observe what is going on in the bath is lost.

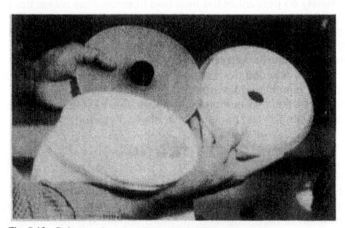

Fig. 5.13—Polypropylene bag filter with locking filter heads.

Fig. 5.14—Filter bag immersed halfway into plating solution.

Sleeves

For the operator who wants to take advantage of the ease of polypropylene felt, but wants more surface area for filtration, there are larger chambers available that use polypropylene felt sleeves. This method more evenly distributes the solution and pressure, thereby optimizing filtration.

Cores

While bags and sleeves provide adequate filtration for most jobs, there are jobs (e.g., laser mirrors, computer discs) that require absolute filtration. For those applications, wound cores (cotton over polypropylene, polypropylene over stainless steel, or polypropylene over polypropylene) are used (Fig. 5.16). Changes in flow and pressure occur when core filtration is used. Pump choice for an adequate turnover rate through filter becomes critical.

For after-the-fact monitoring of the bath operation, it is a good idea to periodically cut the used filter core in half before discarding. Like watching the bag filter, such inspections can often provide indications of potentially serious problems.

Carbon Filtration

Carbon filtration may be required on certain special purpose baths, which usually use fairly simple chemistry. Carbon filtration should be used only when recommended by the supplier of the electroless nickel solution. Most formulations utilize additives that may be removed by carbon filtration, thereby deteriorating the bath. If a decision is made to carbon treat, then care must be given to the types of carbon. Some carbon powders may contain sulfur or other contaminants. Extended exposure of electroless nickel solutions to carbon is not recommended.

Fig. 5.15—Exterior filter chamber.

Most filtration is done either in-tank or with sealed chambers. While there are instances where attached chambers of different designs have been used, they are not generally cost effective unless there is a specific reason for their existence.

PUMP MATERIALS

Most electroless nickel pumping systems are currently either CPVC (Fig. 5.17) or stainless steel. There are, of course, new materials being tried constantly. CPVC has proven itself to be an adequate material in most low volume/low pressure applications—a category that would probably include most of the electroless nickel baths in the country. A good limit for CPVC impellers is 200 L/min (50 gal/min). The range from 200 to 500 L/min (50 to 120 gal/min) is an area in which to consider some of the more resistant plastics (e.g., Kynar), at least for the impellers. Pump bodies may still be made of easily molded CPVC.

When flow rates surpass 400 L/min (100 gal/min), pressure becomes a definite factor. Few, if any, plastics can give adequate, cost effective service under these conditions and therefore stainless steel must be used (Fig. 5.18). There are many options as to what type of stainless steel should be required. Although it is well accepted that 316 will stay passive longer, it is not obvious that it is worth the expenditure. No matter what type of stainless steel is used, it will plate. Therefore, periodic passivation with nitric acid is required, preferably every day. If this procedure is followed, the additional costs associated with 316 become unjustified. If passivation does not occur daily, then 316 is a contender; then, of course, the plumbing becomes a consideration.

Sump Pumps
These pumps are vertically mounted, and whether in or out of the tank, require no seals other than lip seals in some instances. This is a distinct advantage in electroless nickel operations. The open slurry design of most small to medium size models is adequate to provide the open flow required for bag filters. Machined helix-designed impellers found in larger, or more expensive, models will withstand a limited amount of head pressure.

Pumping electroless nickel solution at its normal operating temperature of 190° F (90° C) provides different conditions than what most sump pumps are designed for. Testing has shown that many pumps recommended for use with electroless nickel solutions will either fracture the impeller or pump solution up the column (sometimes even into the motor) after long exposure to the hot solution. Long term test results are available from electroless nickel equipment suppliers recommending pumps for their systems.

Direct Mechanical Drive
These pumps are used when high pressures and flow rates are required (Fig. 5.19). Even though the seals require constant maintenance, certain applications require the output available only from this type of pump. *Dry seals* will have to be replaced regularly. Small particles of electroless nickel will continually provide problems. *Upstream* filtration could prove beneficial. The exact effect of static drag on the pump under those conditions should be tested.

Wet seals are those that are continuously flooded with a deionized water source. A pressure gage will assure that there is always flow to the pump. This system provides reasonable service, although seals will still require constant maintenance. Full-strength plating solution at operating temperature should not be allowed to stand in this or any mechanical pump.

Magnetic Impellers
This pump is usually combined with a filter core chamber mounted on the top (Fig. 5.20). It is ideal for laboratory electroless nickel applications as it is portable

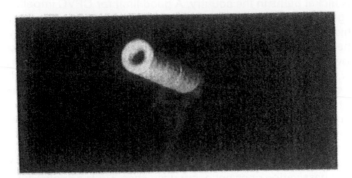

Fig. 5.16—Wound filter core.

Fig. 5.17—CPVC pump.

and can be used as required in multiple test baths. This pump is not recommended for production operations.

The impeller magnet should be Teflon coated. To extend the life of these coatings and minimize particle problems, the draft angle molded into the pump body should be opened all the way to the back. In pumps, as supplied, particles have a tendency to feed to the rear where clearance is minimal, then either jam

Fig. 5.18—Stainless steel pumps.

the pump, chew up the impeller, or both. While there are many pump designs available to the plater, each installation is different and may require modifications to commonly accepted pumping practice.

Pumps should be chosen to provide a minimum of 10 turnovers/hr of the tank capacity. Back pressures, head pressures, particular suction requirements, as well as distance to and height of the holding tanks are important. A knowledgeable equipment manufacturer knows how to obtain maximum performance from a plumbing system, always considering such things as "static drag" created by pipe and fittings. Other important, but seldom considered, problems are altitude and atmospheric pressure. Pumps will not perform the same in Denver as in Los Angeles.

RESIDUAL CONTAMINATION SOURCES

There are (other than parts) three things constantly being introduced into the bath: air, water, and replenisher. The sources and transfer of these items, therefore, becomes an intimate part of the equipment required for a trouble-free electroless nickel operation. Problems in these areas can be particularly insidious because they may provide contamination at imperceptible levels that build up in the bath over time.

Agitation
When designing a tank for electroless nickel, agitation is a prime consideration, as it must set up a flow pattern in the bath that will direct particles to the pump

Fig. 5.19—Direct mechanical drive pump with 400 gal/min capability.

intake, minimize hydrogen bubble retention on the part surface, and provide sufficient solution movement past immersion heaters, if applicable (Fig. 5.21).

While air agitation may affect some baths, it is the easiest to use and most widely accepted method of bath agitation. Low pressure oil-free positive displacement blowers are the best source of air for electroless nickel. The air inlet should be located where the temperature is constant and there are no airborne contaminants, such as solvents, present. Inlets must have filter units that are changed on a regular basis. As these filters are often neglected, it is a good idea to place a used clean bag filter or pillow case over the whole assembly. This will provide a highly visible indication as to what may be going into the bath.

In areas where outside inlets introduce cold air to the bath, a regenerative-type blower (Fig. 5.22) is suggested. The design of the blower provides hot (over 150° F) air, which both supplements the heating of the bath and eliminates problems during the winter. The air coming from this blower is so hot that metal plumbing must be used to dissipate some heat for a distance from the blower before connecting to PVC pipe.

A combination water feed/air sparger (Fig. 5.23) is an ideal source of air agitation for rinse tanks, as it raises water temperature and rinsing efficiency.

Solution agitation using high flow pumps is perhaps the best method of setting up the optimum flow pattern in the tank. More importantly, this methods

Fig. 5.20—Typical laboratory pump with top-mounted filter chamber.

eliminates the question of air source. The major problem to overcome is availability and the cost of pumping systems when larger tanks are contemplated. Installation should include both bottom and side spargers (Fig. 5.24). Not only will directionalized side spargers enhance the flow pattern, but parts hung in this side pattern will be extremely smooth and pit free for heavy build up or critical deposits. This method of agitation is also the choice when using special solutions such as (a) ammoniacal baths, where air agitation would fill the area with ammonia fumes; (b) composites, where particle suspension and equilibrium must be maintained; and (c) memory disc baths, where special problems exist relating to surface and deposit characteristics.

Part Agitation

While seldom used as a primary source of agitation, supplemental part agitation provides an undeniable advantage, especially where complex shapes or part configuration lends itself to hydrogen pitting. Part agitation may be horizontal or vertical, but either should incorporate a method to jar the piece slightly, thereby enhancing the removal of gas bubbles. Offset or notched cam drives may be used.

Mechanical stirrer mixing as an agitation source should only be used as a last resort. Flow patterns derived from mixers are easily broken by parts or racks and become ineffective. Parts hung on wires can sometimes be ruined or become tangled in the mixer apparatus.

Water Deionizers

In general, a purification system for water used in the electroless nickel bath (Fig. 5.25) should include the following:

A *1 μm pre-filter*, which will minimize problems with particles created in the water main.

A *carbon pack*, since certain chlorides and chlorinated hydrocarbons are a potential problem. These, as well as organic contaminants are present in some local water systems.

A *mixed bed deionizer*—Either a mixed bed or separate anionic and cationic resin beds are necessary. Hospital water supplies might be considered over industrial suppliers, as the quality control procedures associated with recharging resin beds are more closely monitored.

In new installations, a water supplier may prove more cost effective in the long run than an in-house DI unit, when taking into consideration the pollution control aspects of backwashing these units.

A *final filter*—All DI water going into an electroless nickel bath should pass through a final 0.5-μm filtration. Sub-μm silicate particles have proved to be one source of roughness and deposit stress problems.

This final filter, as well as a timer valve, may be mounted with your electroless nickel system. The see-through filter chamber is now in constant sight of the operator and the timer minimizes overflowing the tank when making water additions.

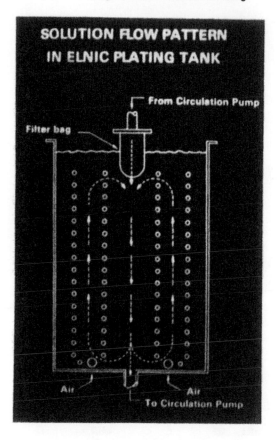

Fig. 5.21—Ideal solution flow using agitation.

Automatic Level Control Devices

The evaporation rate of electroless nickel solutions operated at 190° F (87° C) exceeds 1 in./hr in an air-agitated tank. Under these conditions, it might seem wise to use an automatic level control device to periodically add water to maintain solution level. However, these devices must be used with extreme caution, since there are a number of possible problems associated with liquid level controls for electroless nickel tanks.

Float control valves are used to add DI water to the bath as the level drops. The mechanical arm between the float and the valve is usually plastic and bends when exposed to the high temperatures above the electroless nickel solution.

Fig. 5.22—Regenerative blower.

The vapor above an electroless nickel tank tends to make the plastic parts gummy, thereby contributing to the possibility of the mechanism sticking.

Electronic sensors are devices mounted directly in the solution to sense the liquid level or may be mounted outside and therefore through the opaque (plastic) tank wall. Irrespective of the methodology, we are usually dealing with high and/or low liquid level cutoffs. Problems occur when the solution level fluctuates because of air agitation or when large-volume parts are placed in and removed from the tank. The sensor activates the contactor on the heater controller and, in some instances of fluctuating level, may send intermittent on/off signals, causing the contactor points to burn open or closed. This causes either solution overflow, or in the other case, solution evaporation to a dangerous level that can result in tank fires.

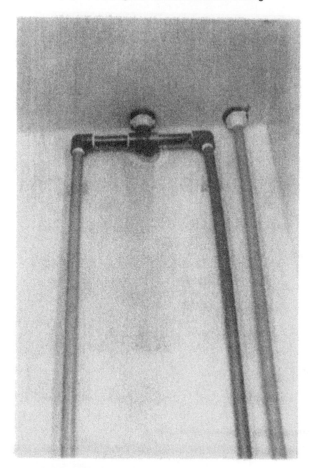

Fig. 5.23—Combination water feed/air sparger for rinse tanks.

Fuse link heaters—As no manufacturer will guarantee that the heaters will shut off 100 percent of the time, the choice and liability fall to the user.

In short, there are no fail-safe methods to automatically shut off electric immersion heaters.

Solution Replenishment
In the majority of installations, it is practical and economical for the operator to handle both the control and replenishment of the bath manually (Fig. 5.26). The operator has constant knowledge of what is happening with the chemistry. While most operations simply add the replenishers "over the side", it is advisable to have replenishment tanks that "bleed" in the solution at a constant rate,

Fig. 5.24—Side sparger for directional solution flow.

thereby minimizing the potential of shocking the bath. The operator can keep such a manual system feeding the bath at a constant rate while periodically making up the shortfall. Such a system of control, along with proper record keeping, can give all the benefits of automatic bath control, except in the case of computer-generated records.

Automatic Analyzers and Replenishers

These units are suggested in high production installations or where the human element must be removed. State-of-the-art controllers (Fig. 5.27) are available that will provide minimal downtime, absolute bath control, and date-time records, which are invaluable for both QC and vendor qualification. Computerized controllers may also control other functions in the plating line (Fig. 5.28).

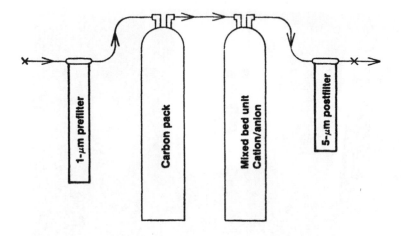

Fig. 5.25—Typical water deionization process.

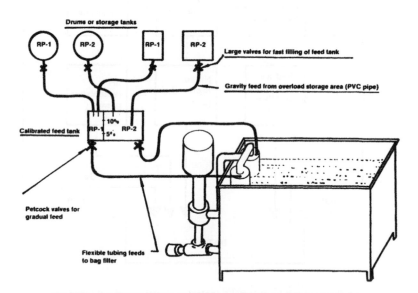

Fig. 5.26—Simple manual replenishment of electroless nickel solution components.

Fig. 5.27—Computer-controlled automatic solution analyzer.

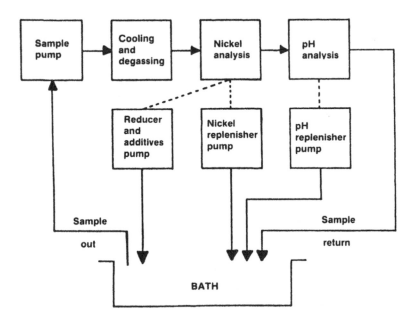

Fig. 5.28—Typical flow diagram for automatic control of electroless nickel processes.

With automatic control, a quick response time guarantees that the chemical components will remain very close to their optimum concentrations. The analysis of the bath is performed continuously, and the pumps are signalled within seconds to add replenisher. Nickel concentration can be controlled to within$\pm$2 percent, while pH is held to$\pm$0.05 units. Since the chemistry is held relatively constant, the result is a reproducible deposit.

Automatic control can also result in a lower operating cost. First, the consistent quality of the nickel deposit can eliminate many costly plating rejects. Second, since optimum concentrations are maintained, the bath stability and operating life are improved. A constant deposition rate is now assured. Automation also removes the possibility of costly errors during analysis and replenishment. Finally, the plater no longer must spend time adding the replenishers; instead he is free to perform other tasks.

One of the most important advantages of an automatic controller is its ability to keep accurate records of what has occurred during the life of the electroless nickel bath. Most large contracts will have QA inspections of the plating vendor facilities. Bath records are a top priority, which the controller record will immediately satisfy.

While the advantages of an automatic controller are numerous, it is necessary to remember that the controller is a machine subject to the calibration and atmospheric limitations that apply to any electronic equipment in a metal finishing facility.

Chapter 6
Quality Control
of Electroless Nickel Deposits

Phillip Stapleton

Electroless nickel deposits are smooth, hard nickel-phosphorus alloy coatings produced by a chemical reduction reaction. The properties of these coatings are dependent upon the substrate, pretreatment, plating process, composition of deposits, and post treatments. This interdependency of processes and materials on the performance of the coating makes quality control and process control key aspects of the manufacturing system.

In most applications, electroless nickel is used because of aggressive corrosion or wear conditions on the base material. In these applications, the coating performance can be critical to the overall system, and the potential loss to society if the coating fails significant.

Changes in coating performance, like surface finish and porosity, are affected by the manufacturing process before plating and bath formulation, while thickness, adhesion, and hardness are dependent on the plating process. Many times a combination of effects will influence the performance of the coating.

Electroless nickel deposit properties can also vary significantly between types of solutions. This is a result of the electropotential in the application environment and differences in the coating structure.

These chemical and physical changes in the deposit can be related to the phosphorus content and the number of lattice defects in the structure. By design, some types of solutions have greater numbers of lattice defects than others. Generally, as the lattice defects in the alloy increase, the deposit becomes more brittle and harder with a reduction in the modulus and increase in tensile strength.

The control of these properties is accomplished by maintaining the plating solution within a narrow operating range and controlling trace elements such as sulfur, antimony, cadmium, and bismuth, which are generally attributed with having the greatest effect on the structure and composition of the deposit.

QUALITY METHOD

Developing a method of controlling the factors that affect quality requires an overall scheme. This scheme must identify the quality level, analyze the process system to identify critical conditions, modify the manufacturing process, and identify the new quality level. Many methods have been developed to describe the relationship of quality to manufacturing process. One that works well for metal finishing is based on the work of Taguchi (Fig. 6.1).

According to his method, the goal is to develop a more robust manufacturing system that has no performance variation of the product, while experiencing a higher level of *noise* or variations within the processes.

The term *noise* is used to describe process variations from a mean. These variations can be from the wear of machines, operator training, contamination of materials, as well as many other factors. The use of the term *inner noise* refers to variations that are being monitored and can be controlled, while *outer noise* refers to variations that are not being controlled.

In Fig. 6.1, the block "Off-line Quality Control" is generally provided by the suppliers of processes, while "Transfer Design" and "On-line Quality Control" are performed by the producing facility.

When these basic concepts are transferred to a shop floor system, several familiar shop functions are revealed. The first item is the *specification*, which describes the requirements by acceptance and qualification tests that must be completed to verify the quality of the coating. In addition to the specification, a *shop traveler* is used to transfer the designs to the processor. This document includes the sequence, conditions, times, and other critical information needed to produce a quality part. A phase of the design includes the frequency of testing of process conditions, which allows the process engineer to determine the *Process Capability Index*.

By following the shop traveler and collecting the process information, studies on the coating performance can be related to process conditions, different operators, base materials, and more. From these studies, the frequency of testing can be modified, new, more robust conditions can be selected, and the quality can be brought under control.

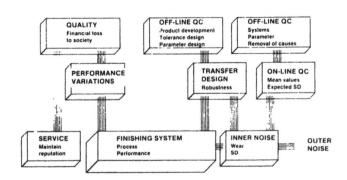

Fig. 6.1—Quality control and process management.

Experience and the pursuit of these methodologies has provided the information to build a series of cause and effect charts covering the electroless nickel process (see Charts 6.1-6.4). These charts describe the relationship of a coating failure to the processes.

Figure 6.2 shows the three areas that require control by operators: (a) process analysis and history; (b) qualification tests; and (c) acceptance tests.

Process Analysis and Management

In order to maintain the quality of the electroless nickel deposit, the electroless solution must be maintained at optimum chemical and physical condition.

Constant filtration of the plating solution is recommended. The solution should be filtered more than 10 times the volume of the process every hour. At this rate of filtration, the filter will be able to remove most of the solids and prevent the formation of roughness on the parts.

An important aspect in the design of the filter system is the pump dynamics and how the system will operate with a high specific gravity and temperature. The system should be sized for an operating temperature of 195° F and a specific gravity of 1.310.

The maintenance of a relatively narrow pH range will insure a constant plating rate and phosphorus content in the alloy. A pH change of 0.1 points may cause a variation of 0.1 mil/hr and affect the final product performance and cost. For some solutions, a change in pH of 0.5 points will cause a 1 percent change in phosphorus content and significantly affect the properties of the subsequent deposit.

Maintenance of the chemical composition of the plating solution in general, and the nickel/hypophosphite ratio in particular, will ensure a uniform plating rate and phosphorus content. Air agitation and low loading will cause oxidation of the available hypophosphite and may lead to a lower phosphorus content in the deposit and slower plating rates. To overcome this, periodic titrations for sodium hypophosphite and additions in the form of replenisher should be made to the plating solution.

The concentration of trace metals within the plating solution will also affect the deposit properties. There are several elements that will directly affect the performance in all solutions. These include sulfur, iron, cadmium, bismuth, antimony, mercury, lead, and zinc. There are several others that may affect the deposit properties, and which are primarly controlled by atomic adsorption, inductively coupled plasma, or poloragraphy. The presence of some of these materials may cause porosity in thin deposits and high stress. Some trace metals increase the propensity to pitting in the environment by setting up active cells on the surface between nickel and the trace element. Excessive concentration of trace elements in the solution may also produce a condition called *step plating*, in which edges are not plated or areas have low thicknesses.

The control of trace elements starts with keeping them out the solution. If they are present, several techniques can be tried to reduce their effect. Dilution is the first choice, followed by dummying, and addition of secondary reducing agents. Generally a combination of treatments will reduce the effect of the trace elements and bring the solution back to a productive condition.

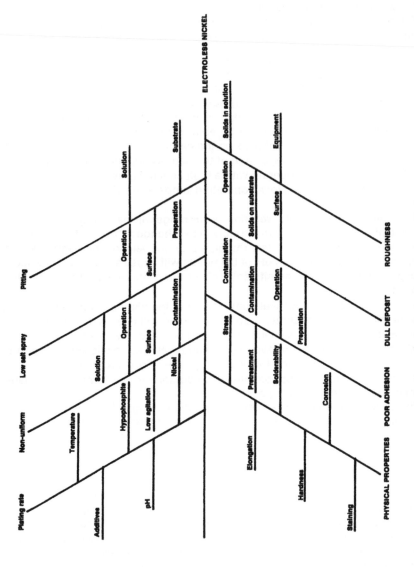

Chart 6.1—Cause-and-effect chart for electroless nickel, showing major defects and possible causes.

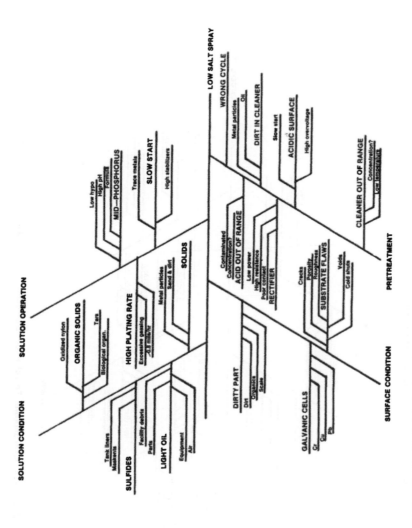

Chart 6.2—Cause-and-effect chart for the effect of electroless nickel with low salt spray resistance, showing some of the causes for such a condition.

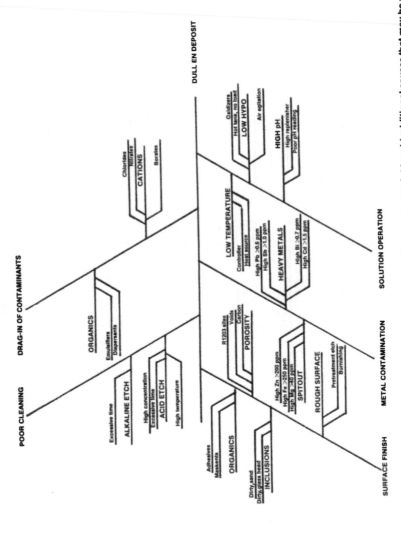

Chart 6.3—Causes of dull electroless nickel deposits. This chart could be used as a starting point to add additional causes that may be unique to the metal finishing facility.

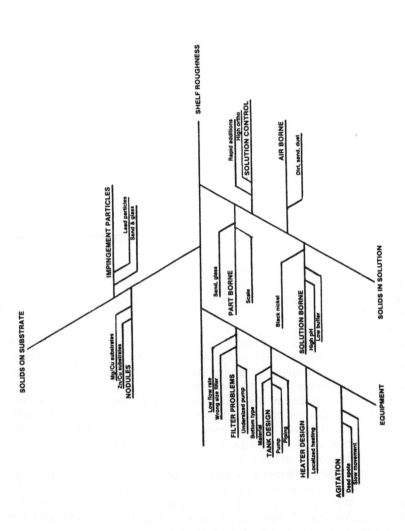

Chart 6.4.—Some causes of shelf roughness of electroless nickel deposits. Additional sources of shelf roughness may be present and should be considered in any quality improvement program.

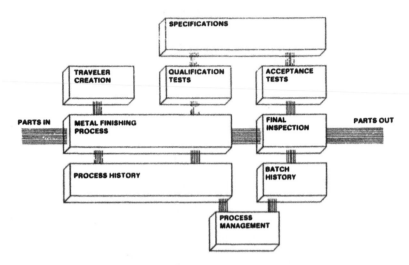

Fig. 6.2—Putting a quality control system together.

Organic contamination within the solution will also cause deposit properties to be compromised. The source of the organic materials may be from the part, facilities and tank, or masking. The problem is manifested by lack of adhesion, porosity, and stress.

Some common sources of organic contamination include maskants, organic materials, oils from within the substrate, plasticizers from hoses and liners, airborne organics, drippage from overhead, and contaminated make-up water. Silicates, though not organic in nature, can adversely contaminate the bath through drag-in from the pretreatment cycle.

Activated carbon treatment of the plating solution at operating temperature can often help reduce or eliminate some types of contaminants. If carbon treatment is used on the solution to remove organic contaminants, care should be taken to replace any desired organic control additives and that the carbon is clean and is not a source of contamination.

Another type of contaminant is *oxidizers*. These materials, such as hydrogen peroxide and nitric acid, will change the deposition potential within the solution and cause black or streaked deposits to be produced. With low levels of volatile oxidizers, heating the solution may help. With higher levels, and with non-volatile oxidizers, the solution must be discarded.

Other conditions that require control are *agitation* and *loading*. These two factors affect the diffusion of nickel ions in the reduction reaction. High agitation and either high or low loading can cause step plating and a low plating rate. Generally, the plating solution will produce coatings when the velocity of the solution is less than 4 ft/sec. This value will be dependent on the solution chemistry and solution operating conditions. At extremely low loads, the deposition potential may be low enough to prevent the reduction reaction from proceeding. This problem may be observed on sharp edges, such as needles, where the sharp point will not plate. In most cases, this problem can be corrected by selection of solutions that can operate at higher velocities (7 to 10 ft/sec).

Testing of Deposit Properties

The selection of requirements for the electroless nickel deposit are generally made by the purchaser of the coating. These requirements are established in *specifications*. Based on the sampling requirements, a part or specimen must be tested to identify any variation in performance.

The choice of tests that will characterize performance variation specific to the application of the part is important. Two types of tests are employed: acceptance tests and qualification tests. These categories can be used to distinguish tests that will be performed on an individual lot, and those that might be run on a regular basis, such as weekly or monthly.

Acceptance tests include:

- Thickness
- Appearance
- Tolerance
- Adhesion
- Porosity

Qualification tests include:

- Corrosion resistance
- Wear resistance
- Alloy composition
- Internal stress
- Hydrogen embrittlement
- Microhardness

The following list of requirements and test methods have been provided to offer the user of electroless nickel coatings an overview of the options available. The actual organization of requirements and test methods have been published in MIL-C-26074C and ASTM B733 and should be used to maintain a national standard for ordering and performance.

TEST METHODS

Appearance

The coating surface shall have a uniform, metallic appearance without visible defects such as blisters, pits, pimples, and cracks.

Imperfections that arise from surface conditions of the substrate and persist in the coating shall not be cause for rejection. Also, discoloration that results from heat treatment shall not be cause for rejection.

Thickness Measurement

The thickness shall be measured at any place on the significant surface designated by the purchaser, and the measurement shall be made with an accuracy of better than 5 percent by a method selected by the purchaser. Examples of common measuring devices are shown in Figs. 6.3 and 6.4.

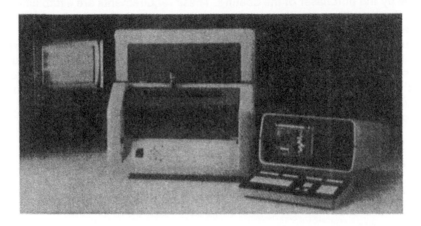

Fig. 6.3—X-ray fluorescence device measures thickness by analyzing the mass per unit area of the electroless nickel deposit according to ASTM B568.

Fig. 6.4—Beta backscatter device also measures thickness by analyzing the mass per unit area of the electroless nickel deposit. There are some limitations to this method, but the cost and availability of instruments makes it an excellent choice for many applications. ASTM B567 can be used to standardize and perform the measurements with this instrument.

Weigh, Plate, Weigh Method

Using a similar substrate material, weigh to the nearest milligram before and after plating, ensuring that the part is at the same temperature for each measurement. Calculate the thickness from the increase in weight, surface area, and density of the coating.

NOTE: The density of the coating will vary with the weight percentage of phosphorus in the coating. For a 9-percent-P alloy, the density is 8 g/cm³.

Example—A coupon made of mild steel has a weight of 3198 mg with an area of 19.736 cm² before plating. After plating with electroless nickel, the coupon weighs 3583 mg. Calculation for thickness is as follows:

$$T = \frac{3583 \text{ mg (after)} - 3198 \text{ mg (before)}}{8.01 \text{ g/cm}^3 \times 1000 \text{ mg/g}} \div 19.736 \text{ cm}^2$$

T = 0.00244 cm x 10,000 μm/cm

T = 24.4 μm

Table 6.1
Substrate Densities
For Weigh, Plate, Weigh Method

Steel, mild 1020	7.86
Stainless steel 316	8.02
Aluminum 2024	2.79
Aluminum 6061	2.70
Copper	8.91

Metallographic Sectioning

Plate a specimen of similar composition and metallurgical condition to the article being plated, or use a sample from the lot; mount and polish at 90° to the surface. Using a Vernier Calibrated Microscope, examine the thickness of the deposit and average over 10 readings.

NOTE: Accurate microscopic metallographic sectioning is very dependent on the sample preparation. Backing springs are recommended to reduce the smearing effects of the polishing step.

Micrometer Method
Measure a part of the test coupon in a specific spot before and after plating, using a suitable micrometer. Ensure that the part is at the same temperature for each measurement and that the surface measured is smooth.

Beta Backscatter Method
The coating thickness can be measured by the use of a beta backscatter device. The use of beta backscatter is restricted to base metals that have an atomic number of less than 18 or greater than 40. The actual phosphorus content of the coating shall be taken into consideration; consequently, the measuring device shall be calibrated using specimens of the same substrate having the same phosphorus content as the articles to be tested.

Magnetic Method
A magnetic thickness detector is applicable to magnetic substrates plated with autocatalytic nickel deposits that contain more than 10 weight percent phosphorus (non-magnetic) and that have not been heat treated. The instrument must be calibrated with deposits plated in the same bath on steel and whose thickness has been determined by the microscopic method.

X-ray Spectrometry
The coating thickness can be measured by X-ray spectrometry. This technique will measure the mass per unit area of the coating applied over the substrate. X-ray spectrometry equipment should be calibrated according to ASTM B568 with standards of known phosphorus and thickness. This method is non-destructive and will produce rapid and accurate results.

Coulometric Method
Measure the coating thickness in accordance with ASTM B504. The solution to be used shall be in accordance with manufacturer's recommendations. The surface of the coating shall be cleaned prior to testing.

Standard thickness specimens shall be calibrated with deposits plated in the same solution under the same conditions.

NOTE: This method is only recommended for deposits in the as-plated condition. The phosphorus content of the coating must be known in order to calculate the thickness of the deposit.

Adhesion Measurement
Bend Test
The sample specimen shall be bent through 180° over a minimum mandrel of 12 mm in diameter or four times the thickness of the specimen and examined at 4X magnification. No detachment of the coating shall occur. Fine cracks in the coating on the tension side of the bend are not an indication of poor adhesion.

Quench Test
Heat a plated article for 1 hr in an oven in accordance with Table 6.2 for the appropriate basis metals within ±10° C. Then quench in room temperature water. The appearance of blisters or peeling is evidence of inadequate adhesion.

NOTE: This test procedure may have an adverse effect on the mechanical properties of the articles tested.

Table 6.2
Substrate Heat-Treatment
Temperatures for Quench Test

Steel	300° C
Zinc	150° C
Copper or copper alloy	250° C
Aluminum or aluminum alloy	250° C

Punch Test
Make several indentations (approximately 5 mm apart) in the coating, using a spring-loaded center punch on which the point has been ground to a 2-mm radius. Blistering or flaking indicates poor adhesion.

File Test
By agreement with the purchaser, a file may be applied to the coated article. A non-significant area shall be filed at an angle of 45° to the coating, so that the base metal/coating interface is exposed. No lifting of the coating shall be observed.

Microhardness
ASTM B578 shall be used for Knoop hardness with a test load of 100 g. The instrument shall be verified on calibrated standard test blocks having a hardness similar to that of the deposit under test.

NOTE: For thin (less than 25 μm) deposits using less than 100 g loads, the standard commercial hardness tester produces varied results. This is due to the plastic deformation of the coating and the optical qualities of the instrument.

On thick (75+ μm) deposits, a surface microhardness determination using ASTM E384 is permissible.
Conversion of microhardness (Knoop or Vickers) to Rockwell scale is inaccurate and therefore inappropriate (see ASTM E140).

Hydrogen Embrittlement

ASTM F519 shall be used once a month to evaluate the plating process for relief of hydrogen embrittlement. A minimum of three V-notch tensile specimens made of AISI 4340 heat treated to a strength of 260 to 280 ksi shall be plated and loaded at 75 percent or greater of their ultimate notch tensile strength and held for 200 hours. No evidence of fracture or cracks shall exist.

Alloy Determination

There are generally three phosphorus ranges for electroless nickel deposits. Specific types of solution formulations will provide the selection of range of phosphorus in the alloy. Specific confirmation of the alloy can be accomplished by analyzing for nickel and phosphorus by one of the methods described below.

Most applications have been developed using a mid-range of 3 to 9.5 percent phosphorus and are considered typical. Confirmation of the alloy for these coatings can be by nickel or phosphorus, while coatings of less than 3 percent P or greater than 9.5 percent P should use both the nickel and phosphorus analysis to determine the alloy.

Preparation of Test Specimens

There are two general methods of preparing a foil specimen for this test. The most efficient technique is to plate a stainless steel panel with a 25- to 50-μm-thick deposit, cut the edges and peel the deposit off the panel.

Another way to produce an autocatalytic nickel phosphorus foil is to deposit a 25- to 50-μm-thick coating onto a masked aluminum panel. Then remove the maskant and remove the aluminum by immersing in 10 percent sodium hydroxide solution. When finished, a foil will have been produced that is acceptable for analysis. Although better adhesion is obtained using a zincate treatment, a coherent plate may be obtained by immersing clean aluminum foil in the autocatalytic nickel solution.

Determination of Nickel Content—Dimethylglyoxime Method

Reagents:

   1:1 v/v concentrated nitric acid (specific gravity 1.42)
   1 percent solution of dimethylglyoxime

Procedure:

Accurately weigh 0.1 g of autocatalytic nickel deposit and transfer to a 400 mL beaker. Dissolve in 20 mL of 1:1 nitric acid, boil to expel nitrous oxide fumes, then cool and dilute to 150 mL with distilled water. Add approximately 1 g of citric or tartaric acid to complex any ion that may be present and neutralize with ammonium hydroxide to pH 8 to 9.

Heat gently to 60 to 70° C, and while stirring, add 30 mL of dimethylglyoxime reagent. Allow to stand at 60 to 70° C for 1 hr, cool to below 20° C, and filter through a clean sintered glass crucible of No. 4 porosity. Wash the precipitate well with distilled water, dry in an oven at 110° C for 1 hr, cool, and weigh the precipitate as nickel dimethylglyoxime.

% Ni = (weight of precipitate x 0.2032 x 100)/sample weight

Determination of Phosphorus Content

The percentage of phosphorus is determined after dissolution of the deposit in acid, either colorimetrically or volumetrically.

Reagents:
For dissolution and oxidation
40 percent v/v concentrated nitric acid (specific gravity 1.42)
2 percent sodium nitrate solution
Approximately 0.1N potassium permanganate solution

For colorimetric phosphorus analysis
Molybdate vanadate reagent—Dissolve separately in hot water, 20 g ammonium molybdate and 1 g ammonium vanadate, then mix the two solutions. Add 200 mL of concentrated nitric acid (specific gravity 1.42) and dilute to 1 L.

For volumetric phosphorus analysis
Solution A: Dissolve 15 g of ammonium molybdate in 80 mL distilled water. Add 6 mL ammonium solution (specific gravity 0.880) and dilute to 100 mL.
Solution B: Dissolve 24 g ammonium nitrate in 60 mL distilled water. Add 33 mL concentrated nitric acid and dilute to 100 mL.

Procedure for dissolution and oxidation:
1. Dissolve 0.19 to 0.21 g (weigh accurately) of autocatalytic nickel deposit in 50 mL of 40 percent v/v nitric acid solution.
2. Heat gently until the deposit is fully dissolved. Then boil to remove brown fumes.
3. Dilute to approximately 100 mL, bring to the boil, and add 20 mL of approximately 0.1N potassium permanganate solution.
4. Boil for 5 minutes.
5. Add 2 percent solution of sodium nitrite dropwise until the precipitated manganese dioxide is dissolved.
6. Boil for 5 minutes, then cool to room temperature.
7. Dilute the solution in a volumetric flask to 250 mL and mix well.
At this stage, the phosphorus content may be estimated either colorimetrically or volumetrically, as described below.

Procedure for colorimetric analysis:
1. Transfer 10 mL of the solution from step 7 above to a 100-mL standard flask, add 50 mL distilled water, 25 mL molybdate-vanadate reagent, dilute to the mark with water, and mix well.
2. Read the absorption at 420 nm after 5 minutes, using 1 cm glass cells with water in the reference cell. Read off the concentration from a previously prepared calibration curve.

% P = mg P from graph/weight of sample (mg)

Procedure for preparation of calibration curve:
1. Dry potassium dihydrogen orthophosphate at 115° C for 1 hour.
2. Weigh out 0.4392 g, dissolve in water, and make up to 1 L of solution (1 mL = 0.1 mg phosphorus).
3. Prepare a calibration curve by adding 25 mL of reagent to 2 mL, 4 mL, 6 mL, 8 mL, and 10 mL liquids of this standard solution in 100 mL standard flasks and diluting to the mark. Read the absorption of these solutions exactly as for the estimated reading at 420 nm, as described above. Plot a calibration curve of absorption against mg phosphorus in samples of 0.2 mg, 0.4 mg, etc., up to 1.0 mg when prepared as above.

Procedure for volumetric analysis:
1. Transfer 10 mL of the solution from step 7 above to a stoppered flask and dilute to 100 mL with distilled water.
2. Warm to 40 to 50° C (*do not exceed this temperature*) and slowly add 50 mL of ammonium molybdate reagent while stirring.
3. Stopper the flask.
4. Agitate the flask vigorously for 10 minutes.
5. Allow the flask to stand for 30 minutes and filter through a Whatman No. 542 filter paper.
6. Wash the flask and precipitate with 1 percent potassium nitrate until the filtrate will not decolorize 1 mL of water containing 1 drop of 0.1N sodium hydroxide and 1 drop of phenolphthalein. This will require about 100 mL of the washing liquid.
7. Place the paper and precipitate in the original flask, add 50 mL water, and shake well.
8. Add 10 mL of 0.1N sodium hydroxide solution and shake well to dissolve the precipitate.
9. Add phenolphthalein indicator and back-titrate with 0.1N hydrochloric acid. Let "X" mL be the titration.

% P = [25 x (10-X) x 0.01349]/weight of sample

Spectra Analysis
A suitable method using emission spectra produced by Inductively Coupled Plasma (ICP) would be acceptable for analysis in nickel, phosphorus, and trace elements.
The following lines have been found to have low interferences when using argon ICP techniques. AA standards should be used for this analysis. Phosphorus standards should be made weekly to ensure accuracy.

Ni 216.10 nm	Al 202.55 nm	Cr 284.32 nm	Pb 283.30 nm
P 215.40 nm	Cd 214.44 nm	Cu 324.75 nm	Sn 189.94 nm
P 213.62 nm	Co 238.34 nm	Fe 238.20 nm	Zn 206.20 nm

Porosity Measurement

The test to be applied shall be decided by the purchaser in agreement with the plater.

Porosity Tests for Ferrous Substrates
Ferroxyl Test

The test solution is prepared by dissolving 25 g of potassium ferrocyanide and 15 g of sodium chloride in 1 L solution. The part is cleaned and immersed for 5 sec in the test solution at 25° C, followed by water rinsing and air drying. Blue spots visible to the unaided eye will form at pore sites. Their allowable number should be specified.

Alternately, strips of suitable paper (e.g., "wet strength" filter paper) are first immersed in a warm (about 35° C) solution containing 50 g/L of sodium chloride and 50 g/L of white gelatin and then allowed to dry. Just before use, they are immersed in a solution containing 50 g/L sodium chloride and 1 g/L of a non-ionic wetting agent, and pressed firmly to make satisfactory contrast onto the cleaned nickel surface to be tested and allowed to remain for 30 minutes.

If papers should become dry during the test, they should be moistened again, in place, with the sodium chloride solution. The papers are then removed and introduced at once into a solution containing 10 g/L potassium ferrocyanide to produce sharply defined blue markings on the papers, wherever the basis metal was exposed by discontinuities in the coating, leading to attack by the sodium chloride and transference of the ion components to the paper. If necessary, the same area may be retested.

Hot Water Test

Immerse the part to be tested in a beaker filled with aerated water at room temperature. Apply heat to the beaker at such a rate that the water begins to boil in not less than 15 minutes, and not more than 20 minutes. Then remove the part from the water and air dry. Examine the part for rust spots, which will indicate pores.

Aerated water is prepared by bubbling clean compressed air through a reservoir of distilled water by means of a glass diffusion disc for at least 24 hours. The pH of the aerated water should be 6.7 ±0.5. The test parts should be covered by at least 30 ±5 mm of aerated water.

Neutral Salt Spray

Testing in accordance with ASTM B117 shall be conducted monthly on the plating process. Coat a 4 x 6 x 0.20 AISI 4130 steel panel with 0.0015 in. of deposit. Wash edges and expose in salt spray chamber for a minimum of 240 hours. This panel shall have a rating of 9.5 or better in accordance with ASTM B537.

NOTE: This test can be made more aggressive by reducing the plating thickness and increasing the exposure time (e.g., 0.0005 in. for 1000 hours).
page 186

Hot Chloride Porosity Test

Immerse a steel part in 50 percent reagent HCl for 3 hours at 83° C. After testing, the acid shall not be significantly discolored or the part will have exfoliation or blisters. Equipment needed to perform this test: a glass beaker, hot plate, thermometer, and a timer.

The purpose of this test is to locate pore sites in the coating when the pore cell potential is greater than 150 mV. Rapid corrosion will occur, causing significant failure of the coating and substrate. A polyethylene cover affixed to the beaker with a large rubber band can be used to reduce the vapor and evaporation of the acid. This test should be performed under a fume hood to remove acid vapor.

Porosity Test for Aluminum Substrates

Wipe the specimens with a 10 percent solution of sodium hydroxide. After 3 min, rinse and apply a solution of sodium alizarin sulfonate (9,10-anthraquinon-1,2-dihydroxy-3-sulfonate, sodium salt) to the specimen. This solution is prepared by dissolving 1.5 g methyl cellulose in 90 mL boiling water to which, after cooling, a solution of 0.1 g sodium alizarin sulfonate dissolved in 5 mL ethanol is added.

After 4 min, apply glacial acetic acid at ambient temperature until the violet color disappears. Any red spots remaining indicate pores.

Porosity Test for Copper Substrates

Wipe the specimen with glacial acetic acid at ambient temperature. After 3 min, apply to the specimen surface a solution of potassium ferricyanide, prepared by dissolving 1 g of potassium ferricyanide and 1.5 g of methyl cellulose in 90 mL of boiling deionized water. After 2 to 3 min, the appearance of brown spots will indicate pores.

Standard Method for Measuring Corrosion Rate
Scope

This method establishes the apparatus, specimen preparation, test procedure, evaluation, and reporting of the corrosion rate of autocatalytic nickel deposits. The results are determined by using linear polarization and potentiodynamic techniques, and can be used to rank the coatings in order of corrosion resistance.

This test method is applicable to electroless nickel deposits applied to specimens of G5 specification and corroded in an artificial environment.

This test method is provided for use in an interlaboratory corrosion procedure for evaluating electroless nickel deposits.

Applicable Documents

ASTM standards:

G3 Standard Practice for Conventions Applicable to Electrochemical Measurements in Corrosion Testing

G5 Standard Practice for Standard Reference Method for Making Potentiostatic and Potentiodynamic Anodic Polarization Measurements

G15 Definition of Terms Relating to Corrosion and Corrosion Testing

G59 Practice for Conducting Potentiodynamic Polarization Resistance Measurements

Summary of Method

The specimen (working electrode) is prepared by plating in an electroless nickel solution a steel G5 plug. The plug is then cleaned and mounted in an electrochemical cell of G5 design. The cell is charged with a test solution at ambient temperature and purged with nitrogen. The cell is then pressurized with the gas in question and heated to the desired temperature. Measurements are then taken after 1000 sec to find the corrosion potential (E_{corr}). Polarization measurements are then made to determine the instantaneous corrosion rate for the applied pressure and temperature. The curves from various conditions and alloys can then be compared and the relative corrosion resistance can be established.

Significance and Use

The significance of these tests is that they can be used to rank electroless nickel deposits in order of their corrosion resistance. These tests can then be used to develop a comprehensive corrosion monitoring program to evaluate the substrate pretreatment and electroless nickel deposit.

Results from these tests produce quantitative values as to the corrosive nature of the environments and the protection afforded by specific electroless nickel deposits under laboratory conditions.

Apparatus

The following required equipment is described in ASTM G5:
1. Potentiostat calibrated according to ASTM G5
2. Potential measuring device
3. Current measuring device
4. Saturated calomel electrode
5. Salt bridge probe

The specimen (working electrode) to be tested is plated with electroless nickel and post-treated with the desired process. The specimen can be made from any material to the following physical dimensions: 1/2 in. (12.7 mm) long, 3.8 in. (9.5 mm) diameter, with a 3-84 tapped hole at one end. The entire surface shall be 16 rms or better.

NOTE: The specimens used in this ASTM program are made from C1215 steel.

The polarization cell shall be made to the following ASTM G5 specifications:
1. Carbon or platinum counter electrodes shall be used in the cell.
2. The salt bridge shall be adjustable and able to be located near the tip of the working electrode.
3. The electrolytes shall be agitated constantly with a magnetic stirrer.
4. The cell shall be purged constantly with N_2 gas.
5. The cell shall be able to operate at 1 atmosphere.
6. The cell shall be able to operate at a temperature of 22° C.
7. The specimen holder shall be designed and maintained to form a tight connection between the specimen and the teflon gasket of the holder.

Reagents

Corroding Electrolyte I is prepared by dissolving 12.5 g of reagent grade sodium chloride (NaCl) into 900 mL of Type I reagent water.* Adjust pH to 7.8 ±0.1 with sodium hydroxide. Fill to 1000 mL with Type I reagent water.

Corroding Electrolyte II is prepared by dissolving 40 g of reagent grade sodium chloride (NaCl) into 1000 mL of Type I reagent water.

Specimen Preparation

Prepare the specimen by the following steps:

1. Vapor degrease or solvent wash.
2. Caustic clean with appropriate cleaner for steel or aluminum substrate. Note mounting hole will expose substrate alloy to cleaning solution.
3. Rinse with Type I reagent water.
4. Wash with 25 percent sulfuric acid for 60 sec at 25° C.
5. Rinse with Type I reagent water.
6. Mount the specimen on the electrode holder. Care should be exercised not to contaminate or disturb the cleaned surface.

Test Parameters

Potentiodynamic polarization:

1. Initial potential -100 mV from E_{corr}
2. Final potential -600 mV
3. Vertex potential +1000 mV
4. Input anodic tafel constant (ATC) 0.145; cathodic tafel constant (CTC) 0.150
5. Scan rate 2.0 mV/sec

Linear polarization:

1. Initial potential -30 mV E_{corr}
2. Final potential 30 mV E_{corr}
3. Anodic and cathodic tafel constant from potentiodynamic
4. Scan rate 0.2 mV/sec

Specimen physical parameters:

1. Area 4.285 cm^2
2. Density 7.95 g/cm^3
3. Equivalent weight 28.51 mols/electron

Wear Analysis

Abrasion Resistance

Abrasion resistance can be measured by a Taber Abrader using a CS-10 wheel. Wear specimens should be dressed for the first 1000 cycles and then weighed. The CS-10 wheels should be redressed for 50 cycles for each 1000 cycles of wear on the specimen. Typical tests are taken to 10,000 cycles with a range of 15 to 30

---

*Type I reagent water is defined in ASTM D1193 as 16.67 M ohm-cm resistivity.

mg/1000 cycles for as-plated deposits, and 6 to 18 mg/1000 cycles for precipitation-hardened deposits. The test is not precise, but can be used to characterize the differences between coatings when several specimens are tested.

Adhesive Wear Resistance

Adhesive wear resistance can be tested by several means. These include pin on disc, block on ring, and pin on notch. Each of these types of adhesive wear tests simulate slight differences in metal-to-metal wear. To evaluate the specific wear for a particular application, the correct wear test method must be selected. Conditions such as temperature, lubrication, load, vibration, wear scar, velocity, and others will affect the results requiring careful study of the test methods before a test can be selected.

The Alpha-LFW1 is a block-on-ring test that has been found to provide wear information on electroless nickel coatings. The Falex tester is a pin-on-notch device that provides information on high load wear. Both of these tests can be run dry or lubricated, and can be used to characterize the differences between coatings.

Stress Analysis

The intrinsic stress of the deposit can be determined using ASTM B636. This test uses a spiral contractometer, plating the spiral and measuring the amount of movement in the helix while the spiral is plating. The amount of intrinsic stress in the deposit is determined by the rate of swing in a needle attached to the spiral. The movement is magnified by a factor of 10, with readings being taken while the process is operating. After the measurement of intrinsic stress has been taken at plating temperature, the helix can be removed and cooled. A second reading can be taken, providing the total stress, and then the thermal component can be calculated:

Spiral total stress = intrinsic stress + thermal stress

The results can be used to predict when adhesion may be compromised on aluminum and other alloys, as well as the preference for certain types of corrosion.

Other methods of analysis for total stress have been developed using a rigid strip. Strips are first plated on both sides, then one side is removed with a stripper. The thickness of the strip and the coating are measured, and then the amount of bow in the strip is measured. Calculations can be made from this information, and the total stress in the part can be selected.

POST-TREATMENTS

The quality control of an electroless nickel deposit implies that certain post-treatments have been completed. These are generally used to improve the deposit adhesion or increase the hardness of the deposit by precipitation of phosphorus to nickel phosphide.

In most specifications, the requirements for heat treatment are described. Generally there are no tests for these processes, and self-certification must be used. In some cases, a test may exist that is destructive, and therefore not usable.

MIL-C-26074C and ASTM B733 both use the same classification system to describe the post-heat treatment of the deposit. These documents use the term *class* and establish the steps to reduce the potential for hydrogen embrittlement, increase the adhesion of the coating, improve the fatigue properties of the part(s), and increase the wear resistance and hardness of the coating.

Classes of Electroless Nickel Coatings
Class 1
As plated, no treatment other than hydrogen embrittlement relief at 190 ±10° C for steels with a tensile strength of 1051 MPa or greater.

Class 2
Heat treated for ·hardness, the coating shall be heat treated to a minimum hardness of 850 Knoop (100 g load). This hardness can be produced by heat treating the coating at 400° C for 60 minutes. Higher temperatures for shorter times may be used.

Class 3
Heat treatment at 180 to 200° C for 2 to 4 hours to improve coating adhesion on aluminum.

Class 4
Heat treatment at 120 to 130° C for a minimum of 1 hour to improve adhesion on heat-treatable (age-hardened) aluminum alloys and carburized steels.

Class 5
Heat treatment at 140 to 150° C for a minimum of 1 hour to improve adhesion on non-age-hardened aluminum and beryllium alloys.

The use of the shop traveler to record the completion of the heat treatment, as well as a notation on the chart record of the oven, is a standard practice.

Additional post-treatments such as silicates, water glasses, waxes, and chromates, are sometimes applied to prevent staining and oxidation of the nickel. The analysis of these treatments are complicated and seldom tested. Contact resistance tests are generally used if the treatment requires testing.

QUALITY CONTROL

To conclude, the electroless nickel facility must have a quality control system. The platers need to identify how the parts are to be processed. In addition, they must maintain the chemistry, qualify the processes, and complete the acceptance tests as required by the specifications.

The system can be operated from a single notebook or from a multiuser computer system. The quality control program can be highly complex, or just cover the essentials.

In the end, it is dedicated people who must build and follow the quality control system. It is people who will produce higher levels of quality and make it possible to extend the performance of electroless nickel deposits into new markets and applications.

Chapter 7
Surface Preparation
For Electroless Nickel Plating

Juan Hajdu

Electroless nickel is plated over a large number of metallic and nonmetallic substrates of very different compositions and properties. For these reasons, it is impossible to select a single general approach to surface preparation. Specific procedures are required for each type of substrate.

In broad terms, we can classify the substrates as metallic and nonmetallic. Most electroless nickel plating is done on metal parts, and in this chapter we will discuss mainly the preparation of metal surfaces.

An important characteristic of plating metals is the strength of the bond that can develop between the base metal and the coating. Metal-to-metal bonds with high adhesion values require thorough surface preparation—removing from the base metal surface foreign contaminants (soil, dirt, corrosion products, oxides, tarnish, and others), and eliminating mechanically distorted surface layers—to present a clean, healthy surface structure.

The removal of foreign contaminants is generally accomplished by using commercial alkaline cleaners. The selection is based on the nature of the contaminants and the type of substrate. The suppliers of cleaners will assist in the selection of the right material; there are also numerous literature sources that discuss metal surface cleaning (1-5).

The removal of surface oxidation and unwanted metal is accomplished by chemical attack. Acid pickling solutions and alkaline deoxidizing materials, similar to those used in electroplating, are also effective for electroless nickel plating. In some cases, mechanical surface treatments, such as shot peening or sandblasting, are used in surface finishing prior to chemical treatment, especially with large, expensive parts that allow the use of manual processing.

Electroless plating differs from other metal coating techniques (electroplating, vacuum metallizing, hot dip galvanizing) in that the substrate initiates the autocatalytic chemical reduction process. Some of the metals, such as nickel, are catalytic when immersed in an electroless plating solution, while steel and aluminum become catalytic by the formation of an immersion nickel deposit in the electroless nickel bath. Other metals, such as copper, are passive and require activating steps in order to initiate the electroless nickel plating process. It should be noted that even naturally active surfaces can become passive when contaminated by foreign residues or oxide layers. For this reason, surface preparation for electroless nickel plating requires the highest degree of care and control of all metal finishing procedures.

Another reason for the need of a very careful selection of the preparation process is that it can significantly affect the porosity of the metal deposit. This is particularly true for electroless nickel. Residues from cleaners and deoxidizers can increase the porosity of electroless nickel, creating passive spots that will not initiate electroless plating (6). Thin (less than 5 μm) electroless nickel deposits are more porous than electrodeposited nickel of comparable thickness. During electroless nickel plating, the deposition process initiates at discrete sites and the substrate will become fully covered through the lateral growth of these sites. Short plating times will not allow full coverage of the base material, causing porous electroless deposits.

IRON AND FERROUS ALLOYS

Iron and its alloys are the most frequently plated substrates. From a surface preparation standpoint, they can be grouped as low alloy and carbon steels; cast iron; and high alloy steels. The principles of preparing ferrous surfaces are based on successive steps for soil removal, deoxidation, and surface activation. The manner in which these steps are carried out will depend on the type of alloy processed.

Steel and cast iron parts are cleaned using conventional techniques such as alkaline cleaning, solvent cleaning, or electrocleaning similar to the processes used in electroplating or other metal finishing procedures. In some cases, where the surface has been severely corroded or is covered by scale or other strongly adherent residues, mechanical pretreatments such as sandblasting, shot peening, or wheelabrading may be necessary.

Alkaline soak cleaning with commercial cleaners are frequently preferred. These materials contain a combination of alkaline sodium compounds such as hydroxide, carbonate, silicates, phosphates, and organic surfactants. The selection of the cleaner will depend on the nature of the surface contamination and substrate, and the assistance of the supplier can be very helpful in achieving effective and economical cleaning. In some cases, solvent or vapor degreasing may be used prior to the alkaline cleaning steps. A good summary of cleaning practices for steel can be found in the ASTM Standard Practices (5).

Carbon and low-alloy steels are deoxidized using either acid pickling solutions or alkaline deoxidizers. Pickling solutions, in general, use hydrochloric or sulfuric acids, combinations of these mineral acids, or salts of the acids. On carbon steels, hydrochloric acid is used at 10 to 50 percent concentration while sulfuric acid (2 to 10 percent) is also used. Sulfamic acid pickles are recommended for leaded steels. It is convenient to avoid the use of pickling inhibitors in this process, since they may interfere in the initiation and activation necessary for electroless plating.

Alkaline deoxidizers containing organic chelating agents and/or sodium cyanide are frequently used in surface preparation, since they offer specific advantages over acid deoxidizers. Alkaline deoxidants will remove oxides without attacking the substrate metal and are less prone to cause hydrogen embrittlement than acid solutions. Alkaline deoxidizers are used electrolytically, either in the anodic mode (the parts become the anode), or using periodically reversed (PR) currents where the parts are sequentially the anode or the cathode in the process. The use of PR with alkaline deoxidizers is especially useful in combination with acid pickling of steels, where carbon and other alloying constituents can form loosely adherent layers on the substrate. These surface products must be removed effectively prior to plating, in order to obtain good bonding of the nickel to the substrate.

The procedures required to obtain initiation and adhesion of the coating to the substrate (substrate activation) will depend mainly on the composition of the substrate. In the case of mild steels and carbon steels, good cleaning and deoxidation is in most cases sufficient for good bonding. In the case of cast iron parts and high alloy steels, special techniques are needed for obtaining good coverage and adhesion.

Cast iron parts are normally porous and may require a number of special alkaline and acid steps to clean the pores and to avoid bleeding-out of rust through the nickel coating. Cast iron parts should be immersed in weak acid pickling solutions only briefly, to avoid exposing and loosening the graphite present in the casting, as well as to minimize acid entrapment.

High alloy steels, such as the 300 and 400 series of stainless steel, require special treatments to activate the surface for fast initiation and good bonding. In general, hydrochloric acid at relatively high concentrations is used for activating stainless steel. Alkaline deoxidizers used with periodic reverse current are also employed for surface preparation. When this type of deoxidation is used on high alloy steels, the last cycle should be run with the parts on the cathodic side to avoid formation of passive oxide layers under anodic conditions.

One approach that has been used very successfully for obtaining active surfaces is to strike the parts electrolytically in a chloride nickel-hydrochloric acid bath ("Woods strike"). Sulfamate nickel- and nickel glycolate-based strikes are used sometimes for activation (5). These electrolytic nickel strikes act by chemical deoxidation, and by depositing a thin layer of active nickel to overcome passivity and to provide a catalytic surface for the preparation of the electroless nickel process.

Typical cycles for ferrous surfaces are:

Low Carbon Steel
  Solvent preclean (if necessary)
  Water rinse
  Alkaline soak clean or electroclean
  Water rinse
  Activate in 40 percent by vol. HCl
  Water rinse
  Electroless nickel plate

High Carbon Steel or Heat Treated Steel
  Solvent preclean (if necessary)
  Water rinse

Alkaline soak clean or electroclean
Water rinse
Deoxidize in 40 percent HCl (and/or alkaline electrolytic deoxidize)
Water rinse
Electroless nickel plate

Stainless Steel or Nickel Alloys
Alkaline clean
Water rinse
Alkaline deoxidizer with periodic reverse current
Water rinse
Activate with 40 percent HCl
Woods strike—cathodic 2 A/dm^2
Electroless nickel plate

Cast Iron
Alkaline clean
Water rinse
Alkaline deoxidizer with periodic reverse current
Water rinse
Electroless nickel plate

Woods Strike (4)
Nickel chloride	240 g/L
Hydrochloric acid (32 percent)	320 mL/L
Current density:	
anodic	1 A/dm^2, 30 to 60 sec.
cathodic	2 A/dm^2, 2 to 6 min.

Sulfamate Strike (4)
Nickel sulfamate	320 g/L
Boric acid	30 g/L
Hydrochloric acid	12 g/L
Sulfamic acid	20 g/L
Current density	1 to 10 A/dm^2

Glycolate Strike (4)
Nickel acetate	65 g/L
Boric acid	45 g/L
Glycolic acid (70 percent)	60 mL/L
Sodium saccharin	1.5 g/L
Sodium acetate	50 g/L
pH	6.0
Current density	2.7 A/dm^2

ALUMINUM AND ALUMINUM ALLOYS

After steel and ferrous alloys, aluminum alloys constitute the largest group of substrates for electroless nickel plating. Electroless nickel imparts hardness, wear, abrasion resistance, solderability, and corrosion resistance to aluminum substrates. The preparation of aluminum and its alloys is quite specific to this metal, since aluminum exposed to air is always covered by a dense oxide coating that must be removed before the parts can be plated. Furthermore, the deoxidized parts must be protected during the transfer, to avoid reoxidation of the highly active aluminum surfaces.

Another characteristic of aluminum is its potential reactivity with electroless nickel solutions. Most of the phosphorus nickel baths are acidic, and with nickel being very positive to aluminum on the electromotive scale, immersion deposits of nickel and chemical attack of the aluminum substrate will occur, interfering with the good adhesion of the electroless nickel coating. To protect the aluminum surfaces during processing, zinc immersion deposits ("zincates") are used. The zinc deposit protects aluminum against reoxidation from atmospheric exposure and redissolves in the electroless nickel solution, exposing an oxide-free aluminum substrate, upon which nickel deposits form adherent coatings. Other methods based on immersion deposits or oxide films (7-9) have been proposed for this application, but are seldom used in electroless nickel plating. For this reason, only processes based on zincating will be reviewed here. Aluminum parts are generally prepared by submitting them to alkaline cleaning, acid or alkaline etching, deoxidizing (conditioning), and zincating. Cleaning is done with mild alkaline cleaners that will not attack (or only attack slightly) the substrate, while still removing superficial organic contaminants. Strongly alkaline cleaners will aggressively attack aluminum, interfering with appearance and adhesion of the electroless nickel deposit.

Conditioning is needed to remove alloying elements from the surface and to prepare it for uniform zincating. This step is generally done by immersion in strong nitric acid-based solutions. Nitric acid forms a thin, light, uniform oxide film on aluminum that protects it from further attack by the acid. The composition of the conditioning solution will depend on the type of alloy being treated. For pure commercial aluminum and aluminum-magnesium wrought alloys, simple nitric acid solutions (50 percent by vol.) can be used. For treating alloys with a higher percentage of metallic constituents, two-step conditioning processes are used, in which the first step consists of removing interfering constituents with mineral acids (such as sulfuric or phosphoric acid mixtures) or alkaline etchants and then treating the parts with 50 percent nitric acid. For casting alloys that contain a high percentage of silicon, hydrofluoric acid or fluorides are added to the nitric acid to dissolve silicon.

Zincating was developed originally for the electroplating of aluminum and has been widely reviewed in technical and patent literature (8,10-12). The principle of the process is the formation of an immersion deposit of zinc on the aluminum surface by displacement from an alkaline zincate solution. The basic formulation of a zincate bath consists of zinc oxide (50 to 100 g/L) and sodium hydroxide (250 to 500 g/L). Proprietary zincate solutions contain other ingredients, such as copper and iron salts and organic chelating agents, to produce more uniform coatings and to permit the use of lower concentrations of zinc ("dilute zincates"). Many formulations for this type of material can be found in the patent literature (13,14). Typical formulations for dilute zincate solutions contain 5 to 10 g of zinc oxide and 50 to 120 g/L of sodium hydroxide. Acid zinc immersion systems have also been reported (16) using zinc sulfate (720 g/L) and hydrofluoric acid (17.5 mL/L).

The mechanism of zincating and the reaction of the zinc immersion deposits in the electroless plating bath have been reported in several papers (8,12,15). The basic mechanism of zincating is the displacement of zinc from an alkaline zincate solution by aluminum:

$$3ZnO_2^= + 2Al° + 2H_2O = 3Zn° + 4OH^- + 2AlO_2^-$$

The rate of this reaction is determined by the composition of the zincate solution and the type of aluminum alloy. There is a strong relation between the structure, uniformity, and thickness of the zinc film and the adhesion and protective value of the electroless nickel coating. In general, fine-grained, thin, and tight zinc films produce the best results. Thus, the zincating bath's composition, temperature, and immersion times must be controlled and adjusted to the nature of the aluminum alloy to be treated. A zincating process that performs well for one type of aluminum alloy may not be effective if the substrate alloy is changed (17,19,20). Figures 7.1 and 7.2 illustrate the influence of temperature and alloy composition on the thickness of the zinc immersion deposit.

One important aspect of the zincating process to be considered is the mechanism of the initiation of the electroless deposition on zincated aluminum surfaces. The bulk of the zinc immersion deposit is dissolved in acid electroless nickel plating solutions, exposing the aluminum, which is covered by nickel by immersion, initiating the autocatalytic deposition process. Still, this dissolution process is not complete, and in general, residual zinc is found under the nickel layer. The influence of this zinc-rich layer has been investigated by Mallory (18), and its presence can be considered beneficial to the performance of the electroless nickel plating.

An approach that has been widely used for cast and wrought aluminum alloys is the double zincating procedure. In this process, the initial zinc layer is removed by a dip in 50 percent nitric acid, and a second zinc layer is deposited by a short immersion in the zincate solution. The benefits of this treatment come from the use of a less critical first zincating step to deoxidize the substrate and remove alloying inclusions, while the second zincate treatment can be adjusted to produce a thin, tight zinc deposit on the uniformly conditioned surface (11,12,17).

An important factor to consider here is change in composition of electroless nickel solutions as they age, by accumulation of byproducts of the chemical

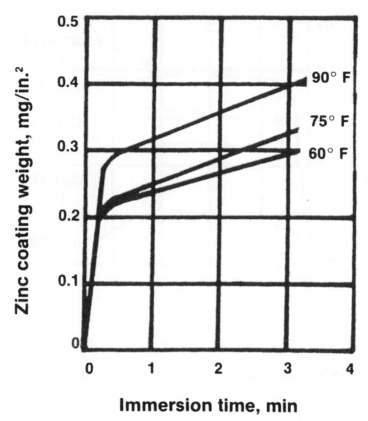

Fig. 7.1—Effect of temperature of zincate solution on weight of zinc film deposited on aluminum sheet.

reduction of nickel (21). Older solutions become more aggressive and give less adherent nickel coatings, and for this reason, surface preparation procedures that produce good results with relatively fresh electroless nickel plating baths may yield nonadherent or blistered deposits as the bath ages. This is the main reason why electroless nickel baths have a shorter life when used for plating aluminum than when used on steel. The use of a non-aggressive electroless nickel strike has been proposed to remedy this problem (22) and extend the useful life of the electroless plating bath.

While the plating of aluminum may appear too complex for consistent processing, industry has been very successful in resolving the requirements of depositing electroless nickel on a broad variety of alloys and satisfying many end use requirements. Experience is the key to success in the plating of aluminum and it cannot be substituted by standard pre-established procedures. The following cycles are presented to illustrate industry practice:

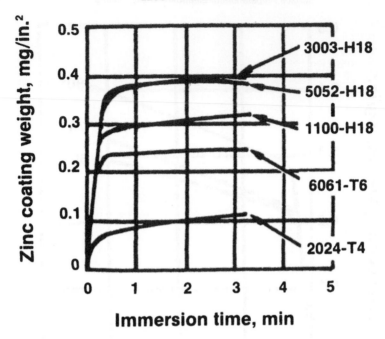

Fig. 7.2—Weight of zinc film obtained from the alkaline zinc immersion process with several aluminum alloys.

Wrought Aluminum Alloys
Solvent preclean (if necessary)
Water rinse
Alkaline soak clean
Water rinse
Etch in 15 percent by vol. sulfuric acid (80° C) or 50 g/L sodium hydroxide (50° C) for 1 to 3 min.

Water rinse
Nitric acid (50 percent) 30 to 60 sec
Water rinse
Second zincate treatment 10 to 20 sec
Water rinse
Electroless nickel plate

NOTE: For aluminum alloys containing 99 percent Al (1000 series), etching and the second zincate treatment may be omitted.

Aluminum Castings
Solvent preclean (if necessary)
Water rinse

Alkaline soak clean
Water rinse
Acid etch 30 sec in nitric acid (50 percent), sulfuric acid (25 percent), and ammonium bifluoride (120 g/L)
Water rinse
Zincate treatment 20 to 30 sec
Water rinse
Electroless nickel plate

COPPER AND COPPER ALLOYS

As mentioned earlier, copper is one of the metals that will not catalytically initiate hypophosphite electroless nickel solutions. Copper can initiate aminoborane and borohydride-reduced baths.

To render copper and copper alloys catalytic, the following approaches are used in commercial operations:

1. Electrochemically initiating by contacting the parts with active metals (steel or aluminum), or by striking the parts electrolytically in a dilute nickel chloride strike, or in the electroless nickel bath.

2. Treat the surface with a strong reducer, such as dimethylamino borane (22,23).

3. Catalyzing the surface by immersion in dilute palladium chloride solutions (not recommended).

A typical sequence could be:

Soak clean
Rinse
Electroclean anodically
Rinse
Acid dip in 1 to 2 percent H_2SO rinse
Activate, using either method 1 or 2 above.

A special hypophosphite-reduced bath using nickel chloride was proposed for direct plating (24), but is not used commercially. The cleaning and conditioning of copper and copper alloys is done with commercial alkaline cleaners and mild acids.

MAGNESIUM

There is a great interest in using magnesium alloys for weight reduction, and electroless nickel offers good protection for magnesium-based parts. While used for commercial applications, plating magnesium alloys is not an easy task. Magnesium is a highly reactive metal and conventional cleaning and deoxidizing solutions are not adequate for preparing magnesium. A general discussion of preparation of magnesium can be found in the literature of electroplating (1,25).

The two recommended approaches are based on zincating or fluoride predips after alkaline cleaning and pickling with chromic acid solutions. The most widely used processes are based on fluoride predips and use special electroless nickel plating solutions that contain fluorides to improve adhesion (26). A typical cycle is:

    Alkaline rack clean
    Rinse
    Activate in mineral acids containing fluorides
    Rinse
    Cyanide copper plate
    Rinse
    Proceed as with copper parts

BERYLLIUM

Beryllium surfaces require protective, corrosion-resistant coatings for nuclear energy and aerospace applications. Since beryllium exposed to air is covered by an oxide layer, zincate treatment is used to remove the oxide and protect the base metal (1,27). A typical surface preparation uses double zincating after alkaline cleaning and acid deoxidation (28).

TITANIUM

Titanium surfaces are always covered by a tight oxide film that interferes with the good bonding of plated metal coatings. The oxide layer must be removed, preferably by mechanical and chemical attack. The pickling solutions are, in general, based on nitric and hydrofluoric acids (1,29).

After removing the oxide, the surface is protected by a conversion coating using fluorides (30), chromates (31), or zinc immersion deposits (1). The most widely used processes are based on fluoride-containing acid solutions, applied either by immersion or anodically prior to electroless nickel plating. It should be noted that both electroless and electrolytic plating of titanium are plagued by adhesion problems. A possible cycle could be:

    Mechanically treat surface
    Activate in 50 percent HCl solution
    Woods nickel strike
    Electroless nickel plate

ZINC

At the present time, a number of commercial zinc die casting operations involve direct plating of electroless nickel for enhanced appearance and protection. The largest single use of electroless nickel-plated zinc die castings is for carburetors used with alcohol fuel, especially in Brazil.

In many operations, the parts are striked first with electrolytic cyanide copper prior to electroless nickel plating. The copper strike baths typically contain 20 to 45 g/L cuprous cyanide, 10 to 20 g/L free sodium cyanide, and 15 to 75 g/L sodium carbonate. Copper strike baths have the ability to cover complex shapes and recessed areas when used under the right operating conditions. Most textbooks on electroplating extensively describe the operation of copper strike baths and the cleaning of parts prior to copper plating (1,32).

While this approach gives excellent results, there is a continuous interest in finding a less costly process that requires no electroplating and avoids the use of cyanide compounds. Several methods have been proposed, and proprietary materials are available commercially that allow direct electroless plating of zinc die castings.

MOLYBDENUM AND TUNGSTEN

Molybdenum and tungsten have been reported to require contact or electrolytic activation prior to electroless nickel plating. Mixtures of nitric, sulfuric, and chromic acids are used to deoxidize the surfaces. A typical cycle for tungsten uses 50 percent nitric acid followed by hot potassium hydroxide (100 g/L) etch. For molybdenum, parts are activated in a 10 percent nitric and 12 percent sulfuric acid mixture after cleaning. Both metals require activation with palladium or with a Woods nickel strike.

A typical cycle for plating on molybdenum or tungsten is as follows:

Alkaline potassium ferrocyanide 215 g/L in 75 g/L KOH), 1 min at room temperature
Water rinse
KOH (10 percent), 10 min at boiling
Water rinse
HCl (10 percent), 30 sec at room temperature
Rinse
Activate with $PdCl_2$ (0.1 to 0.5 g/L $PdCl_2$ in 1 to 3 mL/L HCL)
Rinse
Electroless plate

NONMETALLIC SUBSTRATES

The main difference between metallic and nonmetallic surfaces resides in the nature of the bond between substrate and coating. While adhesion to metal is of an atomic nature, the adhesion to organic and inorganic substrates is only mechanical. The basis for obtaining adhesion to these materials is to develop the right topography on the surface by means of chemical or mechanical treatment.

Nonmetallic surfaces lack catalytic properties and therefore require activating treatments that will render them catalytic. In general, this activation is done by seeding the surface with a catalytically active metal.

Parts are plated with the electroless nickel for several reasons. One of the major uses for electroless nickel is to form a conductive base for electroplating on plastic or ceramic substrates. Ceramic and glass parts are electroless nickel plated for bonding or soldering applications. Diamond and other abrasive particles are encapsulated in electroless nickel for manufacturing plastic-bonded cutting tools.

The use of electroless nickel for metallizing plastics is discussed in Chapter 14, "Plating on Plastics." There are only limited uses of electroless nickel on plastics for mechanical or decorative applications. A novel application of electroless nickel is to protect electroless copper deposits used in electronic devices, such as shields for electromagnetic interference. This application is also discussed in a separate chapter.

Electroless nickel has been applied extensively to ceramic, glass, and silicon parts used mainly in the electronics industry. Typical uses are to metallize conductors, capacitors, transducers, silicon devices, hybrid circuits, and other electronic components. The processing steps generally include roughening by chemical attack (if the surface requires it), activating with a catalytic metal, and then electroless nickel plating. The composition of the chemical etch will depend on the nature of the substrate. Some typical etch applications are as follows:

Alumina Ceramics
Soak clean with alkaline cleaner
Rinse
Etch in 25 vol. percent hydrofluoric acid (50 percent) for 5 min at room temperature

Barium and Zirconium Titanates
Soak clean with alkaline cleaner
Rinse
Etch in 10 vol. percent fluoboric acid (48 percent) for 10 min at room temperature

Glass and Ceramics
Soak clean with detergent
Rinse
Etch with hydrofluoric acid (40 percent)   40 mL/g
Ammonium fluoride   18 g/L

Silicon Devices
Hydrofluoric acid solutions with an addition of ammonium fluoride of varying concentration, depending on the type of etch desired, are used in the metallization of silicon wafers.

The roughened parts are rinsed and then rendered catalytic either by the two-step sensitizing and activating process or by using commercial activators. Sensitizing is accomplished by immersing the parts in an acid stannous chloride solution (10 to 100 g/L of $SnCl_2$, $2H_2O$), then rinsing and activating with a solution of palladium chloride (0.1 to 0.5 g/L PdCl in 1 to 3 mL/L HCl). Commercial activators containing mixed palladium-tin compounds are also used at 1 to 5 percent dilution. After activation, the parts should be ready for electroless plating either with hydrophosphite- or amino borane-reduced electroless nickel solutions. It must be remembered that the final adherence of the electroless nickel deposit will depend not only on the pretreatment, but also on the characteristics of the plating process. Aggressive plating baths or deposits with high internal stresses should be avoided, and thick deposits are very difficult to produce on this type of substrate.

MASKING

Many parts may require only partial coverage with electroless nickel, and portions of the substrate must be protected by masking. Many stop-off materials (tapes and coatings) used for electroplating can be applied for masking parts for electroless plating. Since most electroless nickel solutions operate at high temperatures, and the parts may stay in the bath for a long time, the masking materials should be tested thoroughly before using them on expensive parts. Masking materials may also release organic or metallic contaminants that may harm the nickel deposits.

A problem commonly encountered when plating small areas in large parts is how to adjust the plating process to very low surface-to-volume ratios. All electroless plating solutions will perform best above minimum loading levels, which may be difficult to reach with masked parts.

REFERENCES

1. F.A. Lowenheim, *Modern Electroplating*, third edition, John Wiley and Sons, New York, 1974
2. F.A. Lowenheim, *Electroplating*, McGraw-Hill Book Co., New York, 1978.
3. L.J. Durney, *Electroplating Engineering Handbook*, fourth edition, Van Nostrand Reinholt, New York, 1984.
4. ASTM B656-86, "Standard Guide for Autocatalytic Nickel-Phosphorus Deposits on Metals for Engineering Use."
5. ASTM B322-68, "Standard Practice for Cleaning Metals Prior to Electroplating."
6. E.B. Saubestre and J. Haydu, *Oberflaeche-Surface*, **9**(3), 53 (1968).
7. G.S. Petit et al., *Plating*, **59**, 567 (1972).

8. D.S. Lashmore, *Plat. and Surf. Fin.*, **65**(4), 74 (1978).
9. D.S. Lashmore, ibid., **68**(4), 48 (1981).
10. E.W. Hewiston, U.S. patent 1,627,900 (1927).
11. J. Korpinn, U.S. patent 2,142,564 (1939).
12. D.S. Lashmore, *Plat. and Surf. Fin.*, **67**(1), 37 (1980).
13. W.G. Zelley, U.S. patent 2,650,886 (1953).
14. E.B. Saubestre and L.J. Durney, U.S. patent 3,329,522 (1967).
15. E.B. Saubestre and J.L. Morrico, *Plating*, **53**, 899 (1966).
16. J. Heiman, *J. Electrochem. Soc.*, **95**, 205 (1949).
17. I. Keller and W.G. Zelley, ibid., **97**, 145 (1950).
18. G. Mallory, *Proc. 70th AES Annual Technical Conference*, 1983.
19. D.S. Lashmore, *J. Electrochem, Soc.*, **127**, 543 (1980).
20. .A.W. Blackwood, E.F. Yarkosky and J. Boupo, *Proc. 68th AES Annual Technical Conference*, 1981.
21. E.F. Yarkosky, M. Alexsinas and A. Marzak, *Proc. 66th AES Annual Technical Conference*, 1979.
22. J. Haydu, E.F. Yarkosky and P. Schultz, *Proc. 3rd AES Electroless Plating Symp.*, 1986.
23. T. Brezins, U.S. patent 3,096,182 (1963).
24. N. Feldstein, U.S. patent 3,993,380 (1976); U.S. patent 4,305,997 (1981).
25. D.G. McBride, *Plating*, **59**(9), 858 (1972).
26. ASTM B480-68, "Recommended Practices for Preparation of Magnesium and Magnesium Alloys for Electroplating."
27. H.K. DeLong, U.S. patent 3,152,209 (1967).
28. K. Parker and H. Shah, *J. Electrochem. Soc.*, **117**, 1091 (1970).
29. W.H. Roberts, "Coating Beryllium with Electroless Nickel," U.S. Atomic Energy Commission Contract AT(29-1) 1106, Ber RFP-478.
30. ASTM B418-69, "Recommended Practices for Preparation of Titanium and Titanium Alloys."
31. D.S. Harshorn, U.S. patent 3,725,217 (1973).
32. M. Thoma, *Plat. and Surf. Fin.*, **59**(6), 76 (1983).
33. ASTM B252, "Practice for Plating Zinc Alloy Die Castings for Electroplating and Conversion Coatings."

Chapter 8
Engineering Applications of Electroless Nickel

Joseph Colaruotolo and Diane Tramontana

Engineering applications for electroless nickel can be found in virtually every industry. Various physical characteristics of electroless nickel coatings, such as hardness, wear resistance, coating uniformity, and corrosion resistance, as well as the ability to plate non-conductive surfaces make this a coating of choice for many engineering applications.

The as-plated hardness of electroless nickel-phosphorus coatings ranges from 500 to 650 VHN_{100} although recent advances in proprietary formulations are generating an as-plated hardness of up to 700 VHN_{100}. Some electroless nickel-boron coatings have an as-plated hardness of 600 to 700 VHN_{100}. Increased hardness of chemically-deposited nickel can be achieved using elevated temperatures over a period of time. This heat treatment process encourages the separation of nickel phosphide (Ni_3P) or nickel boride (Ni_3B) (1). The resulting coating approaches, and in some cases surpasses, the hardness of hard chrome (1000 VHN_{100} for Ni_3P and 1100 VHN_{100} for Ni_3B). Consequently, electroless nickel is used as a barrier against wear and abrasion, specifically in applications where a uniform deposit is to be achieved without grinding. Even in severe wear situations, such as abrasion coupled with high temperatures (>500° C), where hard chrome is still the correct choice, electroless nickel plating can be used in conjunction with the hard chrome. The electroless nickel coating, under the chromium coating, is used to inhibit the corrosion of the base metal through the characteristic cracks of hard chrome plating.

Corrosion is one of the major problems facing engineers in industrial situations today. Its effects can range from a simple loss of appearance, which can lead to unsalable merchandise, on up to increased operating costs. Unfortunately, many metallic coatings used as barriers against corrosion have been inherently porous and therefore have provided poor protection for the base metal. In general, a high-phosphorus, low-stress electroless nickel coating largely eliminates the porosity problem.

Table 8.1 reviews the physical characteristics of electroless nickel. In addition to its engineering features, the unique ability to plate non-conductors such as glass, ceramics, plastics and graphite has made electroless nickel a very useful coating that can be regarded as a problem-solving finish offering a cost-effective alternative with equivalent performance to more expensive metal alloys.

Table 8.1
Properties of Electroless Nickel Coatings

Property	Data as a function of phosphorus or boron content			
	5% B	3% P	5-6% P	8-9% P
Density, gm/cm³ᵃ	8.25	8.52	8.25	7.85-8.1
Coefficient of thermal expansion, μm/m/°Cᵇ	12.1	nd	nd	13-14.5
Electrical resistivity, μohm-cmᶜ	89.1	30	72	50-110
Thermal conductivity, cal/cm/sec/°Cᵈ	nd	nd	nd	0.0105-0.0135
Melting point, °Cᵉ	1080	nd	nd	890
Magnetism, oerstedsᶠ	weakly ferromagnetic	ferromagnetic	weakly ferromagnetic	nonmagnetic

---

ᵃ The lower density of electroless nickel as compared to pure metallurgical nickel (8.91 g/cm³) is attributed to the presence of phosphorus or boron. Analysis has also shown minor levels of the following elements that may also affect density: H (0.0016%), N_2 (0.0005%), O_2 (0.0023%), and C (0.04%).
ᵇ Values for electrodeposited nickel are from 14 to 17 μm/m/°C. Heat treatment has been shown to decrease these values to an average of 10.8 μm/m/°C.
ᶜ Thinner films of 0.04 to 0.08 mils generated a resistivity of 55-60 μohm-cm, as compared to 0.4 mils of the same 7-10% P coating, which exhibited resistivity of 30-55 μohm-cm.
ᵈ The same value for metallurgical nickel (99.94% purity) at 100° C is 0.198.
ᵉ A 4.3% boron coating has also been reported with a 1350° C melting point. The melting point of the 8-9% P alloy corresponds to the melting point of nickel phosphide (NiP_3).
ᶠ Heat treatment of both Ni-B and Ni-P alloys has been shown to increase the coefficient of magnetism.

AEROSPACE APPLICATIONS

Electroless nickel has been used extensively over the years in the aerospace industry (2-5). All of the properties of electroless nickel are used to advantage in this industry. In aircraft engines, turbine or compressor blades are plated with electroless nickel as protection against the corrosive environment they are exposed to (6). The coating thickness in this application is generally 1 to 3 mils of high-P electroless nickel. When the blades are plated with electroless nickel, there is about 25 percent less loss of fatigue strength than when plated with hard chrome.

Aluminum is frequently used by engineers in aerospace applications because of its density and, consequently, its light weight. Electroless nickel coatings compliment aluminum's inherent characteristics by adding hardness, wear resistance, corrosion protection, and solderability. Piston heads are a good example of the successful combination of aluminum and electroless nickel in the aerospace industry. The light weight of aluminum allows the piston to work more efficiently, while the electroless nickel provides wear resistance that extends the useful life of the piston. The main shafts of aircraft engines are plated with electroless nickel to provide good bearing surfaces. An additional advantage of electroless nickel is realized when rebuilding of the shafts is required during maintenance overhauls (10-12). The remaining electroless nickel can be stripped off and replated to the required thickness. This contrasts favorably with the more expensive machining required for chromium-plated shafts. The rear compressor hub sleeves and bearing liners of the TF30 jet engine are reconditioned and replated with electroless nickel at a cost savings of several thousand dollars over the purchase of new components. The components are made of a titanium alloy containing 6 percent vanadium and 4 percent aluminum. Two mils of electroless nickel are used as the finish.

In addition to engine-related applications, electroless nickel finds many other uses in the aerospace industry. The relatively low coefficient of friction of electroless nickel, coupled with its corrosion resistance, makes it useful in plating servo valves. Landing gear components are plated with 1 to 2 mils of electroless nickel to build up mismatched surfaces, as well as to provide corrosion resistance.

Metallic optics are becoming widely used in spaceborne systems. In these applications, a strong coating must be deposited over a light, strong metal such as beryllium or aluminum. Special high-phosphorus electroless nickel deposits have been polished to 9 Å (13), providing superior performance in space applications where high G forces are present and low inertia is required.

A deposit of 3 to 5 mils of electroless nickel is applied to the beryllium or aluminum substrate. The coating is controlled to produce a compressive strength of 7000 to 10,000 psi. After heat treatment, the coating has nearly zero stress, which provides a stable optical system for extended periods. The phosphorus content for these deposits is in the range of 12.2 to 12.7 percent, and the solutions must be free of stress-inducing elements. Boron-electroless nickel coatings are not suitable for this type of application because they tend to have high tensile stress.

Table 8.2 summarizes the major uses of electroless nickel in the aerospace industry.

AUTOMOTIVE APPLICATIONS

The automotive industry has been moving toward the use of alcohol/gasoline fuel mixtures. The use of alcohol, aside from some performance problems, also creates corrosion problems in the fuel systems. In Brazil, where ethanol is used as an auto fuel, zinc-die-cast-components, such as carburetors, are routinely plated with electroless nickel to prevent corrosion. Production of automobiles in

Table 8.2
Major Uses of Electroless Nickel
In the Aerospace Industry

Component	Basis metal	Phosphorus, %*	Coating thickness, mils	Property of interest**
Bearing journals	Al	L, M	1-2	WR, U
Servo valves	Steel	M, H	1	CR, LU, U
Compressor blades	Alloy steels	M, H	1	CR, WR
Hot zone hardware	Alloy steels	M, H	1	WR
Piston heads	Al	M, H	1	WR
Engine shafts	Steel	L, M	1-2	WR, Buildup
Hydraulic actuator splines	Steel	L, M	1	WR
Seal snaps & spacers	Steel	M, H	0.5-1	WR, CR
Landing gear components	Al	M, H	1-2	WR, Buildup
Struts	SS	M, H	1-2	WR, Buildup
Pitot tubes	Brass/SS	M, H	0.5	CR, WR
Gyro components	Steel	L, M	0.5	WR, LU
Engine mounts	Alloy steels	M, H	1	WR, CR
Oil nozzles	Steel	M, H	1	CR, U
Optics	Al	H	3-5	

*Phosphorus content: H = 9 to 12; M = 5 to 8; L = 1 to 2.
**CR = corrosion resistance; WR = wear resistance; U = uniformity; LU = lubricity.

the U.S. should eventually be using electroless nickel on carburetors, fuel sending systems, and fuel pumps as ethanol and methanol gasoline mixtures gain wide acceptance and use.

Differential pinion ball shafts are plated with 1 mil of electroless nickel for wear resistance. Some automobile manufacturers use electroless nickel/teflon to plate shafts. The added lubricity of the teflon, coupled with the hardness of the electroless nickel, provide the added service life required for extended warranties.

The auto industry takes particular advantage of the uniformity of the electroless nickel deposit when it is used to plate components such as gears, heat sinks, and fuel injectors. Heat sinks, such as the one shown in Fig. 8.1 are easily plated with electroless nickel for heat protection, while it would be impossible to obtain good coverage on internal surfaces by plating electrolytically. Retaining tolerances of 10 to 20 μin. on gears is easily accomplished

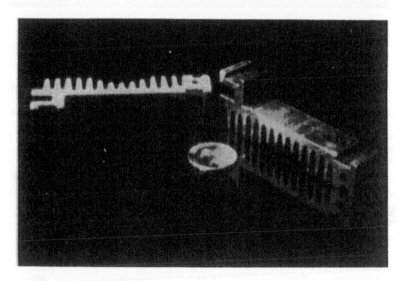

Fig. 8.1—Electroless nickel plated heat sink.

with electroless nickel, whereas electroplated nickel would require further processing to assure proper tolerances. Electroless nickel is also used on fuel injectors, providing important wear resistance against fuel flows, and to a certain extent, corrosion. The result of this wearing and corrosion over time can cause the widening of the injector orifice. As this widening occurs, the horsepower of the engine will increase beyond its design criteria, and therefore increase the warranty liability of the manufacturer. Electroless nickel prevents this wear and corrosion, thus considerably reducing warranty liability.

Table 8.3 summarizes the typical uses of electroless nickel in the automotive industry.

APPLICATIONS IN THE CHEMICAL PROCESSING INDUSTRY

Materials engineers in the chemical process industries are using electroless nickel to solve many unresolved corrosion problems. Use of electroless nickel also decreases the capital costs required for the construction of certain types of chemical plants. This results in more favorable economic conditions for producing chemicals more competitively. Potential gains also exist for improvement in product purity, environmental safeguards, safety of operations, and reliability in manufacturing and transportation of chemical products.

Table 8.3
Major Uses of Electroless Nickel
In the Automotive Industry

Component	Basis metal	Phosphorus, %*	Coating thickness, mils	Property of interest**
Heat sinks	Al	M, H	0.4	CR, S, U
Carburetor components	Steel	M, H	0.6	CR
Fuel injectors	Steel	M, H	1.0	CR, WR
Ball studs	Steel	M, H	1.0	WR
Dif. pinon ball shafts	Steel	M, H	1.0	WR
Disc brake pistons & pad holders	Steel	M, H	1.0	WR
Transmission thrust washers	Steel	M, H	1.0	WR
Synchromesh gears	Brass	M, H	1.2	WR
Knuckle pins	Steel	M, H	1.0	WR
Exhaust manifolds & pipes & mufflers	Steel	M, H	1.0	CR
Shock absorbers	Steel	M, H	0.4	CR, LU
Lock components	Steel	M, H	0.4	CR, WR, LU
Hose couplings	Steel	M, H	0.2	CR, WR
Gears & gear assemblies	Carburized	M, H	1.0	WR, Buildup

*Phosphorus percent: H = 9 to 12; M = 5 to 8; L = 1 to 2.
**CR = corrosion resistance; WR = wear resistance; U = uniformity; LU = lubricity; S = solderability.

The capability to line large process size vessels was developed and patented in the 1950s. At that time, tank cars were being plated with electroless nickel for use in shipping caustic soda. Process controls were not developed sufficiently to produce a consistently good quality product; consequently, adhesion and porosity problems developed. Recent improvements in electroless nickel chemistry have shown the coating to be reliable in many chemical environments.

Steel valves of the ball, gate, plug, check, and butterfly types are widely employed in the chemical industry. Generally, a coating of 0.5 to 1 mil of high-P electroless nickel can more than double the life of such valves in corrosive environments. One exception is that of valves and other components used in caustic soda service. Corrosion data have shown that low-P (1 to 2 percent) electroless nickel provides superior performance to the higher-phosphorus nickels (14,15). The inherently higher hardness of the low-P electroless nickel, along with its tendency to form strongly adherent nickel oxide films, allows this coating to minimize corrosion rates in the caustic soda to about 0.1 mils/year. It is also possible that the solubility of phosphorus in caustic soda affects the corrosion rate of various phosphorus content coatings. Components in caustic soda service can be exposed to corrosion conditions such as 14 percent salt, 15 percent caustic soda at 285° F, and flow velocities of 5 to 10 ft/sec.

Electroless nickel does not perform well in highly acidic environments. Application in areas where hydrolysis of chemical compounds produces acids is also not recommended. For example, in a reactor where thionyl chloride is used as a chlorination agent, the release of hydrochloric acid would produce unacceptable corrosion rates in the electroless nickel coating.

Fasteners made of austenitic stainless steel undergo stress corrosion cracking in chlorine service as well as general corrosion and pitting. One to two mils of high-P electroless nickel over steel, using oversized nuts, provides a cost-effective solution to the problems encountered with stainless steel. In the production of low density polyethylene, 1 mil of high-P electroless nickel is plated over pressure-vessel-quality steel. This prevents unlined vessels from becoming contaminated and discolored by iron. The use of stainless steel was precluded because of its higher cost, roughly twice that of electroless nickel over steel.

Pumps used in mining sulfur undergo severe abrasion and corrosion of their impellers and housings, resulting in a service life of only days. A 1-mil coating of high-P electroless nickel increased the service life of the pumps to months.

Table 8.4 compares the corrosion rates of electroless nickel coatings in caustic soda with other commonly used materials. The low-P electroless nickel is comparable to the more expensive nickel 200, and performs better than stainless steel.

Table 8.5 compares the cost of alternative materials and coatings for 5/8 in. x 3-1/2 in. hex bolts with nuts. Again, the cost/performance of the electroless nickel-coated bolts provides the best overall value.

In Table 8.6, the cost of electroless nickel lined vessels is compared with FRP unlined steel and various lined steel tanks. Table 8.7 summarizes some of the typical applications of electroless nickel in the chemical process industries.

OIL AND GAS PRODUCTION

Oil and gas production is an important market for electroless nickel coatings. Typical environments encountered by tools and equipment used in oil and gas

Table 8.4
**Comparison of the Corrosion Rates of Electroless Nickel Coatings
In Caustic Solutions with Other Commonly Used Materials***

Corrodent	Nickel 200	LP	EN coatings MP	HP	Mild steel	316 SS
45% NaOH + 5% NaCl @ 40 ±2° C	2.5	0.3	0.3	0.8	35.6	6.4
45% NaOH + 5% NaCl @ 140 ±2° C	80.0	5.3	11.9	F	nd	27.9
35% NaOH @ 93 ±2° C	5.1	5.3	17.8	13.2	94.0	52.0
50% NaOH @ 93 ±2° C	5.1	6.1	4.8	9.4	533.4	83.8
73% NaOH @ 120 ±2° C	5.1	2.3	7.4	F	1448.0	332.7

*All corrosion rates in μm/year, 100 days exposure. nd = no data available; F = failure.

production includes brines, CO_2, and H_2S at temperatures of up to 350 to 400° F. Sand and other grit can also be encountered, compounding the severe corrosion problems that can develop. Figure 8.2 is a schematic diagram of a typical oil well, indicating the various major components subject to corrosion and wear.

Tubulars used in oil and gas production are an excellent application for electroless nickel (17-19). The tubes are made of mild steel and if left unprotected, might last only a few months under severe conditions. Coated with 2 to 4 mils of high-P electroless nickel, corrosion rates are reduced to levels comparable with Hastelloy. Figures 8.3 and 8.4 compare corrosion rates of electroless nickel with several other tubular materials. When compared on a cost/performance basis, electroless nickel provides superior protection for tubular goods.

Figure 8.5 shows the components that are used with electrical submersible pumping equipment. Electroless nickel is widely used for protecting the various pumping components. Pump housings, impellers, and pump discharge barrels are plated with 1 to 3 mils of electroless nickel, depending on the corrosion environment encountered. Figure 8.6 shows another pumping system used for low volume producing wells. The sucker rod joint shown in the figure is an ideal electroless nickel application. The uniformity of the coating and the precision with which it can be applied will maintain the integrity of the threads. Tubing packers are used in the tubing string to seal off the space between the tubing and casing. The packers also help support the weight of the tubing string, and

Table 8.5
Economics of Electroless Nickel Coatings:
Cost Comparison for Corrosion Resistant
Bolting Applications for Use in the
Chemical Process Industry*

Material	Cost each/$	Cost ratio
ASTM A307 Low carbon steel bolt	0.44	1.0
ASTM A193 B-7 Alloy steel bolt	0.77	1.8
ASTM A193 B-7 **Fluorocarbon coated	1.41	3.2
ASTM A320 B-7 ***EN coated	0.99	2.3
ASTM A320 L-7 Alloy steel bolt	2.75	6.3
ASTM A320 L-7 **Fluorocarbon coated	3.16	7.2
ASTM A320 L-7 ***EN coated	2.97	6.8
ASTM A193 B8M (316 SS) bolt	3.09	7.0
Nickel 200	17.80	40.5

*Based on 5/8-in. x 3-1/2-in. hex bolt with nut in 500 pieces quantity.
**20 to 25 μm proprietary coating not recommended for immersion.
***15 to 20 μm coating as-plated high phosphorus.

anchor it to prevent shifting. To prevent corrosion of packers, 1 to 2 mils of electroless nickel is employed.

At the surface, electroless nickel is widely used in the heat treater systems used to separate the water-oil mixtures produced from most wells. Large U-shaped tubes, used to heat the water-oil mixtures, are routinely coated with 1 to 3 mils of high-P electroless nickel to prevent corrosion. Valves and other fittings and fixtures used on the surface to collect and transfer the produced oil are also plated with electroless nickel. Table 8.8 summarizes the typical uses of electroless nickel in the oil and gas industry.

Table 8.6
Economics of Electroless Nickel Coatings:
Comparison of Standard Wall Vessels*

Capacity, gal (m³)	FRP	Mild steel	Steel-EN-lined 50 μm	316 SS	Nickel 200	Steel-glass-lined 1270 μm	Steel-teflon-lined 1000 μm
100 (0.32)	1.0	1.2	1.6	1.7	4.5	5.5	7.0
200 (0.76)	1.5	1.6	2.4	2.3	7.1	6.5	9.0
500 (1.9)	2.0	2.0	3.4	3.0	9.0	10.5	13.3
1000 (3.8)	3.0	2.9	5.5	4.1	14.0	13.0	18.7
5000 (18.9)	5.0	4.5	10.6	8.0	30.0	35.7	42.5
10,000 (37.9)	10.0	8.5	21.3	15.0	56.0	59.0	64.0
16,000 (60.0)	16.0	12.0	32.1	22.0	78.0	81.0	75.3

Vessels designed for use with liquid with 1.2 specific gravity. Each contains two nozzles (top and bottom), closed top and vented. Values are thousands of dollars.

FOOD PROCESSING INDUSTRY

The food processing industry offers a tremendous potential for the use of electroless nickel. There are, however, barriers to its widespread application (20-22). First and foremost, there has been a lack of regulatory standards for electroless nickel from the USDA and FDA. Generally, the agencies have approved the use of electroless nickel in a food contact application on a case-by-case basis. In the related area of food packaging, electroless nickel finds easy acceptance for its wear and corrosion resistance properties. In these applications, there is no direct contact of the electroless nickel with food. Typical applications include bearings, rollers, conveyors, hydraulics, and gears, and all components materials normally associated with materials handling.

There are some notable food contact applications. The processing conditions for meat products involve substances such as sodium chloride solution, nitrite,

Table 8.7
Applications of Electroless Nickel Plating
In the Chemical Industry

Component	Basis metal	Phosphorus, percent*	Coating thickness, mils	Property of interest**
Pressure vessels	Steel	H	2.0	CR
Reactors	Steel	H	4.0	CR, PP
Mixer shafts	Steel	L,M,H	1.5	CR
Pumps and impellers	Cast iron/steel	L,M,H	3.0	CR
Heat exchangers	Steel	H	3.0	CR, ER
Filters and components	Steel	H	1.0	CR, ER
Turbine blades and rotor assemblies	Steel	H	3.0	CR, ER
Compressor blades and impellers	Steel/Al	H	5.0	CR, ER
Spray nozzles	Brass/steel	H	0.5	CR, WR
Ball, gate, plug, check, and butterfly valves	Steel	L,M,H	3.0	CR, LU
Valves	SS	L,M,H	1.0	WR, GR, protection against stress and corrosion cracking

*Phosphorus percent: H = 9 to 12; M = 5 to 8; L = 1 to 2.
**CR = corrosion resistance; WR = wear resistance; U = uniformity; LU = lubricity; PP = product purity; ER = erosion resistance; GR = galling resistance.

citric or acetic acid, natural wood smoke, acid-added liquid smoke, volatile organic acids, and others. The process temperature ranges from 60° C to 200° C, and the process air is usually laden with moisture. Under these conditions, corrosion or metal fatigue has been encountered in most of the processing equipment. Screws for extruding meat products are subject to abrasion and corrosion in service. Chromium plate has been a traditional coating for this application. Electroless nickel is a cost effective alternative to hard chrome because of its superior corrosion resistance and its ability to coat complex parts without the need for fixturing anodes to achieve a uniform deposit.

Coating dough kneaders used in the baking industry is an excellent application for electroless nickel contact applications. Dough kneaders are

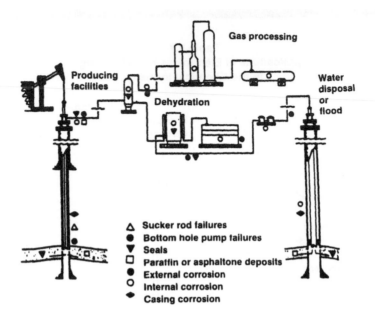

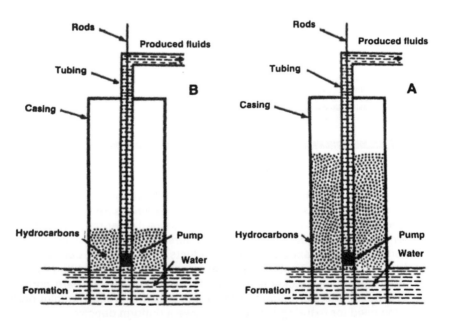

Fig. 8.2—Schematic diagram of a typical oil well, showing common corrosion sites.

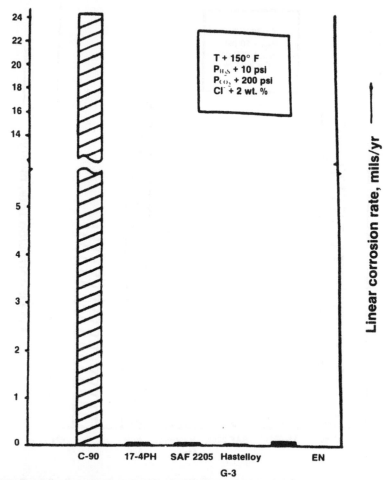

Fig. 8.3—Corrosion rates of various tubing materials used in oil and gas applications. Corrosion environment consists of mixed H_2S, CO_2, and Cl at temperature of 150° F.

rather convoluted components, so the uniformity of the electroless nickel coating is an advantage. The wear resistance, along with the excellent release properties of electroless nickel make it superior to hard chrome for this application.

MINING AND MATERIALS HANDLING APPLICATIONS

In the mining and associated materials handling industries, a wide range of operating conditions can be encountered that require surface finishes. In coal

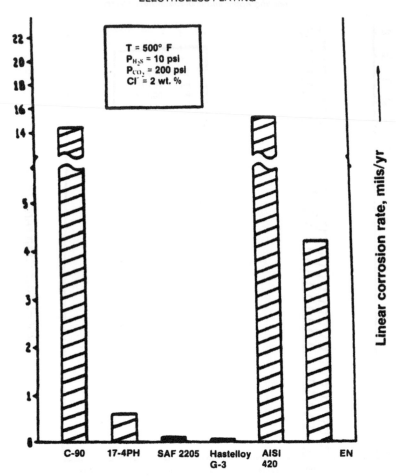

Fig. 8.4—Corrosion rates of various tubing materials used in oil and gas applications. Corrosion environment consists of mixed H_2S, CO_2 and Cl^- at 500° F.

mining operations, below-ground mining components have to endure contact with brines and acid waters, high temperatures, and very abrasive conditions.

Hard chromium has been applied extensively in mining as a corrosion and wear resistant coating for the hydraulic cylinders used in roof support systems. Hard chromium deposits, however, are porous, as a result of their cracked structure. In service, these coatings developed significant corrosion, to the point where these cylinders can seize and fail to function. This problem is accentuated by the high pressures at which these cylinders operate (e.g., 6000 to 7000 psi), causing coating elongation to occur, which increases cracking of the highly

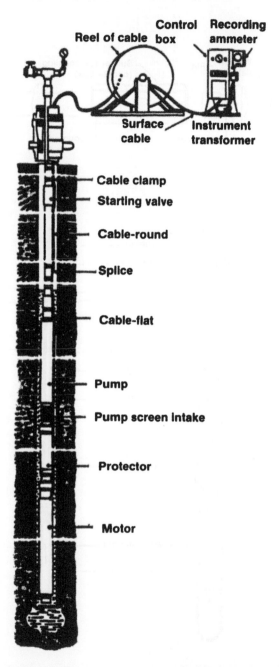

Fig. 8.5—Schematic diagram of submersible centrifugal pump. Electroless nickel plating on pump housing, blades and impellers provides excellent corrosion protection.

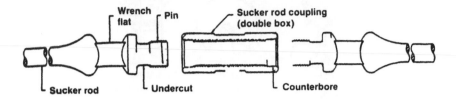

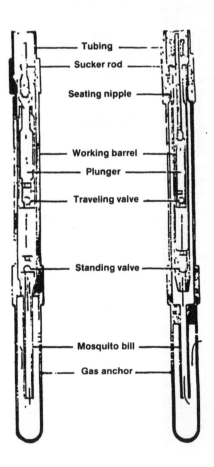

Fig. 8.6—Schematic diagram of pumping system for low-pressure wells. The sucker rod (inset) is plated with electroless nickel for corrosion protection.

Table 8.8
Applications of Electroless Nickel
In the Oil and Gas Industries

Component	Basis metal	Phosphorus, percent*	Coating thickness, mils	Property of interest**
Tubes	Steel	H	2-4	CR, WR, U
Pump housing	Steel	H	2-3	CR, WR
Sucker rods	Steel	H	1-3	CR, WR, U
Ball valves	Steel	H	1-3	CR, WR
Rod boxes	Steel	H	1-3	CR, WR, U
Packers	Steel	H	1-3	CR, WR
Mud pumps	Steel	H	1-3	CR, WR
Blow out preventors	Steel	H	1-3	CR, WR
Fire tubes	Steel	H	1-3	CR

*Phosphorus percent. H = 9 to 12; M = 5 to 8; L = 1 to 2.
**CR = corrosion resistance; WR = wear resistance; U = uniformity.

stressed chromium deposits. A 1-mil coating of high-P electroless nickel with good compressive stress properties is used in this application. The electroless nickel deposits do not crack after elongation of the cylinder, and can withstand the corrosive and abrasive environments found in coal mines.

In certain open pit mineral mining operations, such as phosphate rock for use in fertilizers, high pressure pumps are used to suspend the mineral/soil mixtures as the first step in processing. Jetting pump heads also find wide use in mining operations. In this case, corrosion and erosion of materials is a problem. The corrosion resistance and hardness of electroless nickel is used over a steel substrate to prevent premature replacement of components.

Table 8.9 summarizes the typical mining and materials handling applications of electroless nickel.

MILITARY APPLICATIONS

The military presents numerous unique problems that were solved by the corrosion and wear resistant properties of electroless nickel, as well as taking advantages of other characteristics, such as its optical properties (13,23,24).

Table 8.9
Applications of Electroless Nickel Plating
in the Materials Handling and Mining Industries

Component	Basis metal	Phosphorus, percent*	Coating thickness, mils	Property of interest**
Materials handling				
Hydraulic cylinders and shafts	Steel	H	1.0	CR, WR, LU
Extruders	Alloy steel	M, H	3.0	CR, WR
Link drive belts	Steel	M, H	0.5	CR, WR, LU
Gears and clutches	Steel	M, H	1.0	WR, buildup
Mining				
Hydraulic systems	Steel	H	2.4	CR, AR
Jetting pump heads	Steel	L, M, H	2.4	CR, ER
Mine engine components	Steel/cast iron	M, H	1.2	CR, WR
Piping connections	Steel	M, H	2.4	CR
Framing hardware	Steel	M, H	1.2	CR

*Phosphorus percent: H = 9 to 12; M = 5 to 8; L = 1 to 2.
**CR = corrosion resistance; WR = wear resistance; LU = lubricity; ER = erosion resistance; AR = abrasion resistance.

Catapult covers and tracks for aircraft carriers have recently been coated with electroless nickel. In this application, both wear and corrosion cause the failure of the catapult system within 6 to 12 months in the uncoated state. The catapult environment is very harsh. When the planes are launched, high temperature steam (450° F) is released and rushes by the track. At the same time, up to 1,000,000 lb of force is applied to the track, pulling down and putting pressure on the keyway. As the plane becomes airborne, the track is pulled up, creating additional wear. Salt water and corrosion products from the planes are always present and collect in the keyway, which is a galvanic cell between the dissimilar metals of the cover track.

Many other coatings were tried, with electroless nickel being the most effective in preventing the fretting wear and loss of keyway material through galvanic corrosion. The covers are first repaired and returned to size. After cleaning, a 1.2-mil coating of electrolytic glycolate nickel is applied, followed by a 4-mil coating of electroless nickel, and then 0.5 mils of cadmium, which is subsequently chromated. This combination of treatments provides fretting, wear, and corrosion protection, and chloride in the aggressive marine environment.

After coating the catapult covers with electroless nickel and cadmium, the tracks and covers have a 14- to 18-year service life. This increase in life means that the carriers can remain operational for up to 5 years without having to return to base for a four-month-long overhaul.

Trunions for military vehicles have been plated with electroless nickel for many years. The trunions are located on the side of the vehicle and provide a bearing for the track splines. In this environment, salt and mud are always present and would cause rapid corrosion of the steel bearing surface if it were not protected.

The rear view mirrors of tanks are made of highly polished aluminum and plated with electroless nickel to provide a wear- and corrosion-resistant coating. This application requires that the reflectivity of the electroless nickel be 80 percent over the visible light spectrum. Electroless nickel easily meets these requirements. Aluminum radar wave guides are plated with 1 mil of electroless nickel to protect them against the elements either on land or at sea. Electroless nickel easily provides a uniform coating over the various shapes of the wave guides.

MISCELLANEOUS APPLICATIONS

Molds and Dyes
When electroplating complex molds, careful and complicated anode fixturing is required to achieve coverage in every recess of the mold. The time required for such anode fixturing, and the post-plating machining required to achieve uniformity, adds significantly to the cost. Plating molds in an electroless nickel bath provides the uniformity required, plus the inherent lubricity and excellent release properties that combine to make electroless nickel the most cost effective way to coat molds and dyes.

Foundry Tooling
Electroless nickel is successfully used to protect foundry tooling against pattern wear (25). Aluminum and cast iron tooling can be upgraded by applying 3 to 5 mils of a high-P electroless nickel coating, which increases the surface hardness of these metals and extends their useful life.

Printing Industry
In this industry, both printing rolls and press beds are plated with 1 to 2 mils of high-P electroless nickel (26). These components are generally made of steel or cast iron, and need to be protected from the corrosive inks. In the case of printing rolls, the uniformity of the electroless nickel coating eliminates the post-plating machining to achieve roundness.

Textiles

Wear is the single most important problem solved by electroless nickel in the textile industry. As literally hundreds of miles of thread pass through, around, and over components, wear becomes a significant maintenance problem. Steel thread feeds and guides are plated with 2 mils of heat-treated electroless nickel. Other electroless nickel-coated components include spinnettes, loom ratchets, and knitting needles.

Medical

Surgical staples have been plated with 0.1 to 0.2 mils of electroless nickel. Dental drills use hardened electroless nickel. Various hospital equipment moldings and rails are using electroless nickel in place of chrome or nickel-chrome coatings.

REFERENCES

1. G.G. Gawrilov, *Chemical (Electroless) Nickel Plating*, Portcullis Press Ltd., Queensway House, Redhill, Surrey, RH1 1QS, England, 1979; p. 68-72.
2. D.W. Baudrand, "Aircraft Applications for Electroless Nickel Plate," Business Aircraft Meeting, Society of Automotive Engineers (Apr. 1974).
3. G. Reinhardt, *Electroless Nickel Applications in Aircraft Maintenance*, Society of Automotive Engineers, Inc., Airline Plating and Metal Finishing Forum, Mar. 1982; p. 27-29.
4. M. Levy and J.L. Morrossi, "Wear and Erosion Resistant Coatings for Titanium Alloys in Army Aircraft," Army Materials and Mechanics Research Center, Watertown, MA (Dec. 1970).
5. M. Levy et al., "Erosion and Fatigue Behavior of Coated Titanium Alloys for Gas Turbine Engine Compressor Applications," Army Materials and Mechanics Research Center, Watertown, MA (Feb. 1976).
6. P. Ahuja, "Electroless Nickel Coating for Jet Engine Turbine Blades," *Proc. Electroless Nickel Conference, Products Finishing Magazine* (1982).
7. K.M. Gorbunova and A.A. Nikiforova, *Physiochemical Principles of Nickel Plating*, Translated from Russian by the Israel Program for Scientific Translations, Jerusalem, 1963.
8. H. Wiegand, G. Heinke and K. Schwitzgebel, *Metalloberflaeche*, **22**(10), 304 (1968).
9. *The Engineering Properties of Electroless Nickel Deposits*, International Nickel Co. Inc., New York, NY, 1971.
10. F. Bryant, "Use of Electroless Nickel to Salvage Steel and Aluminum Aircraft Parts," *Proc. Electroless Nickel Conference, Products Finishing Magazine* (1979).
11. J. LeGrange, "Electroless Nickel Application at a Jet Engine Overhaul Plating Facility," ibid. (Nov. 1979).

12. G. Reinhardt, "Potential Applications of Electroless Nickel in Airline Maintenance Operations," ibid. (Nov. 1979).
13. J.W. Dini et al., "Influence of Phosphorus Content and Heat Treatment on the Machinability of Electroless Nickel Deposits," ibid. (1985).
14. R.P. Tracy and B.R. Chuba, "Corrosion Resistance, Applications, and Economics of Electroless Nickel Coatings in Sodium Hydroxide Production," *Proc. Corrosion '87* (Mar. 1987).
15. R.P. Tracy et al., *Materials Performance*, p. 21 (Aug. 1986).
16. P. Talmey et al., U.S. Patent 2,717,218 (Sep. 1955).
17. J.F. Colaruotolo, B.V. Tilak and R.S. Jasinski, "Corrosion Characteristics of Electroless Nickel Coatings in Oil Field Environments," *Proc. Electroless Nickel Conference IV, Products Finishing Magazine* (1985).
18. R.D. Mack and M.W. Bayes, "The Performance of Electroless Nickel Deposits in Oil Field Environments," *Proc. Corrosion '84* (Apr. 1984).
19. R.N. Duncan, "Corrosion Resistance of High Phosphorus Electroless Nickel Coatings," *Proc. AESF Third Electroless Plating Symp.* (Jan. 1986).
20. G.L. McCowin, "The Food Additive Amendments: Electroless Nickel for Food Application," *Proc. Electroless Nickel Conference III, Products Finishing Magazine* (Mar. 1983).
21. C.J. Graham, "Electroless Nickel: Its Role in Food Processing," ibid. (Mar. 1983).
22. R.N. Duncan and T.L. Arney, "Performance of Electroless Nickel Coatings in Food Processing," *Proc. Corrosion '84* (Apr. 1984).
23. J.S. Taylor et al., *Optical Engineering*, **25**(9), 1013 (1986).
24. J.M. Casstevens and C.E. Daugherty, "Diamond Turning Optical Surfaces on Electroless Nickel," Society of Photo-Optical Instrumentation Engineers, 22nd International Symposium (Aug. 1978).
25. J. Henry, "Electroless Nickel Successful Application in Protecting Foundry Tooling," *Proc. Electroless Nickel Conference III, Products Finishing Magazine* (Mar. 1983).
26. J.V. Paliotta, "Functional and Economic Impact of Electroless Nickel on the Printing Press Industry," ibid. (Mar. 1981).

Chapter 9
Electroless Nickel Applications In Electronics

E.F. Duffek, D.W. Baudrand, CEF and J.G. Donaldson, CEF

Electroless nickel baths have been developed to produce a wide range of deposit characteristics that are desirable for electronic applications. Deposits that exhibit excellent solderability, conductivity, corrosion protection, receptivity to brazing, wire and die bonding, and act to retard precious metal migration are an integral part of the fabrication and functional finishing of electronic devices. Typical areas of use include electronic components and packages, printed wiring boards, ceramic hybrid circuits, metal and plastic connectors, and computer memory discs. Tables 9.1 and 9.2 show typical applications of electroless nickel, deposit characteristics required, and alloys used.

ELECTRONIC COMPONENTS AND PACKAGES

Transistor and diode package bases, caps, and pins are plated with electroless nickel to provide corrosion protection, solderability, and to allow the brazing of the caps to the bases. Figure 9.1 shows TO-5 transistor packages that were plated with 1.5 to 2.0 μm of nickel-boron instead of gold, because Ni-B is solderable using RMA fluxes.

Electroless nickel is also used in the manufacture of capacitors. When the conductor area of the capacitor is made of silver paste, electroless nickel can be plated onto the fired silver to provide a solderable connection. Electroless nickel can be plated as the conductive area itself. Nickel-boron alloys are preferred over nickel-phosphorus deposits on capacitors, because of its low electrical resistivity (7 to 15 ohm-cm) and ease of soldering. Figure 9.2 shows a typical barium titanate "feed-through" capacitor plated with nickel-boron.

Heat sinks, as seen in Fig. 9.3, are devices used to cool semiconductors during operation, another example of electronic components that employ electroless nickel plating. They are constructed of aluminum or copper for heat conductivity, with a large number of closely-spaced fins to facilitate heat dissipation. Electroless nickel provides corrosion protection for the basis metals and provides a surface that is receptive to soldering, brazing, or welding the semiconductor in place. It would be impossible to achieve the same results by electroplating because of the limits imposed by current distribution. *Electrodeposited* nickel would not reach into the many recessed areas of such a complex part. *Electroless* nickel deposits a uniform coating thickness on all

Table 9.1
Applications and Characteristics of Electroless Nickel Deposits

Application	Type of electroless nickel	Required deposit characteristics
"Burn-in" contacts	Ni-B (1-3% B) plus gold, or Ni-P (9-12% P) plus gold	Diffusion barrier, hardness
Capacitors	Ni-B (1-2% B)	Solderability, low electrical resistance
Contacts (surface, sliding, pins & sockets)	Ni-B (1-3% B), or Ni-P (2-8% P) plus gold	Hardness, smoothness, diffusion barrier
Ceramic substrates, reducing/replacing gold	Ni-B (0.5-3% B), or Ni-P (1-5% P) plus overcoating if attachment is required	Low electrical resistance, solderability, wire bond-ability, brazability
Headers	Ni-P (9-13% P)	Corrosion resistance, appearance
Heat sinks	Ni-P (9-13% P)	Corrosion resistance, appearance
Hybrid circuits, reducing/replacing gold	Ni-B (1-3% B)	Solderability, low electrical resistance, corrosion protection for molybdenum or molybdenum/manganese metallization on ceramics
High energy microwave	Ni-P (2-10% P), Ni-B (0.5-2% B)	Conductance, corrosion resistance, brazability
Lead frames	Ni-P (2-8% P), Ni-B (1-3% B)	Wire bondability, die bond-ability, solderability
Memory disks	High-P EN (11-13% P)	Non-magnetic, corrosion resistance, smoothness, hardness
"Metallized" ceramics, reducing/replacing gold	Ni-B (1-3% B)	Low electrical resistance, solderability, die bondability, wire bondability, brazability
Printed wiring boards, edge card connectors	Ni-P plus gold	Hardness, diffusion barrier
PWB-connectors	Ni-P plus gold, or Ni-B alone or with thin gold (0.125 μm), or ternary alloy deposits	Diffusion barrier, corrosion resistance
PWB surface mount	Ni-B (1-3% B), Ni-P (1-8% P)	Low electrical resistance, wire and die bonding, diffusion barrier
Transistor packages (TO-3, TO-5, etc.)	Ni-P (8-13% P)	Corrosion protection, solderability, appearance
IC, transistors, diode wafers	Alkaline electroless Ni-P (1-5% P)	Low electrical resistance, solderability

Table 9.2
Electroless Nickel Selector Chart

Deposit characteristics	Electroless nickel alloy							
	Low-P P*		Med.-P P	High-P P	Ni-B		Polyalloys	
					P	B**	P	B
	1-5%		5-9.5%	10-13%	0.2-1%	2-3%	3-10%	0.2-2%
	pH							
	8-10	5-7	4.5-5.5	4.5-5.5	5-7	5-7	6-7	6-7
Brazing	X	X	X	X	X	X	X	X
Compressive strength	—	—	—	X	—	—	—	—
High hardness	X	X	—	—	X	X	X	X
High temperature resistance	X	X	X	—	X	X	X	X
Non-magnetic	—	—	—	X	—	—	X	X
Solderability (RMA flux)	X	X	X	X	X	X	X	X
Wire bonding (thermosonic/ultrasonic)	X	X	—	—	X	X	X	X
Wire bonding (thermocompression)	—	—	—	—	—	—	—	—
Wear resistance	X	X	—	X	—	X	X	X
Welding	X	X	X	X	X	X	X	X
Corrosion resistance	—	—	X	X	—	—	—	X

*Phosphorus content.
**Boron content.

areas of the part that are in contact with the solution, regardless of part shape or complexity, and is the only practical method of providing corrosion protection, solderability, hardness, and durability required on heat sinks.

Transistor chips, made from silicon wafers, are difficult to solder, and often use electroless nickel deposits to form a solderable, adherent, and ohmic contact on the back side of the wafer, which is used as a common ground. Plating electroless nickel directly onto silicon wafers requires an alkaline plating solution, with a deposit containing 1 to 5 percent phosphorus (1,2). Alkaline nickel-boron solutions can also be used. An initial deposit of 0.1 to 0.2 μm is thermally diffused into the surface of the silicon to "getter" contaminants and enhance adhesion. The part is then activated and replated with about 0.25 to 0.35

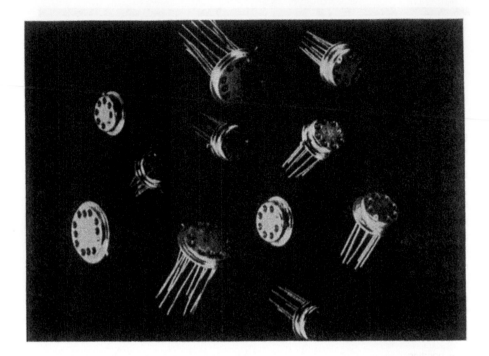

Fig. 9.1—TO-5 transistor packages plated with 60 to 90 μin. (1.5 to 2.25 μm) of nickel-boron.

μm of electroless nickel to provide a solderable coating. Activation is achieved by removing the oxide layer in diluted hydrofluoric acid or ammonium bifluoride. Solar cell production utilizes electroless nickel plating methods as described above. The electroless nickel coating is used to produce the current-carrying circuits shown in Fig. 9.4. When nickel-boron plating is used, excellent wire bonding characteristics are obtained in addition to good solderability. Ultrasonic techniques are used for wire bonding aluminum wire to the deposit; thermocompression methods are required for wire bonding gold wire.

High-energy microwave devices, including wave guides and circuits, often use electroless nickel because it is readily brazed or welded, as well as providing corrosion protection.

Copper lead frames are plated with electroless nickel to provide a diffusion barrier between the copper substrate and gold or silver deposits.

Typical Process Cycle (3,4)
Transistor bases consisting of mild steel body, glass sealed pins made of "52 alloy" (52 percent nickel, 48 percent iron) with oxygen free high conductivity (OFHC) copper heat sink soldered or brazed to the steel base (see Figs. 9.5 and 9.6) are plated as follows:

Fig. 9.2—Feed-through capacitor plated with nickel boron.

- Alkaline clean and rinse.
- Descale in inhibited hydrochloric acid to remove any oxides from the pins. Rinse thoroughly. Alternative: alkaline permanganate treatment.
- Bright dip, 8 to 12 sec. Typical bright dip compositions: sulfuric acid (66° Be), 44 percent by vol.; nitric acid (40° Be), 22 percent; water, 33 percent; sodium chloride, 0.25 oz/gal; phosphoric acid (85 percent), 55 percent by vol.; nitric acid (40° Be), 20 percent; acetic acid (98 percent solution), 25 percent; temperature 130 to 175° F.
- Rinse.
- Acid dip in either HCl 20 percent by vol., H_2SO_4 20 percent by vol., or equivalent dry acid salts.
- Rinse.
- Nickel strike (Wood's strike or low-pH sulfamate nickel) for maximum adhesion to the 52 alloy pins. (Note—Kovar is sometimes used for pins and also requires a nickel strike for best adhesion).
- Rinse.
- Electroless nickel plate.
- Rinse and dry.

Aluminum components are plated as follows (5,6):

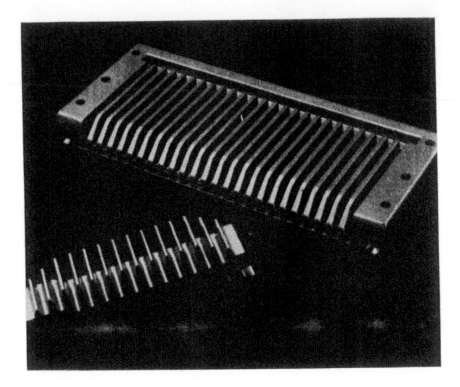

Fig. 9.3—Heat sinks plated with electroless nickel.

• Clean in a mildly alkaline solution compounded for aluminum (3 to 6 min at 150 to 160° F). For best results, a very slight etch should result.
• Rinse.
• Deoxidize in nitric acid (50 to 60 percent by vol.) or proprietary mixtures. This step removes oxide films of unknown composition and thickness, and leaves a thin, more uniformly-controlled oxide film. In addition, some alloying constituents, such as copper, are partially removed from the surface. Deoxidizing aids in producing a more uniform surface condition. A typical deoxidizing time at room temperature would be 30 sec.
• Rinse.
• Immersion zinc (30 to 60 sec at 70 to 80° F). Proprietary solutions that have been developed especially for subsequent electroless nickel plating are readily available.

Fig. 9.4—Solar power cell plated with electroless nickel.

- Rinse.
- Strip the zinc in 50 to 60 percent nitric acid for 30 sec to 1 minute.
- Rinse.
- Immersion zinc (20 to 25 sec at 70 to 80° F), same as the immersion zinc step above. A separate zincate solution is preferred.
- Rinse (double rinse).
- Electroless nickel plate.
- Rinse and dry.

Printed Wiring Circuit Boards

Printed wiring boards (PWBs) typically consist of discrete electronic components mounted onto a substrate with copper plated on its surface to form wiring interconnections (Fig. 9.7). The principal uses of electroless nickel on PWBs are for wear resistance and diffusion barrier characteristics on plug-in contacts and circuit conductors.

Plug-in or "edge" connectors allow PWBs to be easily removed and replaced by mounting them in spring-loaded contacts, but this subjects the contacts to abrasive wear. The service life of the contacts is greatly enhanced when electroless nickel is plated onto the copper because the hardness of the nickel

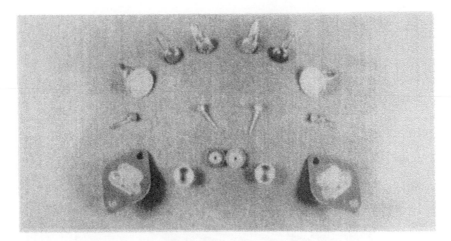

Fig. 9.5—Examples of electroless-nickel-plated transistor bases.

Fig. 9.6—Examples of copper transistor bases and a final assembly plated with electroless nickel-phosphorus.

Fig. 9.7—Printed circuit boards in various stages of fabrication, along with miscellaneous other electronic components: LED, lead frames, transistor bases, and tube sockets.

improves wear resistance. In many applications, gold is plated onto the PWB contacts and circuit traces to provide conductivity and oxidation resistance. If the gold is plated directly onto the copper, however, migration and diffusion of the copper into the gold can produce a surface that is easily oxidized, causing high-resistance electrical connections or loss of contact. A deposit of electroless nickel between the gold and the copper substrate forms a diffusion barrier that retards the diffusion of copper into the gold (7). Although nickel will diffuse through gold at elevated temperatures, electroless nickel has less tendency to diffuse, resulting in longer life without significant loss of contact conductance. Electroless nickel provides a superior diffusion barrier to that of electrolytic nickel, which makes it more suitable for protecting all of the conductors of PWBs, especially those which must operate at higher than room temperature and, therefore, have increased diffusion rates. "Burn-in" boards (those that are tested by cycling from high to low temperatures) will fail unless electrodeposited nickel, tin-nickel, electroless nickel, or another suitable diffusion barrier layer is used. Electroless nickel is frequently preferred.

In some cases, nickel-boron deposits alone can replace dual coatings of gold and nickel-phosphorus deposits on copper (8). In addition to acting as diffusion barriers, Ni-B deposits possess greater electrical conductivity than Ni-P deposits. Although nickel-boron deposits form an oxide layer, it is usually thin and easily penetrated. Nickel-boron deposits can replace gold plating in those situations where contact pressure is moderately high and resistance values in the range of 10 to 100 μohm-cm can be tolerated.

Low phosphorus ($<$5 percent) or low boron ($<$1 percent) electroless nickel plating solutions can be used for through-hole interconnections in place of electroless copper. Electroless nickel is especially useful for circuit devices operated at high frequency. The signal-to-noise ratio is improved using electroless nickel-boron deposits. The oxygen present in the electroless copper deposit forms copper oxide, a semiconductor, resulting in "noise" in the high frequency ranges.

Electroless nickel solutions can be used for very small hole diameters and for multilayer circuit boards. Holes of 0.010-in. diameter that were over 1/2-in. deep have been successfully plated with electroless nickel. These were used successfully in a high frequency application without any subsequent overplate.

Following is a typical process cycle for plating electroless nickel on PWBs with a simple electroless nickel bath formula. However, proprietary baths are preferred because they provide improved performance and deposit characteristics.

Process Steps for Plating
Edge Card Connectors (Plug-In Fingers)

• Mask off areas not to be plated. High grade tapes are often used, masking just above the line where plating is to take place. Controlled depth immersion in preparation and plating solutions is the usual method of plating the PWB edge connectors.

• Remove or strip the plated resist, such as tin-lead, if any. Scrub the exposed copper connectors with a mild abrasive as required (usually required after stripping tin-lead).

• Acid dip. Sulfamic acid (4 to 6 oz/gal) or 10 percent by volume sulfuric acid are typical solutions.

• Electroless nickel plate. With a bus bar connector in place, touch any part of the area to be plated with a dissimilar metal for a few seconds to galvanically initiate deposition, since copper is not catalytic to electroless nickel. A brief cathodic current will also start the electroless reaction. Alternatively, use an electroless nickel-boron strike in a solution that is catalytic to copper prior to the final electroless nickel deposit.

An example of preparation and electroless nickel plating of copper-clad PWB through-hole interconnects (epoxy-fiberglass, two-sided and multilayer boards) is as follows:

• Drill holes for through-hole connections.

- Remove drilling smear. Alkaline permanganate, concentrated sulfuric acid, or chromic acid are used for this purpose. These procedures involve several steps in each case. When using a proprietary process, follow suppliers' recommendations.
- Rinse.
- Alkaline clean and condition.
- Rinse.
- Microetch copper. Remove small amounts of copper in a persulfate or hydrogen peroxide-sulfuric acid etchant.
- Rinse.
- Acid dip in 10 percent by volume sulfuric acid.
- Rinse.
- Catalyst predip.
- Catalyze using mixed tin/palladium or a non-precious metal system. A catalyst system must be selected which will not lead to an immersion (chemical replacement) deposit on the copper and will effectively catalyze the plastic inside the holes. Sometimes a precatalyst conditioner (surfactant) is used to increase the absorption of catalyst.
- Rinse.
- Activate in 10 to 20 percent HCl, 10 percent KOH, or proprietary solutions.
- Rinse.
- Electroless nickel plate to a thickness of 0.5 to 5 μm. A typical electroless nickel solution formula consists of 25 to 30 g/L nickel sulfate, 30 g/L sodium hypophosphite, 70 to 88 g/L sodium citrate, 40 g/L ammonium chloride, a pH of 9 to 9.5, and a temperature of 125 to 185° F.
- Rinse.
- Electrolytic copper strike for 5 min. The copper strike should minimize the possibility of producing an immersion deposit on the nickel-plated surfaces. One example of a copper strike formula consists of 20 g/L copper pyrophosphate, 45 g/L potassium pyrophosphate, 30 g/L sodium carbonate, 20 g/L sodium citrate, and a pH (adjusted with dilute KOH solution) of 8.0 to 8.5. Enter the solution with current on.
- Rinse.
- Electrolytic copper plate.

NOTE: Panel or pattern plating procedures can be used. Depending upon the usual metal etch resist used, etching to remove unwanted copper is done in peroxide/sulfuric acid, ferric chloride, persulfate, or cupric chloride solutions in order to remove the electroless nickel deposit along with the copper. Alkaline etchants are not effective in removing electroless nickel.

FINE LINE PATTERNING
OF CERAMIC SUBSTRATES FOR ELECTRONICS

Metallized conductor patterns or uniform metal layers on ceramic substrates have been widely used in the electronics industry. For many years, ceramics

Fig. 9.8—IC chip die bonded to substrate and wire bonded to external circuitry.

have been metallized by processes such as the ones using fused metal-glass pastes, or by thin film vacuum deposition techniques. Direct electroless deposition, although a problem in the past because of poor adhesion of metal to ceramic, is now also practiced.

Extensive use is made of the nickel-boron and nickel-phosphorus electroless systems. An important feature is the ability to plate electrically isolated areas and fine lines.

Ceramics were the first type of insulating substrate for printed circuit production, having been used in World War II for miniature radio proximity fuses (9). Although ceramic use has declined since the 1950s, there is renewed interest in precision ceramics for printed circuit applications. Ceramics have important properties such as high temperature strength, inertness, hardness, and high dielectric and thermal conductivity. Applications include hybrid microwave circuits, printed wiring boards, chip carriers, resistors, and capacitors (Figs. 9.8-9.10). Figure 9.8 shows an integrated circuit chip carrier that has been silicon bonded to the metallized substrate using a gold-silicon eutectic preform heated to 850° C. Connection to the external circuitry is by aluminum or gold wire bonding from the chip to metallized pads.

Alumina (Al_2O_3) is the most commonly used material, although others, such as beryllium oxide (BeO) and aluminum nitride (AlN), with higher heat dissipation than alumina, barium titanate ($BaTiO_3$) and silicon nitride (Si_3N_4) are also available. In each case, there is a need for a metallic coating or a conductive pattern on the ceramic to support and interconnect the active and passive

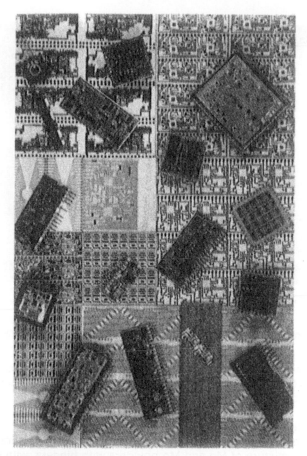

Fig. 9.9—Various hybrid packages.

elements contained in a particular device or system. The conductor pattern has bonding pad areas that provide a means of attaching leads suitable for connector or printed circuit board attachments. Electroless nickel plating is used to protect the metallization, make it more conductive, brazable, hermetic, solderable, and wire bondable.

Four systems for producing patterns on ceramic substrates, using electroless nickel, will be discussed. As shown later, the system selected for fabricating patterns on a substrate depends upon line definition, the electrical requirements, mechanical strength, and cost.

Moly-Manganese Film Metallization
The classic molybdenum-manganese (Mo-Mn) metallization system is selected

Fig. 9.10—Ceramic copper-electroless nickel-gold electroplated alumina circuit boards.

for high strength and for its compatibility with the thermal coefficient of expansion (TCE). Patterns are screened or photo-etched with plating before or after pattern delineation (10).

Mo-Mn Application
The Mo-Mn technique consists of forming a metallic layer on a ceramic surface by applying a mixture of Mo and Mn powders with binders, with subsequent heating in a hydrogen atmosphere (11-14). Bond strengths of 70 to 105 MPa (10,000 to 15,000 psi) are typical.

Plating (15)
Typical process steps are as follows:
- Alkaline cleaning (or vapor degreasing) for removal of soils.
- Activation/microetching of surface to attain microporosity. A typical activating solution consists of 200 to 300 g/L potassium ferricyanide and 100 g/L potassium hydroxide, operated at 25 to 50° C for 30 to 60 sec. This should be followed by deionized (DI) water rinsing.
- Glass removal from Mo-Mn surfaces with a solution of 100 g/L potassium hydroxide at 100° C for 8 to 15 min. Time is critical and must be accurately determined. Rinse with DI water.
- Acid activation with 10 percent HCl at room temperature for 5 to 15 sec. Rinse with DI water.

• Catalyzation with a solution of palladium chloride (0.01 to 0.1 percent) and hydrochloric acid (0.1 percent) at room temperature for 30 to 60 sec. Rinse with DI water.

• Nickel plate with either Ni-P or Ni-B. Follow supplier recommendations to deposit 2.5 to 5.0 μm. DI water rinse.

Direct Plating on Ceramics

After surface conditioning, this method becomes similar to plating on a nonconductor or plated through-hole in PC boards. Background on this process is interesting to note, since optimum metal adhesion and coverage, along with cost effectiveness, may have been achieved at this time. Two key steps are necessary, namely, etching for microporosity of surface layers, and coverage of these pore surfaces with a catalyst-absorption promoter.

Ceramic Conditioning

The need for surface microporosity was first recognized by Stalnecker (16). Processes using fluorides or molten inorganic compounds are now followed to achieve bond strength.

Fluorides

Early work showed minimal metal adhesion on BeO (17) and Al_2O_3 (18). BeO was conditioned with NaOH at 250° C and HBF_4 prior to plating. This process achieved only 1.7 MPa (250 psi) bond strength, or about one-third the preferred value of 5 MPa. Various fluoride solutions, e.g., ammonium fluoride-NaCl, HF-NaCl, and ammonium bifluoride have been used, preferably on 90 to 97 percent alumina for optimum porosity development (15,19,20). A typical process is:

• Alkaline cleaning with ultrasonics, followed by DI water rinsing with ultrasonics.

• Surface conditioning with a solution of 100 g/L ammonium fluoride and 100 g/L sodium chloride at 60° C for 2 to 20 min, followed by a DI water rinse with ultrasonics.

• Sensitization with 0.05 g/L $SnCl_2 \cdot 2H_2O$ and 1 mL/L HCl at 40° C for 5 min, followed by a DI water rinse.

• Electroless nickel plating to a thickness of 2.5 to 5.0 μm. Borohydride solutions are operated at pH 13 at 90° C; aminoborane solution—pH 6 to 7, 65° F; hypophosphite solution—pH 6, 70° C.

• Fine-line patterning, carried out using positive-type photoresist and etching with nitric acid, ferric chloride, CrO_3/H_2SO_4 or H_2SO_4/H_2O_2 (13,22,25).

Molten inorganic compounds

Use of a heated NaOH or NaOH/KOH coating in the molten state plus absorption promoters is claimed to give high adherence of electroless nickel (and copper) directly on ceramics (21-24). Various ceramics, including 89 to 99 percent Al_2O_3 can be processed. The procedure on 90 percent Al_2O_3 is as follows:

- Alkaline cleaning at 60° C for 10 min; water rinse.
- NaOH solution dip and drain, 760 g/L; dry at 175° C.
- Heat substrate to 450° C for 15 min to fuse NaOH. An alternate process uses NaOH/KOH at 210° C, followed by water rinsing.
- Sulfuric acid dip with a 20 percent H_2SO_4 solution at 25° C for 2 min, followed by water rinsing.
- Absorption promoter dip, with surfactants and monoethanolamine or 4 to 12 M HCl at 25° C for 2 min, followed by water rinsing.
- Sensitization with tin chloride and hydrochloric acid solution at 25° C for 10 min; water rinse.
- Catalyzation with 0.01 g/L palladium chloride in 0.01 M HCl at 25° C for 2 min, followed by water rinse.
- Nickel plate with nickel-phosphorus solution, pH 4.5 to 5.2, 90° C for 30 min, to a thickness of 10 μm (0.4 mils). Follow with water rinsing. Adhesion values of 7.5 to 11 MPa were noted. Note that nickel thicknesses over 5 μm can be plated satisfactorily without blistering upon heating (24).

Thin film evaporation and sputtering (26)

These techniques are used extensively in microelectronics and semiconductor technology, including high-frequency and microwave devices often used for high-cost military applications. Evaporated, sputtered and chemical vapor deposition films are used extensively for thin film structures. Deposition usually takes place on the ceramic through a mask or over the entire surface. The process involves two layers: (a) aluminum, chromium, titanium, or molybdenum for adhesion to the substrate; and (b) copper, silver, gold, or other precious metals for conductivity. Thin film resolution of 1 μm (10,000 angstroms) or less is possible. Electroless nickel and gold plating is used to increase electrical conductivity and applications. Photoresist and etching techniques are used to define the pattern.

Screened conductive patterning (15)

Screen-applied pastes of copper, silver, or gold are applied to the ceramic and then fired. A typical process cycle is as follows:

- Alkaline clean, then rinse with DI water.
- Microetch; silver pattern; apply 10 percent HNO_3, then DI rinse; copper pattern; 10 percent ammonium persulfate.
- Dip in 10 percent sulfuric acid; DI rinse.
- Electroless nickel plate directly onto the screened ceramic, as discussed earlier in this section.

SELECTION OF A PATTERNING SYSTEM

The system selected for patterning on a ceramic substrate depends on end use, electrical parameters, mechanical strength, line width, and cost.

The Mo-Mn system with electroless nickel plating, for example, is selected for its resultant high mechanical strength, thermal coefficient of expansion, compatibility of materials, and capability for hermetic packaging. Electronic properties, with 60 percent electrical conductivity of other systems, and line widths become secondary issues in the selection process; costs are moderate. A metal(s) thickness of 0.5 mils on ceramic has been chosen in order to make line width comparisons of these systems equivalent. Practical line widths (for 0.5 mil metal thicknesses) are 7 to 10 mils for screened patterns and 4 to 7 mils for photoetched patterns.

Screened conductor patterns show higher conductivities and lower costs than Mo-Mn systems, although mechanical strength is lower. Typical line widths of 7 to 10 mils are similar to screened patterns.

Direct plating on ceramics, notably the molten inorganic processes discussed above, have application in electronics because of their optimum electronic parameters, metal adhesion, variety of substrate materials, and possible low production costs. Typical line widths at 0.5 mil metal thicknesses are 2 mils.

Thin film systems are used in high-frequency, low-impedance applications. Costs are high because of the need for expensive capital equipment such as sputtering, electron beam, and evaporation machines; these systems also have a low output. Line widths depend upon substrate smoothness and metal thickness. Thus, at 0.08 mils (2 μm) and 0.5 mils (12.5 μm) thicknesses, practical line widths are 1 to 5 μm and 1/2 to 1 mil, respectively.

PLATING CIRCULAR CONNECTORS WITH ELECTROLESS NICKEL

Aluminum is used to fabricate the barrels, shells, coupling nuts, endbells, and other metal components of electrical connectors (see Figs. 9.11 and 9.12) because of its physical and metallurgical properties. Electroless nickel plating is specified for a large percentage of these aluminum connector components.

The properties of the electroless nickel deposit have facilitated its widespread use in many electronics applications. Its usage in the manufacture of connectors is no exception. The electroless nickel properties cited by design engineers as being the most important are:

- Corrosion resistance
- Hardness
- Wear resistance
- Lubricity and non-galling
- Weldability and solderability
- Relative ductility
- Uniform deposit thickness
- Non-magnetic

Fig. 9.11—Aluminum circular connector parts plated with electroless nickel.

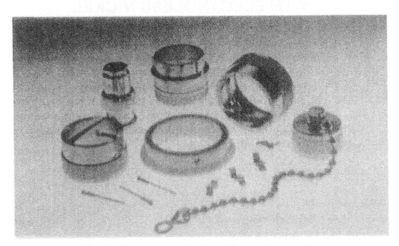

Fig. 9.12—More examples of EN-plated circular connector parts.

Although the first six items on the list are most often cited, the last two are also extremely important properties. Since electroless nickel deposition is uniform on all surfaces with which the plating solution comes in contact, the integrity of all dimensions are maintained both on I.D. and O.D. The non-magnetic property of electroless nickel eliminates any interference with signal transmissions.

ELECTROLESS NICKEL SPECIFICATIONS

Military specification MIL-C-26074 ("Coatings, Electroless Nickel, Requirements for") is the main document governing electroless nickel deposition in the connector industry. The Society of Automotive Engineers (SAE) has issued the following aerospace material specifications:

AMS 2399—Electroless Nickel/Boron Plating
AMS 2404B—Electroless Nickel Plating
AMS 2405A—Electroless Nickel Plating, Low Phosphorus

The American Society for the Testing of Materials (ASTM) has issued a general specification—B733-86, "Standard Specification for Autocatalytic Nickel-Phosphorus Coatings on Metals". This document specifies coating thicknesses for meeting wear and corrosion requirements and defines test methods for various coating properties.

While not a specification, another important document is ASTM B-654, "Standard Guide to Autocatalytic Nickel-Phosphorus Deposition on Metals for Engineering Use".

PLATING PROCEDURES AND PROCESSES (27)

Obtaining adherent and defect-free electroless nickel deposits on aluminum connector components requires knowing which alloys are being processed and how they were produced, whether by machining bar stock, die casting, extrusion, or a combination thereof. It is also necessary to know that certain parts must be handled carefully to avoid any mechanical damage of the threads, sharp corners, keyways, etc. Care must be exercised to maintain critical after-plate dimensions, yet to plate within the specified thickness tolerances.

Equipment
Plating Systems
Whether designed to produce hundreds of thousands of components each month, or just a few hundred, the required quality of the end product mandates that the electroless nickel plating be performed in a satisfactory plating system.

A properly designed system should be large enough in volume to enable close control over the process chemistry and variables, and will include the following features:

Tanks
Polypropylene, 316 stainless steel, pyrex, or appropriate high-temperature liners, such as 8- to 12-mil-thick PVDC. Agitation, either mechanical or by clean, dry, low-pressure air, through CPVC or polypropylene pipes located on the tank bottom is required, and part agitation using moving work rods is recommended. Tank walls, seams, and welds must be as smooth as possible to reduce the tendency for the solution to plate out.

Heaters
Heaters can be fabricated from teflon, quartz, pyrex, and stainless steel. It is important that they be sized large enough to avoid excessive heat-up time, but have low heat density to prevent localized over-heating of the solution.

Filters
Continuous filtration is important to preclude any chance of any degree of surface roughness. In-tank filters are recommended, particularly CPVC cartridge holders with 0.25 to 1 μm polypropylene cartridges. The size of the filters should be adequate to permit 10 solution turnovers per hour.

Reservoir tanks
Also recommended are "hard-plumbed" reservoir tanks for the nitric acid solution required for stripping the processing tanks after the inevitable, and often times, all too frequent plate-out phenomena. These reservoirs can also hold the plating solution while the stripping process is underway.

The equipment mentioned above is recommended for any good electroless nickel plating system; however, when the deposit is expected to meet the requirements of the electronics industry at the lowest possible cost, then the above recommendations should be considered *mandatory*.

Handling Procedures
The appropriate technique for transporting the parts to be plated through the preplating and plating operations are determined by the number of parts to be processed, their size and shape, and how easily they might be damaged.

Racking
The larger components, and those having complex shapes, or that are easily damaged, must be racked on appropriately designed racks. Spacing between parts must be sufficient to allow good solution flow. Complex shapes, deep bores, and recesses, must be positioned in such a way that the solution impingement on those surfaces is optimized. Racked parts should be at an angle to avoid the chance of trapped gases causing roughness on undersurfaces, and to promote good draining after each processing step. Rack marks may be permitted by specification at sites indicated in engineering drawings. If not permitted, the parts should be suspended loosely and the racks agitated frequently to prevent shielding.

If a copper strike is used before electroless nickel, it will be necessary to initiate the nickel deposit by catalytic activation. One technique is to apply a brief cathodic current to the parts; stainless steel contacts on the racks are recommended when this is necessary.

Wiring

It is unlikely that appropriately-designed racks are feasible or economically advisable for all of the many shapes and sizes of components that must be plated. Affixing them to small diameter copper or aluminum soft wire, either individually or in chains, is a common practice.

The parts must be suspended loosely to prevent shielding by the wire, and each must be separated from its neighbor by 1 to 2 inches to be sure that there will be no interference with solution impingement on a part, and to avoid mechanical damage.

Baskets

Because aluminum components are light, they can be bulk-processed in polypropylene or other heat-resistant, nonconductive dipping baskets. The baskets should be filled to no more than 1/3 to 1/2 their height with parts, and the same baskets can be used throughout the entire preplating sequence.

The baskets are immersed in the electroless nickel solution suspended from a suitable support or crossbar. In a manual installation, the plating operator must carefully agitate and swirl the baskets at regular intervals to assure uniform deposition on all parts.

By using separators in the baskets, it is possible to plate several types of components at the same time, avoiding a post-plating sorting operation.

Barrels

Conventional horizontal plating cylinders, without cathode contacts, can also be successfully used for electroless nickel plating aluminum connector components that will not be damaged by tumbling. The effect of the tumbling action can be minimized by rotating the cylinder just a few times per minute. Rotation speed should be slow—i.e., 3 rpm. The perforations in the barrel should be as large as possible to assure good solution transfer.

The cylinders can be loaded to about the center line with parts that will tumble freely, and which are small and light enough that threads and sharp edges will not be adversely affected. About 1/3 of the barrel should be above the nickel solution so that the gas being generated will escape. Preplating operations are best performed in dipping baskets, with the parts being transferred to the plating barrels before nickel plating.

Preplating 2000 and 6000 Series Aluminum

Aluminum connector components are manufactured by several different methods, usually by machining from bar stock, extrusion, and die casting. This means that several aluminum alloys will be used, and that the condition of the surface can differ significantly, ranging from smooth and highly polished, to coarse with both macro- and micro-surface and subsurface porosity. To further complicate the task of the plater, fabricating techniques often include secondary operations that produce lapped metal, burrs, abraded surfaces, and combinations of cast and machined areas.

Other factors that must be considered when establishing preplating processes include the different types of lubricants used in the various fabricating operations; the need to maintain critical dimensions, especially on threaded parts; the presence of blind threaded holes; the size of the inner diameters that must be plated; and the capability of the parts to trap solutions in small bores and recesses.

Given all of the above complications, it would seem that a satisfactory preplating process would be impossible to develop and maintain. However, it can be accomplished, and surprisingly enough, with but little modifications to conventional techniques that can be found in any text on the subject of plating aluminum and its alloys.

Following is a typical processing sequence used for the 2000 and 6000 series aluminum alloys from which connector components are fabricated, and which may have either a machined or cast surface:

- Solvent vapor degrease in perchloroethylene, 1,1,1-trichloroethane, or equivalent.
- Phosphoric acid etch for 1.5 min in a 40 percent by vol. solution.
- Rinse (two-stage). Good rinsing between operations is very important. For all rinses, a two-stage cascading rinse is advised, and air agitation is recommended. Rinsing time should be from 30 sec to about 1 min. The temperature of the water should not be less than 60° F, preferably at least 70° F.
- Non-etch aluminum cleaner. The use of an inhibited cleaner is recommended. Proprietary compounds containing alkaline silicates that inhibit attack of the aluminum can be used; however, non-silicated cleaners are preferred to avoid subsequent difficulties in removing a residual silicate film that might form.

The solution should be operated according to the supplier's recommendations; however, the operating temperature should be held between 120 and 130° F.

Immersion time should be about 2 to 4 minutes for racked parts, and 5 to 10 minutes for parts being processed in dipping baskets or horizontal barrels. The pH should be maintained between 9.5 and 11.0.

- Mild etch (optional). Alkaline solutions that etch aluminum are not recommended for the purpose of cleaning. If it is desired to etch the surface to improve the appearance of the component by masking machining marks, scratches, or other surface imperfections, this operation should follow a non-etch cleaning.

Both mildly alkaline and acidic proprietary etching solutions can be obtained from the supply houses, and when operated in accordance with the supplier's recommendations will produce a satisfactorily etched surface. The amount of each etch desired will vary, depending on the nature of the parts. The degree of etching is varied by (a) adjusting the immersion time; (b) raising or lowering the solution temperature; or (c) raising or lowering the concentration of the

- Rinse.
- Desmut for 10 to 30 sec. There are a number of proprietary desmutters available; or use a mixture of 75 percent nitric acid and 25 percent hydrofluoric acid; or 65 percent nitric acid, 25 percent sulfuric acid, and 10 percent hydrofluoric acid.
- Rinse.
- Nitric acid (concentrated 42° Be) dip for 1 min at room temperature.
- Rinse.
- Zincate for 1 min. Formulas for zincating solutions can be found in most plating texts. A typical formulation recommended for wrought and diecast alloys consists of (28): 100 g/L zinc oxide, 400 g/L sodium hydroxide, 1 g/L ferrous chloride, 5 g/L Rochelle salts, at a temperature of 60 to 80° F.

A number of especially formulated proprietary zincating solutions are marketed by the many companies that feature processes for electroless nickel plating on aluminum. The use of one of these is recommended rather than a self-made solution in order that the user can benefit from the knowledge and experience of the supplier.

- Rinse.
- Zincate strip with 50 percent nitric acid at room temperature for 30 sec.
- Rinse.
- Zincate for 30 sec. A "double zincate" process is necessary when processing the 6000 series aluminum alloys to assure good adherence of the electroless nickel plating. While it may not be necessary for alloys of the 2000 series, double zincating is recommended as a safety precaution.
- Rinse.
- DI water rinse.
- Copper strike (optional). Two techniques may be used to improve the adherence of nickel plating to aluminum. One is to submit the component to a baking operation of 30 to 60 min at 375° F; the second is to copper strike the zincated parts before immersing in the nickel solution.

If a copper strike is used, it must be one that is especially formulated for plating on zincated aluminum. This is to avoid the possibility of an immersion copper film, which will result in poor adherence and the formation of blisters after nickel plating. This is particularly true when the parts are being plated in baskets or barrels because of the inevitable presence of very low current density "pockets". For the same reason, the zincated parts must be "hot" (cathodically charged) when they enter the copper plating solution. A suggested copper strike solution consists of 5.5 oz/gal copper cyanide, 8.5 oz/gal sodium cyanide, 8.0 oz/gal Rochelle salts, at pH 10.2 to 10.5 and a free cyanide level of 2.0 to 2.8.

- Rinse.
- DI water rinse.
- Electroless nickel plate.

Plating

Acid electroless nickel baths are commonly used for plating aluminum connector components. The phosphorus should be maintained between 11 and

13 percent to sustain the desired properties of the deposit, and to assure that it will be non-magnetic.

The process selected for plating aluminum connector hardware must be one which will produce deposits having the properties specified and/or required for optimum connector operation and reliability. In addition, the plater must consider a number of other factors to assure the continuing production of parts of suitable quality at the lowest possible cost:

Appearance

A semibright coating is the accepted standard of the connector industry.

Ease of control

The chemistry of the process must be maintained within close tolerances to produce deposits of required characteristics, and to govern the deposition rate. Therefore, the fewer additives required to sustain the process, the better.

Stability

A high tolerance for contaminants and high temperature stability are important factors that improve the life of the solution and reduce the tendency for plate-out—a common phenomena in electroless plating processes.

Optimum plating rate

that is, a rate that is both consistent and predictable. Since plating thickness tolerances must be rather tight to assure later matching of opposing connector members, it is important that the plater know what deposition rate to expect.

Cost

Experienced electroless nickel platers know that to accurately determine plating costs, a number of factors must be considered. These are items such as controllability, stability, and plating rate, as well as chemical make-up costs and the cost of replenishers. Obviously a solution that costs less at the start but which is difficult to control, tends to plate-out too frequently, and/or has a relatively short life expectancy is going to ultimately result in higher processing costs.

The chemistry of electroless nickel plating solutions can be found in chapters 1 and 2 of this book, and in any of the conventional texts and guidebooks relating to plating and metal finishing practices. Operation and control of these solutions are also described, and many plating operations have used or are using them successfully—but not necessarily for plating parts for electronic applications.

It is the recommendation of the authors, based on personal experience, that when plating electronic components, particularly aluminum connector hardware, commercially available processes be used. There are many available that have been developed specifically for such applications, which provide all of the characteristics required. It is critical in electronics applications that the electroless nickel deposits consistently have all of the properties and characteristics needed for optimum and reliable performance. To achieve this, it is necessary that all process variables be closely controlled. Good practice includes the use of automatic bath controllers. If they are not used, it is mandatory that the solutions be analyzed frequently so that only small additions of required chemicals are made. Some general operating guidelines include:

Loading
Typically, an electroless nickel solution can be loaded with about 0.25 to 1.0 ft^2/gal of plating surface area per gallon. To attain closer control over the deposition rate, it is recommended that the loading be restricted to 0.4 to 0.6 ft^2/gal.

Temperature
Although the typical operating range of most solutions can vary 15 to 20° F, it is recommended that when used for plating electronic parts, particularly aluminum connectors, that the range be kept smaller, preferably about ±3° F, e.g., from 186 to 192° F. It may be advantageous to operate a new solution about 5 degrees cooler for its first cycle.

Make sure, through good solution agitation, that the temperature of the solution is equal throughout the entire tank, "Hot-spots" around the heaters cannot be allowed to occur.

Filtration
The importance of continuous filtration was discussed earlier in this chapter. For best results, the filtering media should be 0.25 μm.

Agitation
Solution agitation is accomplished mechanically, i.e., pumping and the use of wires, and by air agitation using a low-pressure air blower. Agitating the parts during plating using horizontal work rod movement is recommended, or when this is not possible, the racks and plating baskets should be agitated manually at frequent intervals.

Deposit monitoring
The most important of the characteristics and attributes of the deposit that can be checked at the time of the plating process by the plater are thickness, adherence, and ductility. The plating operator can perform these tests easily by the following method:

With suitable process controls in place, the deposition rate of an electroless nickel solution is quite predictable, and a typical plating specification of 0.0002 to 0.0004 in., or 0.0004 to 0.0007 in. is easy to meet. Thicker coatings of 2 to 3 mils, which are required for special connector applications may prove to be more of a problem, particularly when the specified range may be a seemingly impossible ±0.0001 in. Depending on the age of the plating solution and other conditions, the total plating time will be about 3 to 4 hours, and the plating "window" will be about 8 to 10 minutes.

About the simplest technique used to monitor plating thickness, as well as plating adherence and ductility, is to plate aluminum test panels along with the parts being processed. The panels should be made from the same alloy and be about 1 x 4 in. and 0.03 to 0.04 in. thick. They should be run along with the parts being plated through the complete process.

Several panels should be plated, wired, or racked to avoid shielding problems. Both adherence and ductility can be tested by a standard bend test (see chapter on testing of deposits). The hand-held micrometer can be used effectively by a trained operator to determine how much additional time should be given the parts in process, but if an X-ray unit is readily available for use, it is the preferred and more accurate tool to employ. (It can also be used for checking the parts themselves.) One of the advantages of the post-plating baking operation used to enhance the adherence of the deposit is that blisters will be formed if pre-plating operations have not adequately prepared the surface for plating. Since many of the components will be subjected to higher temperatures during subsequent bonding or marking operations, it is important that blistering be detected immediately after plating. A baking test is recommended—30 minutes at about 400° F on several specimens before the entire lot is subjected to the baking process.

ELECTROLESS NICKEL FOR MEMORY DISKS

For many years, manufacturers of disk drives relied on disks coated with a magnetic iron oxide layer. The emergence of thin film technology, however, has increased the magnetic density of memory disks several orders of magnitude over the older particulate-oxide-coated disks. Figure 9.13 shows electroless nickel plated 3.5 and 5.25 in. memory disks.

Thin-film disks are manufactured in the following manner, using most or all of the following process steps:

Substrate preparation
- Punching and machining aluminum alloy blanks
- Precision grinding
- Stress relief annealing
- Surface polishing and/or diamond turning

Electroless nickel deposition
- Surface cleaning
- Oxide removal
- Immersion zinc
- Electroless nickel plate

Polishing and texturing
- Coarse polishing
- Fine polishing
- Surface texturing to reduce stiction

Magnetic layer deposition
- Sputtered chrome layer
- Magnetic thin film coating (sputter or plate)

Fig. 9.13—3½ and 5¼ in. electroless nickel-plated memory disks in a cassette carrier.

Finishing
- Protective overcoats of chrome alloys and/or carbon
- Antistiction coating
- Burnishing

Details of the above process can be obtained from most proprietary electroless nickel suppliers. The magnetic properties of electroless nickel, definitions and references can be found in the literature (29,30).

The crucial element of thin-film memory disk fabrication is the electroless nickel that is deposited between the aluminum substrate and the thin magnetic recording layer. It is essential that the electroless nickel deposit be non-magnetic in the as-plated state, and remain non-magnetic when heat treated at temperatures up to 300° C for 1 to 2 hours. Ni-P deposits containing at least 9 percent by weight phosphorus are usually non-magnetic in the as-plated state but become magnetic when exposed to temperatures as low as 200° C for short time periods. If the disk has a residual magnetism less than about 0.1 emu/cc after heat treatment, it is considered acceptable. Deposits that contain at least 11 percent phosphorus by weight will, in general, meet the residual magnetism criteria.

The structure of electroless Ni-P deposits undergo changes during heat treatment with the formation of intermetallic nickel-phosphide compounds. It has been observed by X-ray diffraction techniques that when the dominant nickel-phosphide that forms during heat treatment is Ni_3P, the deposit will be ferromagnetic. If, on the other hand, the intermetallic phosphides that precipitate are mixtures of the type Ni_xP_y (usually $Ni_{12}P_5$), with very little Ni_3P present, the deposit will meet the residual magnetism criteria.

The magnetic behavior of the Ni-P deposit after heat treatment not only depends on the phosphorus content of the deposit, but also on the internal intrinsic stress of the deposit (31,32).

Deposits that show compressive or near-zero intrinsic stress and contain 11 percent or more phosphorus will precipitate nickel phosphides of the form Ni_xP_y when heat treated at temperatures up to about 300° C. These deposits will be essentially non-magnetic. However, if the intrinsic stress of the deposit (with P content >11 percent) is tensile, the probability that the dominant intermetallic compound that forms during heat treatment will be Ni_3P is greatly increased, in which case the deposit will fail the residual magnetism test; i.e., there is a correlation between intrinsic tensile stress and ferromagnetic behavior.

Imperfections of any kind in the electroless nickel coating that are not removed by polishing are cause for rejection. Imperfections that only show up on microscopic examination are also cause for rejection. Surface imperfections cause loss of information and noise in the recording media.

In order to achieve high phosphorus (>11 percent by weight), low stress (zero to compressive) Ni-P deposits with a smooth, pore-free surface, special attention must be paid to the chemistry of the electroless nickel plating bath that is used to plate disks. Citric acid and its salts have been the most widely used complexing agents in electroless nickel plating solutions used to obtain high-phosphorus, corrosion-resistant deposits. In 1950, Wesley (33) reported a plating bath containing citrate anions that produced Ni-P deposits that were non-magnetic. An electroless nickel plating bath similar to Wesley's, utilizing citrate, that yields high-phosphorus, compressively-stressed, pore-free deposits consists of 0.1M Ni^{++}, 0.2M $K_3C_6H_5O_2 \cdot H_2O$, 0.28M $NaH_2PO_2 \cdot H_2O$, 1.0 ppm Pb^{++}, at a pH of 4.8 and temperature of 87° C. The plating rate was 7.0 $\mu m/hr$; the resulting deposits had a compressive stress of -500 psi and contained 12.1 percent P.

Typically, complexing agents that form more than one chelate ring with nickel, e.g., citrate and malate anions, will, in properly formulated and operated plating baths, produce non-magnetic Ni-P deposits.

Regardless of bath chemistry, the disk plating bath must be stable and produce consistent deposits over 4 to 5 metal turnovers. Exceptional consideration must be given to bath purity. The raw materials used in these plating solutions are often purified more extensively than in most other electroless nickel baths. Further, special attention must be given to filtration to prevent even the smallest particle from causing deposit defects. Filtering through a 1-μm depth filter, followed by an absolute filter (screen) of 0.25 to 0.45

μm is required. The plating of disks is generally done in "clean-room" conditions.

Plating Electroless Nickel on
5086 Aluminum Alloy

The surface of the finished disk must be nearly perfect; therefore, it is mandatory to use a high-quality aluminum alloy with the highest possible finish, and to prepare the surface in such a way that its finish is not destroyed. The alloys presently used to produce magnetic memory disks are 5086, 5586, CW66, FB-1, CW75, CZ46, PA00, and 5252. Magnesium tends to congregate at the grain boundaries, providing electrochemical cells that could lead to pitting of the aluminum if severe processing steps are used; therefore, a mild non-etching alkaline cleaner is required in the initial stage of the preplate cycle. An extremely mild acidic treatment is then used to remove oxide films from the aluminum disk, so it can properly receive the immersion zinc treatment. A double zinc treatment provides the best base and the smoothest surface for subsequent electroless nickel plating. In terms of adhesion requirements, a properly cleaned and prepared aluminum disk that is plated with 0.0005 to 0.001 in. (12.5 to 25 μm) of electroless nickel should be capable of being bent double with no flaking of nickel. A typical process cycle follows:

- Alkaline clean in mild, non-etch cleaner.
- Rinse.
- Acid clean and deoxidize in a mild, non-etch cleaner/deoxidizer.
- Rinse.
- Immersion zinc 35 to 60 sec at room temperature.
- Rinse thoroughly.
- Strip the zinc deposit in 60 percent by vol. nitric acid (42° Be).
- Rinse.
- Immersion zinc 15 to 20 sec at room temperature. Must be separate immersion zinc solution.
- Rinse thoroughly (double-rinse, counterflow).
- Rinse (DI water).
- Electroless nickel plate.
- Rinse (DI) and dry.

Plating Electroless Cobalt on
Electroless Nickel Deposits

A typical process cycle for plating an electroless cobalt medium on an electroless nickel undercoat is as follows:

- Polish electroless undercoat to a high finish.
- Clean in a mild alkaline cleaner.
- Rinse in DI water.
- Activate. Mild acids are often used (proprietary solutions vary from mild alkaline to acidic solutions).

- Rinse in DI water.
- Electroless cobalt plate for 1 to 2 min at a bath temperature of 150 to 170° F (65 to 76° C).
- Rinse (DI water).
- Dry.
- Apply lubricant overcoating (carbon, silicon dioxide, fluocarbons, oxidizing treatments, and other processes).
- Test for magnetic characteristics.

REFERENCES

1. Symposium on Electroless Nickel Plating, ASTM Special Technical Publication 265, p. 23 (1959).
2. M.V. Sullivan and J.H. Eigler, *J. Electrochem. Soc.*, **104**(4), 226 (1957).
3. W. Karaces, *Metal Finishing* (Jan. 1985).
4. C.D. Brown and G.R. Jarrett, *Trans. Inst. Met. Finish.*, **49**(1), 1 (1971).
5. G.O. Mallory, 17th Connector and Interconnection Technology Symposium, 1984.
6. D.W. Baudrand, *Plat. and Surf. Fin.*, **66**(12), 14 (1979).
7. J.C. Turn and E.L. Owen, *Plating*, **61**(11), 1015 (1974).
8. D.W. Baudrand, *Plat. and Surf. Fin.*, **68**(12) (1981).
9. T.D. Schlabach and D.K. Rider, *Printed and Integrated Circuitry*, McGraw-Hill Book Company, New York, NY, 1963; Chap. 2.
10. E.F. Duffek, *Plating*, **56**, 505 (1969).
11. H. Rochow, *Ber. Dtsch. Keram. Ges.*, **44**, 224 (1967).
12. V. Jirkovsky, J. Mikulickova and K. Balik, *Tesla Electronics*, **4**, 107 (1967); 3, 75 (1977).
13. E.F. Duffek, *Plating*, **57** (1970).
14. H.D. Kaiser, F.J. Pokulski and A.F. Schmeckenbecker, *Solid State Technology*, **15**(4), 39 (1972).
15. D.W. Baudrand, *Plat. and Surf. Fin.*, **68**, 67 (1981); 71, 72 (1984); 74, 115 (1987).
16. S.G. Stalnecker, U.S. patent 3,296,012 (1967).
17. S. Schachameyer, U.S. patent 4,428,986 (1984).
18. Ameen et al., *J. Electrochem. Soc.*, **120**, 1518 (1973).
19. T. Okamura and C. Sasaki, *Jitsumi Hyomen Gijutsu (Metal Finishing Practices)*, **33**(2), 38 (1986).
20. H. Honma and K. Kanemitsu, *Plat. and Surf. Fin.*, **74**, 62 (1987).
21. G.V. Elmore, U.S. patent 3,690,921 (1972).
22. M. DeLuca and J.F. McCormack, U.S. patent 4,574,094 (1986); 4,604,299 (1986); 4,701,352 (1987).
23. M. DeLuca, U.S. patent 4,647,477 (1987).
24. M. DeLuca, J.F. McCormack and P.J. Oleske, U.S. patent 4,666,744 (1987).

25. B.P. Hecht and C. Buckley, *Proc. Hybrid Microelectronics Symposium, International Society for Hybrid Electronics,* Chicago, IL, 1968; p. 397.

26. L.I. Massel and R. Glang, *Handbook of Thin Film Technology,* McGraw-Hill Book Company, New York, NY, 1970.

27. G.O. Mallory, 17th Connectors and Interconnection Technology Symposium, 1984.

28. E.B. Saubestre and J.L. Morico, *Plating,* **53**(7), 899 (1966).

29. A.L. Ruoff, *Materials Science,* Prentice-Hall, NJ, 1973; Chaps. 4 and 32.

30. W.H. Safranek, *The Properties of Electrodeposited Metals and Alloys, 2nd ed.,* American Electroplaters and Surface Finishers Society, Orlando, FL, 1986; Chap. 23.

31. M. Schwartz and G.O. Mallory, *J. Electrochem. Soc.,* **123**(5) (1976).

32. G.O. Mallory and D. Altura, *SAE Technical Paper 830693* (1976).

33. W.A. Wesley, *Plating,* **37**(7) (1950).

Chapter 10
Electroless Deposition of Alloys

By Fred Pearlstein

Electroless deposits may be comprised either of an essentially pure metal element (with less than 1 percent of other elements), or they may be alloys. Examples of the single-element deposits include copper from formaldehyde-based baths, gold from borohydride-based baths, nickel from hydrazine-based solutions, and silver from dimethylamine borane (DMAB)-based electrolytes. Most commonly used electroless plating baths, however, produce binary alloys of a metal such as nickel, cobalt or palladium with phosphorus or boron, the latter elements depending on the reducing agent employed—e.g., hypo-phosphite or DMAB. These alloys make up the most important and well known of the electroless deposits, and are thoroughly discussed elsewhere in this book.

Since the number of metals that can be electrolessly deposited is rather limited compared to the number of metals that can be electrodeposited, the incorporation of additional metal elements into the electroless deposits can be an important means of enlarging the range of chemical, mechanical, physical, magnetic, and other properties attainable. A number of alloys can readily be deposited by combining metals that are independently deposited electrolessly from similar baths; an example is nickel and cobalt from alkaline hypophosphite solutions. Also, and more importantly, certain metals that cannot themselves be deposited by the autocatalytic mechanism can be induced to codeposit with an electrolessly depositing metal. The discussion that follows will concern the binary, ternary, and quaternary alloys that have been deposited electrolessly, some of their unique properties, and potential applications.

ELECTROLESS NICKEL-PHOSPHORUS AND NICKEL-BORON BASED ALLOYS

Since electroless nickel-phosphorus (EN-P) deposits have found the greatest commercial applications, it is not surprising that most of the investigative effort in alloy deposition has involved codeposition of a third element with EN-P. EN-P and cobalt-phosphorus (Co-P) deposits may be produced from similar alkaline ammoniacal baths, and therefore, nickel-cobalt alloys of virtually all proportions are possible with phosphorus as the third element. It is interesting to note that Co-P cannot be deposited from acid-type baths, and cobalt is not even induced

261

to codeposit with nickel when cobalt salts are added to the conventional acid-type EN-P baths. EN-Co-P deposits are more electrochemically active (more negative in mixed potential in neutral salt solutions) than EN-P deposits, and thus, double-layer deposits of conventional EN-P on steel followed by EN-Co-P provides enhanced basis metal corrosion protection in marine environments. The outer EN-Co-P deposit apparently provides sacrificial protection to the inner EN-P layer, thereby preserving the inner layer's protective barrier (1).

Tungsten can readily be codeposited with EN-P. For example, Table 10.1 describes a bath capable of producing EN-P deposits containing up to 20 percent W (2). The influence of citrate concentration on the deposition rate and composition of electroless nickel-tungsten alloys is shown in Fig. 10.1. Lowering the citrate concentration of the bath resulted in lower tungsten and phosphorus contents but higher deposition rates. At pH 8.2, the deposition rate from the bath containing 20 g/L sodium citrate dihydrate was about 0.2 mil/hr, while that containing 40 g/L was less than 0.1 mil/hr.

Table 10.1
Electroless Nickel-Tungsten Alloys
Effect of Solution Makeup* and pH

Deposit Composition, percent by weight

Solution pH	20 g/L sodium citrate			40 g/L sodium citrate		
	Ni	W	P	Ni	W	P
5.5	—	—	—	87.5	2.9	9.5
7.0	87.5	6.7	4.8	74.6	16.6	8.6
8.2	83.8	12.7	4.4	73.5	20.5	6.5
9.5	—	—	—	72.0	18.2	9.8

*7 g/L $NiSO_4 \cdot 6H_2O$, 35 g/L $Na_2WO_4 \cdot 2H_2O$, 10 g/L $NaH_2PO_2 \cdot H_2O$, 20 or 40 g/L $Na_3C_6H_5O_7 \cdot 2H_2O$, temperature 98° C.

These deposits may find applications where resistance to high temperatures is required, as the melting points are substantially increased by the presence of tungsten. Deposits containing about 20 percent tungsten were resistant to attack by concentrated nitric acid and 7N hydrochloric acid. An electroless nickel deposit containing 9.1 percent tungsten and 8.2 percent phosphorus was considerably more protective to steel than comparable EN-P deposits (3).

Mallory and Horhn (4) studied ternary alloys of EN-P and EN-B with codeposited tungsten, molybdenum, or tin from solutions containing gluconate to complex the oxyanion (tungstate, molybdate, stannate). The results are summarized in Table 10.2. Nickel deposits containing up to 10.5 percent W (0.87

percent B), 31 percent Mo (1.67 percent B) or 8.9 percent Sn (3.2 percent P) were obtained.

X-ray diffraction studies (5) showed that molybdenum codeposited with EN-P or EN-B resulted in deposits of larger grain size than the corresponding binary EN deposits. It is noteworthy that an EN-P deposit with 11 percent molybdenum and only 2 percent phosphorus was nonferromagnetic and retained this property even when aged at 400° C for 96 hours.

Small quantities of copper codeposited with EN-P from acid baths can have a profound effect on deposit characteristics. For example, nickel deposits containing 12 percent phosphorus and 1 percent copper are unusually smooth and bright, with improved ductility and corrosion resistance. The deposits are also virtually nonferromagnetic, even when exposed to elevated temperature and, therefore, have found application as an initial coating on aluminum computer memory disks, prior to application of the magnetic deposit.

It is also possible to produce electroless copper-nickel-phosphorus alloys (containing 70 to 99 percent copper) from alkaline non-ammoniated hypophosphite-based baths (6). A quaternary alloy deposit of EN-P containing 1.7 percent copper, 8.2 percent tin, and 16.3 percent phosphorus was found to provide superior protection to aluminum or steel substrates during salt spray exposure (3).

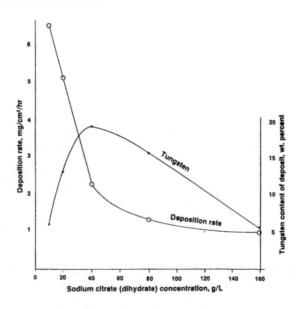

Fig. 10.1—Effect of citrate concentration on electroless deposition of nickel-tungsten alloys. Bath contains 7 g/L nickel sulfate·6H$_2$O, 35 g/L sodium tungstate·2H$_2$O, and 10 g/L sodium hypophosphite H$_2$O; pH = 8.2, temperature = 98° C.

Table 10.2
Effect of Bath Temperature and pH
On Composition of Electroless Nickel Alloy Deposits

	Deposit composition, percent by weight											
	87° C		70° C		87° C		70° C		87° C		70° C	
Solution pH	W	P	W	B	Mo	P	Mo	B	Sn	P	Sn	P
5	2.1	9.1	10.0	5.1	—	—	2.9	3.2	2.5	9.1	8.9	3.2
7	5.5	8.0	10.0	3.5	tr.	6.9	5.8	1.7	2.7	10.3	6.2	2.3
9	8.0	6.3	10.5	0.8	7.0	1.8	31.0	1.6	1.7	1.8	3.5	1.3

Additions of potassium perrhenate to an ammoniacal-alkaline EN-P bath produced alloys containing more than 45 percent rhenium and 2.3 percent phosphorus, though the quantity of rhenium salt in the solution was quite small compared to nickel (2) (see Table 10.3). The deposit had a melting point of about 1700° C, compared to 885° C for comparable EN-P deposits. The addition of zinc sulfate instead of perhennate to the same EN-P bath produced deposits containing up to 15 percent zinc. The deposits are electrochemically active and may be capable of offering sacrificial protection to steel substrates and, therefore, may be potential alternatives to cadmium for uniformly coating complex shapes and threaded parts. The effects of additions of several metals to the ammoniacal alkaline EN-P bath are shown in Fig. 10.2.

Combining hypophosphite and DMAB in an electroless nickel plating bath containing glycolate produced EN-P-B alloys (typically 92.5% Ni-6.8% P-0.7% B) with hardness, wear resistance, and corrosion resistance comparable to EN-B but superior to conventional EN-P deposits from similar baths. The main advantage appears to be one of economics of the mixed EN-P-B system over EN-B (7).

When iron ions were added to an ammoniacal EN-B bath containing tartrate and DMAB, nickel-iron-boron alloys containing up to 30 percent iron were produced. Some of these alloy deposits are considered useful as magnetic memory elements (8). Likewise, iron can be codeposited with EN-P from suitably formulated baths (9).

ELECTROLESS COBALT-PHOSPHORUS BASED ALLOYS

Electroless cobalt-phosphorus deposits find their greatest application by virtue of their useful and unique magnetic properties. However, there are indications that Co-P deposits may be superior, under certain conditions, to EN-P for

Table 10.3
Electroless Nickel-Phosphorus Alloys

Bath constituents	Concentration, g/L	
	EN-P-45% Re	EN-P-12% Zn
$NiSO_4 \cdot 6H_2O$	30	27
NH_4Cl	50	50
$Na_3C_6H_5O_7 \cdot 2H_2O$	85	85
$KReO_4$	1.5	—
$ZnSO_4 \cdot H_2O$	—	7.5
$NaH_2PO_2 \cdot H_2O$	10	10
pH (adjusted with W/NH_4OH)	9	9
Temperature, °C	98	98

protecting steel against corrosive attack. A variety of interesting and useful magnetic properties can be obtained by addition of other metal ions to the typical alkaline-ammoniacal bath to form ternary alloy deposits. Addition of 10 g/L sodium tungstate dihydrate or 0.8 g/L potassium perrhenate to a typical bath at pH 8.9 and 98° C resulted in deposits containing about 9 percent tungsten and 30 percent rhenium. Addition of 10 or 30 g/L nickel sulfate hexahydrate resulted in deposits containing 30 or 75 percent nickel, respectively (10). Electroless Co-P deposits containing up to 45 percent iron (11) or 4 percent zinc (12) have also been produced.

ELECTROLESS COPPER BASED ALLOYS

Electroless copper deposits for most commercial applications (e.g., printed circuits and plated plastics) are produced from highly alkaline formaldehyde-based baths, though DMAB or hypophosphite baths can be used to successfully deposit copper. A bath comprised of 13.8 g/L copper sulfate pentahydrate, 69 g/L rochelle salt tetrahydrate, 20 g/L sodium hydroxide, and 40 mL/L formaldehyde (36 percent HCHO + 12.5 percent methanol) was used as a base to produce various copper alloys by addition of other metal ions. (See Table 10.4 for details and remarks.) Addition of a bath stabilizer such as 2-mercaptobenzothiazole tended to decrease the alloy content of some of the copper deposits, while others were increased (13). Additional studies may be warranted using electroless copper baths containing EDTA or quadrol complexing agent along with stabilizers currently in use. Electroless copper-tin-boron deposits containing more than 10 percent tin and less than 1 percent boron were produced from a DMAB-based bath (14).

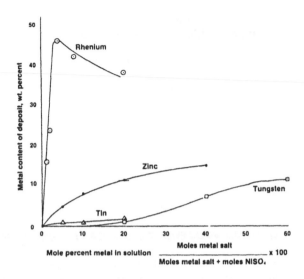

Fig. 10.2—Effect of additions of potassium perhennate, sodium tungstate, sodium stannate, or zinc sulfate to ammoniated alkaline-citrate electroless nickel solution on deposit composition. Metal and nickel salts—0.13 moles/L; ammonium chloride—50 g/L; sodium citrate·2H$_2$O—85 g/L; sodium hypophosphite·H$_2$O—10 g/L; ammonium hydroxide (29%)—50 mL/L (additional 30 mL/L added per hour during deposition); temperature—98° C.

ELECTROLESS PALLADIUM-PHOSPHORUS ALLOYS

An alkaline ammoniacal bath based on a hypophosphite reducing agent produced electroless palladium deposits containing about 1.5 percent phosphorus. Ternary alloy deposits of palladium-phosphorus containing significant quantities of nickel, cobalt, or zinc can be produced by addition of appropriate metal ions (see Table 10.5). Palladium-nickel alloys may have value for application to electronics contacts by increasing hardness and wear resistance, and by moderating the catalytic properties of palladium that can result in formation of resistive films by polymerization of organic vapors in the air.

It is confidently anticipated that new and useful electroless alloy deposits, in addition to those described above, will be developed for specific desired properties along with the superior properties that already characterize electroless deposits.

Table 10.4
Electroless Copper Alloys

Metal salt added	Conc., g/L	Metal in deposit, %	Remarks
$NiSO_4 \cdot 6H_2O$	14.2	3.9 (Ni)	Light copper color
$CoSO_4 \cdot 7H_2O$	15.2	3.5 (Co)	Light copper color; tarnish resistant
Na_2CrO_4	0.35	1.1 (Cr)	Stabilized bath; increased deposition rate
CdCl	5.0	3.2 (Cd)	Increased deposition rate
$CdCl_2$	20.0	18.2 (Cd)	Somewhat brassy color

Table 10.5
Electroless Palladium Alloys

Metal salt addition to bath*		Deposit composition, % by weight		
Salt	Conc., g/L	Metal	P	Pd
None	—	—	1.52	Balance
$NiSO_4 \cdot 6H_2O$	29.6	5.99 Ni	2.68	Balance
$CoSO_4 \cdot 6H_2O$	29.6	9.82 Co	2.77	Balance
$ZnSO_4 \cdot 7H_2O$	36.0	36.0 Zn	1.24	61.4

*2 g/L $PdCl_2$, 4 mL/L HCl (38%), 160 mL/L NH_4OH (28%), 27 g/L NH_4Cl, 10 g/L $NaH_2PO_2 \cdot H_2O$; pH = 9.8 ±0.2; temperature = 60° C.

REFERENCES

1. L. Gruss and F. Pearlstein, *Plat. and Surf. Finish.*, **70,** 47 (Feb. 1983).
2. F. Pearlstein and R.F. Weightman, *Electrochemical Technology*, **6,** 427 (1968).
3. *Finishing Highlights*, **11,** 14 (Jan./Feb. 1979).
4. G.O. Mallory and T.R. Horhn, *Plat. and Surf. Finish.*, **66,** 40 (Apr. 1979).
5. G.O. Mallory, ibid., **63,** 34, (Jun. 1976).
6. U.S. patent 4,482,596 (1984).
7. C.K. Mital and P.B. Shrivastova, *Metal Finishing,* **84,** 67 (Oct. 1986).
8. A.F. Schmeckenbecher, *Plating,* **58,** 905 (1971).
9. A.F. Schmeckenbecher, *J. Electrochem. Soc.,* **113,** 778 (1966).
10. F. Pearlstein and R.F. Weightman, *Plating,* **54,** 714 (1967).
11. J.R. DePew and D.E. Speliotis, ibid., **54,** 705 (1967).
12. M. Soraya, ibid., **54,** 549 (1967).
13. F. Pearlstein, *Plating and Surf. Fin.,* **70,** 43 (Oct. 1983).
14. G.T. Duncan and J.C. Banter, paper submitted to *Plat. and Surf. Finish.*

Chapter 11
Composite Electroless Plating

Nathan Feldstein

The ability to codeposit fine particulate matter within an electroless metal matrix has led to a new generation of composite coatings. Successful codeposition is dependent on various factors including particle catalytic inertness, particle charge, electroless bath composition, bath reactivity, compatibility of the particles with the metallic matrix, plating rate, and particle size distribution.

The mechanics of composite electroless plating are different from prevailing practices for conventional electroless plating. Finely divided, solid particulate material is added to and dispersed throughout the electroless plating bath, even though the plating bath is thermodynamically unstable and is prone to homogeneous decomposition. The dispersed particles are not filtered out. The dispersion of the particulate matter results in a new surface area loading in the range of 100,000 cm^2/L, which is some 800 times greater than the plating load generally acceptable in electroless nickel plating.

Composite electroless plating is still in its infancy; however, future prospects are attractive both to the user as well as to the designer. The adoption of a coating that offers improved wear resistance and/or lubricity can yield ways of conserving both energy and natural resources. The use of a composite electroless coating also offers the benefit of reduced solution handling and waste treatment problems, as well as reduced reliance on strategic materials (e.g., chromium) for wear applications.

Some types of composite plating were demonstrated in earlier works. In pursuing improved corrosion resistance for nickel-chromium electrodeposition, Odekerken (1) interposed the structure with an intermediate layer containing finely divided particles distributed within a metallic matrix. Electroless nickel deposition was demonstrated as a means of depositing the intermediate layer, utilizing finely divided aluminum oxide, polyvinyl chloride resin (PVC), as well as binary mixtures of powders (e.g., PVC and aluminum oxide). Though Odekerken employed a relatively thin intermediate layer, thicker layers can be attained with longer immersion periods because of the autocatalytic nature of electroless plating.

Metzger et al. (2) duplicated an electroless nickel deposit containing alumina particles. Their efforts were then directed towards the codeposition of silicon carbide, with the first commercial application of the coating on the Wankel (rotary) internal combustion engine. These efforts are documented in a German patent application (3) which, although it did not issue as a German patent, was later modified in the United States and resulted in the issuance of several patents.

In typical composite coatings, the fine particulate matter can be selected in the size range of 0.1 μm to about 10 μm, with a loading of up to about 40 percent by volume within the matrix. The ratio of codeposited particles to the metal matrix in composite electroless plating can be adjusted to a fixed and constant ratio.

Most commercial practices, however, appear to focus on 18 to 25 vol. percent of the particle within the matrix. Because of the uniform manner by which the particulate matter is codeposited, these coatings are known as *regenerative*. This type of coating maintains its properties even when portions of the coating are removed.

Though it is possible to generate thicker coatings, deposits of 0.5 to 1 mil have been found to be adequate for most commercial applications.

While it would appear that, potentially, a wide variety of particulate matter can be codeposited, at present commercial composite electroless plating activities have been limited to just a few types of particulate matter: diamond, silicon carbide, aluminum oxide, and PTFE.

Figures 11.1 through 11.4 are photomicrographs of cross-sectional cuts of electroless nickel containing diamond, silicon carbide, aluminum oxide, and PTFE, respectively.

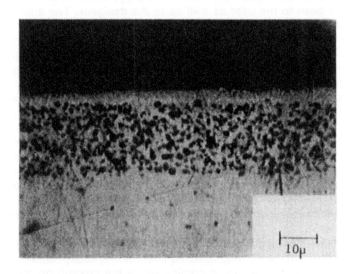

Fig 11.1—SEM photograph showing cross-section of electroless nickel deposit containing diamond particles.

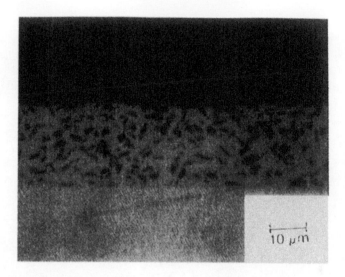

Fig. 11.2—SEM photograph showing cross-section of electroless nickel deposit containing silicon carbide particles.

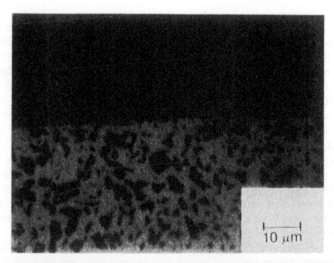

Fig. 11.3—SEM photograph showing cross-section of electroless nickel deposit containing aluminum oxide particles.

Fig. 11.4—SEM photograph showing cross-section of electroless nickel deposit containing PTFE particles.

WEAR RESISTANCE

So far the primary objectives of composite electroless coatings have been to improve the wear resistance and/or corrosion resistance and/or lubricity of machinery parts. Although the basic mechanism for the incorporation of the insoluble particulate matter into the metallic matrix is not fully understood, highlights of the dominant parameters have been characterized in various reports (4-10).

Various test procedures have been employed to evaluate the degree of wear resistance achieved: the Taber Wear Test, the Alfa Wear Test, and the Accelerated Yarnline Wear Test (4).

The Taber Wear Test evaluates the resistance of surfaces to abrasive rubbing produced by the sliding rotation of two unlubricated abrading wheels against the rotating sample. In the Alfa Wear Test, samples are subjected to wear against a hardened steel ring under clean, lubricated conditions. The Accelerated Yarnline Wear Test was designed to simulate typical conditions commonly found in textile machinery.

In all of these laboratory tests, it is generally the scar depth or the worn volume that is measured. It is particularly important in compiling comparative data from these tests that all test specimens be of similar morphology. Variations in particle size, heat treatment cycle, concentration of loading, and surface finish could yield misleading and erroneous wear test results. Surface finish of the coating is especially important in the Taber test.

Table 11.1 provides wear test results for various composite electroless coatings based on the Accelerated Yarnline Wear Test. Wear rate (in μm/hr) is shown in the last column; the higher the number, the greater the deterioration in the wear resistance.

Three different types of diamond particles were tested since they were all commercially available at the time of the Christini et al. investigation (4). The "A" diamond test of Test 1 was a polycrystalline diamond prepared in accordance with Cowan (11). Test 2 used natural diamond. The "B" diamond of Test 3 was a synthetic diamond believed to have been synthesized in accordance with U.S. patents 2,947,608 through 2,947,611.

Table 11.1
Accelerated Yarnline Wear Test Results

Test	Material	Test time, min	Wear rate, μm/hr
1	Electroless Ni-B/9-μm diamond "A" composite coating (polycrystalline)	85	5.1
2	Electroless Ni-B/9-μm natural diamond composite coating	85	10.2
3	Electroless Ni-B/9-μm diamond "B" composite coating	85	13.1
4	Electroless Ni-B/8-μm Al$_2$O$_3$ composite coating	9	109.0
5	Electroless Ni-B/~10-μm SiC composite coating	5	278.0
6	Electroless Ni-B as-plated (with no particles)	1/30	23,000.0

Tests 4 and 5 employed composite coatings bearing aluminum oxide and silicon carbide, respectively, prepared in an electroless nickel-boron bath. Test 6, used as a basis for comparison, has the same composition as the matrix used in Tests 1 through 5, but with no particulate matter codeposited.

The wear rates were determined after specific test time intervals, as seen in column 3. The results demonstrate that diamond "A" is superior to all the other diamond particulate matter evaluated, as well as being the best of all the composite coatings regardless of the particulate matter occluded.

Comparison of the results obtained for the aluminum oxide composite and the silicon carbide composite shows a trend that appears to be contrary to previous reports (8) and, perhaps, to reasonable anticipation. The inclusion of the aluminum oxide appears to provide superior wear even though the testing time was longer (9 min as compared to 5 min) and the particle size was smaller (8 μm as compared to 10 μm).

Greater wear resistance is generally obtained with larger particulate matter. It should be remembered that the hardness of silicon carbide is greater than the hardness of aluminum oxide. This behavior, then, though not fully understood, may be attributable to the manner in which the particles are anchored within the metallic matrix, i.e., the compatibility of the particles within the matrix, as well as their resistance to being pulled out of the matrix during the wear mode (4).

All the composites, regardless of the particulate matter incorporated, performed substantially better than the electroless nickel (Test 6) without any particulate matter.

The merits of composite electroless coatings in comparison with conventional electroless coatings are further illustrated by Parker (12,13) in Table 11.2. Parker measured the wear resistance of miscellaneous composites bearing carbides, borides, diamond, and aluminum oxide. The composites he tested, however, employed particulate matter of different sizes, and neither the concentration of

Table 11.2
Abrasive Wear Data for Electroless Nickel
With Particle Inclusions

Particle	Particle hardness, Knoop	Taber wear index[a]	
		No heat treatment	Heat treated[b]
None	—	18	8
Chromium carbide	1735	8	2
Aluminum oxide	2100	10	5
Titanium carbide	2470	3	2
Silicon carbide	2500	3	2
Boron carbide	2800	2	1
Diamonds	7000	2	2
Hard chromium, 1000 KHN		3	
Aluminum hardcoat		2	

[a]Weight loss in mg/1000 cycles (average of 5000 cycles) with CS 10 wheels and a 1000 g load.
[b]Heated 10 to 16 hr at 290° C.

particles within the matrix nor the surface roughnesses for the coating prior to testing were revealed. Accordingly, based on the data published, no definitive conclusions can be drawn about the performance of one particle over another.

Christini et al. (4) have also demonstrated the role of the metallic matrix binding the insoluble particulate matter in achieving the ultimate wear characteristics. In addition, the following trends were noted:

• Wear resistance is related to particle size. Increases in particle size, however, will yield an optimum point in the obtainable wear resistance values; beyond a particle size of 9 μm, there does not appear to be any discernable gain in wear resistance.

• Wear resistance is related to volume percent (loading) of the codeposited particulate matter in the electroless bath.

In Table 11.3, additional results employing the Taber test (7) are demonstrated. Table 11.4 summarizes additional wear data employing steel pins rotating in V blocks (13). Kenton et al. (14) studied the wear and friction coefficient for a variety of electroless nickel composites. They found that supplementing silicon carbide with the addition of calcium fluoride produced the least wear. Yano et al. (15) studied the codeposition of a binary mixture of particles, and reported that the inclusion of silicon carbide along with a cubical form of boron nitride appears to yield better wear and sliding properties.

Table 11.3
Taber Wear Test Data

	Wear rate	
Wear resistant coating or material	per 1000 cycles (10^{4} mils^{1})	vs. diamond
Polycrystalline diamond*	1.159	1.00
Cemented tungsten carbide Grace C-9 (88WC, 12 Co)	2.746	2.37
Electroplated hard chromium	4.699	4.05
Tool steel, hardened R 62	12.815	13.25

*Composite coating contained 20 to 30 percent of a 3-μm grade diamond in an electroless nickel matrix.

Table 11.4
Wear of Steel Pins in V Blocks[a]

Coating on pin[b]	Coating on V block[b]	Pin wear, mg	Block wear, mg
EN	None	3.7	0.7
EN + WC	None	0.1	4.4
EN + TiC	None	0.4	2.3
EN + SiC	None	0.1	2.6
EN + B₄C	None	0.2	0.7
None	EN	0.1	0.4
None	EN + SiC	0.3	0.7
None	EN + TiC	0.2	0.8
None	EN + B₄C	0.8	0.3
None	EN + WC	2.0	0.3

[a]Data for wear with an applied load of 90 kg for 60 min followed by a load of 182 kg for 40 min, using a Falex lubricant tester.
[b]Coated steel pins and blocks were heat treated 16 hr at 288° C.

FRICTION COEFFICIENT

It has been observed (13) that electroless composites can yield a lower friction coefficient than the same coating without particulate matter. Table 11.5 presents the results obtained for a friction coefficient utilizing an apparatus with a rotating steel ring contacting plated steel blocks (13). In recent years, the incorporation of PTFE into electroless nickel deposits attracted commercial interest, which then led investigators to an exploration of such properties.

The incorporation of PTFE into the electroless nickel composite serves several functions:

- Dry lubrication
- Improved wear resistance
- Improved release properties
- Repellency of contaminants such as water and oil

Most applications employ coating thicknesses of approximately 0.25 to 0.5 mil, with a preferred underlayer of electroless nickel. The presence of the underlying electroless nickel is believed to provide improved corrosion resistance. Typical electroless nickel-PTFE composite coatings incorporate PTFE in the range of 25 percent by volume, with focus on deposits having 18 to

Table 11.5
Friction Coefficients and Wear Data
For Electroless Nickel Composites

Coefficients of friction[a]

Coating on block	Static		Kinetic		Block wear, mg	Ring wear, mg
	initial	final	initial	final		
None	0.233	0.123	0.140	0.123	3.8	+0.3
EN	0.193	0.133	0.165	0.128	9.0	0.6
EN[b]	0.180	0.117	0.178	0.120	2.3	0.5
EN + B$_4$C	0.177	0.142	0.113	0.103	2.0	0.8
EN + SiC	0.180	0.137	0.133	0.128	1.0	1.0
EN + WC	0.167	0.160	0.133	0.133	3.2	3.1
Teflon	0.213	0.140	0.160	0.113	2.8	0.9

[a]Friction coefficients calculated from friction forces measured on a friction and wear machine consisting of a steel ring rotating at 72 rpm on a line contacted against a coated block.
[b]Heat treated at 400° C for 1 hour.

25 vol. percent. Contrary to the inclusion of wear resistant particles (e.g., silicon carbide and diamond), electroless nickel-PTFE composite coatings appear to be limited to particles of 1 μm or smaller.

With the growing commercial interest in electroless nickel-PTFE composite coatings, various investigators have explored the wear properties of such composites utilizing accelerated testing procedures, which may or may not reflect actual field conditions.

Using a rotating ring apparatus, Tulsi (16) investigated the friction coefficients for electroless nickel and composites with PTFE. Table 11.6 summarizes these observations, which suggest that the lowest coefficient of friction is attained when both the pin and the ring are coated with an electroless nickel-PTFE composite coating. Ebdon (17) reported that an electroless nickel-PTFE composite exhibited a friction coefficient of 0.10 in contact with an SAE 300 series stainless steel and lasted much longer (8,000,000 cycles) on pneumatic cylinders than did hard anodized aluminum (10 to 33,000 cycles). After 3,000 cycles, the friction coefficient increased to 0.19; the friction coefficient of the hard anodized aluminum was 0.7.

So far, the reported (14,16-18) results for composites containing Teflon reveal that only particles in the range of 1 μm and less are incorporated, and that the plating rate is 8 to 9 μm/hr.

Table 11.6
Friction Coefficient and Wear Data
For Electroless Nickel-PTFE Composite

Coating on pin	Coating on ring	Coefficient of friction	Relative wear rate[a]
EN	Cr steel	0.6 to 0.7	35
EN + PTFE	Cr steel	0.2 to 0.3	40
EN + PTFE	EN + PTFE	0.1 to 0.2	1
EN + PTFE[b]	Cr steel	0.2 to 0.5	20
EN + PTFE[b]	EN + PTFE	0.1 to 0.7	2

[a]Wear determined in a test machine consisting of a pin and a rotating ring.
[b]Heated 4 hr at 400° C.

Table 11.7 (19) documents the friction coefficient for miscellaneous coatings, of which boron nitride appears to yield the lowest coefficient of friction, especially with increased loads employed in the friction machine. At this point, however, little if any commercial interest has been focused on boron nitride. Kim (20) describes the codeposition of carbon fluoride (CF_x) along with electroless nickel for the purpose of attaining self-lubricating coatings to temperatures as high as 500° C. The success of the Kim process appears to rely upon the inclusion of non-ionic and cationic surfactants where the latter is at a significantly lower concentration. The presence of the cationic surfactant tends to increase the volume percent codeposited. Preferred non-ionic surfactants are those having an HLB value in the range of 10 to 20.

In pursuing the deposition of composites containing PTFE or carbon fluoride, filtration appears to be required in the removal of large agglomerates of particles larger than 50 μm. In addition, the number of metal turnovers is significantly less than that commercially obtained by conventional electroless nickel deposition and the inclusion of hard particles (e.g., silicon carbide and diamond).

SURFACE FINISHING

The degree of surface roughness is dependent upon various parameters such as particle size, concentration of loading, coating thickness, particle size distribution, and smoothness of the starting substrate. Inclusion of particulate matter within the electroless metal matrices tends to increase the surface roughness, especially in the case of hard particles. Since it is important to consider the ultimate application of the plated article, surface roughness must be a factor. In certain circumstances, a surface finishing operation may be

Table 11.7
Friction Coefficients
For Miscellaneous Composites

Coating	Load, kg/cm^2	Friction coefficient
TFE	0.1	0.12
Ni-P-BN	0.1	0.13
Ni-P-SiC	0.1	0.15
Ni-P	0.1	0.18
Cr	0.1	0.25
Ni-P-BN	0.3	0.09
TFE	0.3	0.13
Ni-P-SiC	0.3	0.14
Ni-P	0.3	0.16
Cr	0.3	0.40
Ni-P-BN	0.5	0.08
TFE	0.5	0.13
Ni-P-SiC	0.5	0.14
Ni-P	0.5	0.15
Cr	0.5	150.00

required to modify the as-plated roughness, thereby also controlling the frictional properties of the coating.

Table 11.8 shows the surface roughness of composite coatings containing polycrystalline diamond of different nominal sizes in a nickel-phosphorus matrix. It also shows the final roughness attainable with various newly-developed smoothing techniques.

Composite coatings, especially those containing wear resistant particles, are difficult to smooth. Newly developed methods afford a simplicity and economy of the smoothing operation not previously available. One recent development (21) relies upon the deposition of a second metallic layer that covers all exposed particulate matter, followed by the removal of a portion of the secondary layer, while still covering the exposed particles. In so doing, a smoothness level can be attained in a much shorter period of time (e.g., 30 sec vs 480 sec). Table 11.9 (21) documents the ease of smoothing with the application of the overcoat layer.

The use of an overcoat layer on top of the composite layer has been found beneficial in commercial applications. In the deposition of such a layer free of any particulate matter, smoothing of parts may be carried out more readily without expending significant energy and designing specialized tools for surface polishing hard-to-reach areas.

In another method, Spencer (22) discovered that the addition of comparatively small particles along with larger sizes of the same particulate matter provides a

Table 11.8
Surface Roughness Variations
With Particle Size and Finishing

Diamond size, μm	Roughness values, μin.*	
	initial (as-plated)	final (after smoothing)
6	40 to 45	15 to 18
4	26	13.5
3	19	12.0
1.5	7.2	—

*Substrates used were 1 to 2 μin. in roughness.

way of achieving shorter smoothing times. The disparity in particle size cannot be very large if beneficial effects are to be obtained. In a typical example, as illustrated in Fig. 11.5, the combination of 2- and 5-μm-sized particles produces an equilibrium roughness value that is virtually the same as that for the 5-μm particle alone. In addition, it has been found that a histogram of the particle size distribution (from the deposit) resulting from the admixture of 2- and 5-μm particle size was similar to the histogram of the particle size distribution in the plating bath. Although the exact mechanism for this phenomenon is not fully understood, it is believed that the smaller particles fill the voids between the large particles as they are deposited simultaneously, thereby diminishing the roughness for the as-plated conditions and, consequently, yielding a smoother final product.

In an alternate attempt, 6-μm particles were admixed with 0.5-μm particles. Because of the greater disparity in particle sizes, no beneficial effect was found in the resulting deposit roughness. The smoothing curve for the 6-μm deposit alone was identical to the smoothing curve for the 6-μm plus the 0.5-μm particle size deposit.

PARTICLE-SOLUTION INTERACTIONS

Examination of a composite electroless coating generally reveals a uniform dispersion of the particulate matter within the metallic matrix, with some particles exposed from the outer surface of the composite.

Table 11.9
Surface Roughness Variations
With Overcoat and Smoothing

Test	Brushing time, sec	Roughness[a]
1 (control)	as-plated	60.6
	15	55.2
	30	51.3
	60	48.5
	120	42.5
	240	37.8
2[b]	as-plated	69.4
	15	39.3
	30	36.9
	60	30.4
	120	24.4
	240	19.2
3[b]	as-plated	74.0
	15	43.4
	30	35.6
	60	30.4
	120	25.4
	240	19.4
4[b]	as-plated	80.6
	15	41.9
	30	37.5
	60	32.1
	120	26.2
	240	20.4

[a]Roughness data are as measured using a Gould Surf-Indicator.
[b]Tests 2, 3, and 4 represent the supplemental layers plated for periods of 1.0, 1.75, and 2 hr, respectively. Test 1 (control) is without a supplemental electroless layer.

Figure 11.6 shows a top-view scanning electron micrograph of a composite diamond coating using a particle size of 4 μm. There does not appear to be any specific orientation in the exposure of the particle, nor does there appear to be any orderly fashion as to the degree by which these particles are exposed above the surface.

In an extensive examination of particulate matter for potential composite electroless coatings, Christini et al. (4) noted that diamond, particularly

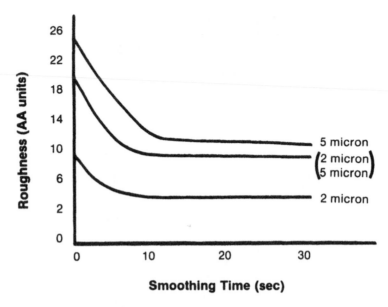

Fig. 11.5—Effect of particle size on surface roughness and time required to reach equilibrium roughness values.

polycrystalline diamond, has a tendency to form metallic sites onto exposed diamond particles in isolated regions. The degree of this tendency was found (4) to be dependent upon plating conditions and the nature of the plating bath employed (e.g., nickel-boron vs nickel-phosphorus). No such observation was reported for other particulate matter which includes, among others, wear resistant particles such as silicon carbide and aluminum oxide.

Although diamond exists in various forms, the polycrystalline form has attracted much interest in commercial composite electroless nickel applications, particularly due to its *non-catalytic* properties and *inertness* when added to electroless plating solutions.

Feldstein et al. (23) investigated the interaction of particles with electroless solutions devoid of particulate matter. Coated substrates were contacted with an electroless nickel-phosphorus plating bath. The degree of overcoat thickness was controlled by varying the plating time, and the resulting surface roughnesses were monitored, as shown in Fig. 11.7. In all cases, the initial roughness

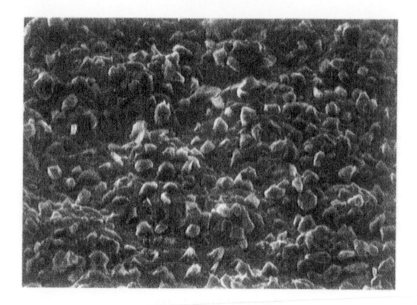

Fig. 11.6—Top view of a SEM photograph showing electroless nickel composite deposit containing diamond particles.

increased, leading to a leveling-off phenomenon. Substitution (24) of a monocrystalline-type diamond for the polycrystalline diamond, in accordance with the procedure in Fig. 11.7, resulted in a gradual decrease in surface roughness with an ultimate leveling effect.

Figure 11.8 represents the variations in roughness resulting from varying the thickness of the overcoat on a composite containing silicon carbide and aluminum oxide particles. It is apparent from Fig. 11.8 that the behavior of aluminum oxide and silicon carbide is significantly different from the behavior of polycrystalline diamond particles. This difference can only be accounted for by the manner in which the outer particles are exposed from the metallic matrix and their interaction with the overcoat plating solution.

The polycrystalline diamond particles provide sites having a catalytic nature onto which electroless plating can be deposited. This is not the case when silicon carbide and aluminum oxide are used. The additional deposition of the overcoat film appears to be selective; thus, the surface roughness is diminished.

Based upon the roughness measurements with varied electroless nickel thicknesses, it has been demonstrated that the exposed particles of polycrystalline diamond provide a catalytic tendency, which results in the

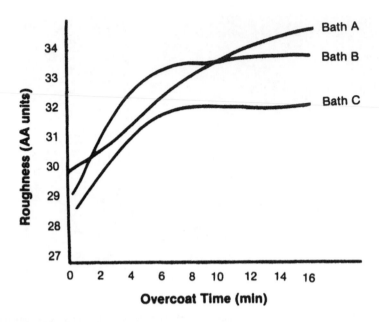

Fig. 11.7—Effect of overcoat thickness on surface roughness for electroless nickel-phosphorus deposits containing polycrystalline diamond.

plating of the diamonds. At times, the plating onto the diamond particles is at a preferential rate to the plating onto the nickel matrix, thereby resulting in an increase in the initial surface roughness. Selective "fill-in" can only take place if the exposed polycrystalline diamond particles are truly non-catalytic and inert. It is conceivable that the catalytic sites associated with the polycrystalline diamond may result from its unique morphology and/or residual transition metals used in its synthesis (4,11,25,26).

In the cases of silicon carbide and aluminum oxide, in contrast to the behavior of the polycrystalline diamond, there is a continuous lowering of the surface roughness, probably because of the selective "fill-in" as the secondary layer is deposited. Thus it appears that though diamond, silicon carbide and aluminum oxide are wear resistant particles, their catalytic inertness differ; their incorporation within the metallic matrix probably varies as well.

Hence, even though Metzger et al. (27) and Parker (28) stipulated otherwise, it would appear that successful composite electroless coatings can be achieved with particles that are not truly inert or non-catalytic. It is anticipated that this type of observation will ultimately help define the codeposition of particulate matter with electroless metal, as well as the growth mechanism of the deposits.

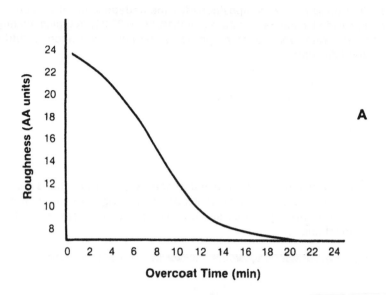

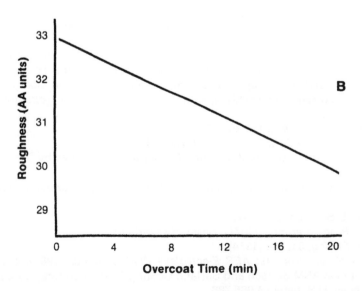

Fig. 11.8—Effect of overcoat thickness on surface roughness for electroless nickel-phosphorus deposits containing (a) silicon carbide and (b) aluminum oxide.

Several electroless plating compositions for the codeposition of particulate matter were provided in earlier literature (2,4,8,16,17,19,27,28). A typical plating solution for use with wear-resistant particles, particularly silicon carbide, consists of the following:

Nickel sulfate—0.08 moles/L
Sodium hypophosphite—0.23 moles/L
Lactic ion—0.30 moles/L
Propionic ion—0.03 moles/L
Lead ion—1 ppm
pH—4.6

Nevertheless, it would presently appear that specialized formulations may be required for industrial applications. Factual data concerning these formulations, which are still in a proprietary status, should advance the understanding of the codeposition mechanism.

REFERENCES

1. Odekerken, British patent 1,041,753 (1966); U.S. patent 3,644,183; W. Ger. patent 41,406.
2. Metzer et al., *Trans. Inst. Met. Finish.,* **54,** 173 (1976).
3. Metzger et al., W. Ger. patent application B-90776, filed 1967.
4. Christini et al., U.S. patent 3,936,577; reissued 29,285.
5. Graham and Gibbs, *Properties of Electrodeposits—Their Measurement and Significance,* The Electrochemical Society, 1974; chapter 16.
6. Sharp, "Properties and Applications of Composite Diamond Coatings," *Proc. 8th Plansee Seminar* (1974); *Proc. Diamond—Partner in Productivity,* p. 121 (1974).
7. Berger, *Machine Design,* (Nov. 1977).
8. Hubbel, *Plat. and Surf. Finish.,* **65,** 58 (Dec. 1978).
9. Wapler et al., Proceedings of Diamond Conference sponsored by DeBeers (May 1979).
10. Feldstein et al., *Prod. Finish.,* p. 65 (Jul. 1980); *Materials Engineering,* (Jul. 1981).
11. Cowan, U.S. patent 3,401,019.
12. Parker, *Proc. Interfinish* (1972), Basel, Switzerland.
13. Parker, *Plating,* **61,** 834 (1974).
14. Kenton et al., *Proc. 1st AES Electroless Plating Symp.* (1982); paper presented at ASM Surtech and Surface Coating Exposition (May 1983).
15. Yano et al., U.S. patent 4,666,786.
16. Tulsi, *Finishing,* **7**(11) (1983); *Trans. Inst. Met. Finishing,* **61**(4), 147 (1983); *Proc. AES 71st Annual Technical Conference,* Session B-4 (1984).

17. Ebdon, ibid; *Materials and Design VI*, No. 1 (1985).
18. Feldstein et al., *Proc. 1st AES Electroless Plating Symp.* (1982).
19. Honnia et al., *Proc. Interfinish* (1980).
20. J.T Kim, U.S. patent 4,716,059.
21. Feldstein, U.S. patents 4,358,922 and 4,358,923.
22. Spencer, U.S. patent 4,547,407.
23. Feldstein et al., *J. Electrochem. Soc.*, 131, 3026 (1984); *Proc. EBRATS 1985.*
24. Feldstein et al., unpublished results.
25. Seal, *Industrial Diamond Review*, p. 104 (Mar. 1968).
26. Komanduri, *Proc. Diamond—Partner in Productivity*, p. 174 (1974).
27. Metzger et al., U.S. patents 3,617,363 and 3,753,667.
28. Parker, U.S. patents 3,562,000 and 3,723,078.

Chapter 12
Fundamental Aspects
of Electroless Copper Plating

Perminder Bindra and James R. White

When an iron substrate is immersed in a solution of copper sulfate or silver nitrate, the iron dissolves while the copper or silver is plated out onto the surface of the substrate. This is called *immersion plating*. The process can be explained scientifically in terms of the electrochemical series, part of which has been reproduced in Table 12.1. The electrochemical series predicts the course of a reaction when two redox systems with different values for E^0, the standard redox potential, are brought into contact. The most electropositive redox system will be reduced (deposition), while the most electronegative will undergo oxidation (dissolution). Table 12.1 indicates that the standard redox potential for silver (E^0 = +0.799 V vs. SHE) or for copper (E^0 = +0.337 V vs. SHE) is more electropositive than that for iron (E^0 = -0.44 V vs. SHE). Therefore, the reactions occurring on the surface of the iron substrate are as follows:

Anodic
$$Fe^0 \rightarrow Fe^{2+} + 2e^- \tag{1}$$

Cathodic
$$Cu^{2+} + 2e^- \rightarrow Cu^0 \tag{2}$$

or
$$2Ag^+ + 2e^- \ 0 \rightarrow 2Ag^0 \tag{3}$$

The surface of the metallic substrate consists of a mosaic of anodic and cathodic sites and the process will continue until almost the entire substrate is covered with copper. At this point, anodic oxidation virtually ceases; hence, plating is stopped. This imposes a limit on film thickness obtained by this technique, with average values of only 1 μm being observed. Other limitations of this method are that the coatings so produced usually lack adhesion and are porous in nature. The solution quickly becomes contaminated with ions of the base metal, and the process is only applicable to obtaining deposits of noble metals.

In order to overcome these limitations and to achieve thicker deposits, an alternative oxidation reaction is essential. This reaction must occur initially on a

Table 12.1
Standard Redox Potential at 25° C

Reaction	Potential (V vs. SHE)
$Ag^+ + e^- \rightleftharpoons Ag^0$	+ 0.799
$Cu^{2+} + e^- \rightleftharpoons Cu^+$	+ 0.158
$Cu^{2+} + 2e^- \rightleftharpoons Cu^0$	+ 0.340
$HCOOH + 2H^+ + 2e^- \rightleftharpoons HCHO + H_2O$ (pH = 0)	+ 0.056
$Ni^{2+} + 2e^- \rightleftharpoons Ni^0$	+ 0.230
$Fe^{2+} + 2e^- \rightleftharpoons Fe^0$	- 0.440
$H_3PO_3 + 2H^+ + 2e^- \rightleftharpoons H_3PO_2 + H_2O$ (pH = 0)	- 0.500
$HCOO^- + 2H_2 + 2e^- \rightleftharpoons HCHO + 3OH^-$ (pH = 14)	- 1.070
$HPO_3^{2-} + 2H_2O + 2e^- \rightleftharpoons H_2PO_2^- + 3OH^-$ (pH = 14)	- 1.650

metallic substrate (activated substrate if the substrate is non-conducting) and subsequently on the deposit itself. From Table 12.1, it is clear that the redox potential for the reducing agent must be more negative than that for the metal being plated. The selection of the reducing agent is further limited in that the electrochemical reactions must only occur on the substrate, and not cause homogeneous reduction (decomposition) of the solution. An example of solution decomposition is the addition of formaldehyde (E_R = +0.056 V vs. SHE at pH = 0) to silver nitrite solution, which causes spontaneous precipitation of metallic silver. A summary of common reducing agents that may be employed for electroless copper plating is given in Table 12.2. Theoretical conditions for the electroless plating of a metal may be determined from the Pourbaix (1) diagrams for the metal and the principal element in the reducing agent. Pourbaix diagrams show the ranges of potential and pH over which various ions, oxides, pure metals, etc., are thermodynamically stable. Potential-pH diagrams for the copper-water system and the carbon-water system are shown in Figs. 12.1 and 12.2, respectively. At potentials greater than +0.337 V vs. SHE, in acid solution, the copper ion is thermodynamically stable. For deposition of the metal to occur, the potential of the substrate must be lowered below this value, into the region where metallic copper is stable. In the carbon-water system, oxidation of formaldehyde to formic acid or formate anion occurs at potentials below +0.377 V vs. SHE over the entire range of pH. Thus the electrochemical processes that could theoretically occur on the metal surface in acid solutions are:

Cathodic
$$Cu^{2+} + 2e^- \rightarrow Cu^0 \quad E^0 = +0.340 \qquad\qquad [4]$$

Table 12.2
Components of Electroless Copper Plating Baths

Reducing agent	Complexant	Stabilizer	Exaltant
Formaldehyde	Sodium potassium tartrate (Rochelle salt)	Oxygen	Cyanide
Dimethylamine borane (DMAB)	Ethylenediamine tetraacetic acid (EDTA)	Thiourea	Proprionitrile
Sodium hypophosphite	Glycolic acid	2-mercaptobenzo-thiazole	O-phenanthroline
	Triethanol amine	Diethyldithio-carbamate	
		Vanadium pentoxide	

Anodic

$$HCHO + H_2O \rightarrow HCOOH + 2H^+ + 2e^- \quad E^0 = +0.0564 \qquad [5]$$

When the two diagrams are superimposed, as shown in Fig. 12.3, the region of potential and pH over which plating is theoretically possible is indicated by the shaded region.

COMPOSITION OF ELECTROLESS COPPER PLATING SOLUTIONS

Thermodynamic conditions for electroless copper plating are described in Fig. 12.3, but successful electroless plating cannot be guaranteed by simply adding a solution of the reducing agent to one containing metal ions. In actuality, local changes in pH can lead to precipitation of the metal in bulk solution. To overcome this difficulty, complexants are added to the plating bath to maintain the metal ion in solution. The complexants depress the free metal ion concentration to a value determined by the dissociation constant of the metal complex. In addition to preventing precipitation within the solution, the complexant also allows the bath to be operated at higher pH values. Figure 12.3 indicates that the thermodynamic driving force for copper deposition becomes greater as the pH increases.

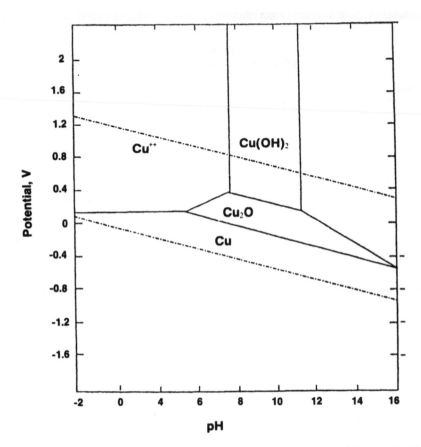

Fig. 12.1—Potential-pH equilibrium diagram for copper-water system at 25° C. The area inside the dotted lines indicates the range of values within which various copper ions are thermodynamically stable.

Table 12.1 shows that the redox potential for formaldehyde, E_R, at pH = 0 and pH = 14 differs by more than 1.0 V. Therefore, copper deposition is thermodynamically more favorable in alkaline solutions. However, copper precipitates at elevated pH values, and to prevent this from happening, a complexant such as sodium potassium tartrate is added to the plating formulation. Some common complexants used in commercial electroless copper processes are also listed in Table 12.2.

The selection and concentration of the complexant must be considered very carefully, because if the metal is too strongly complexed, insufficient free metal

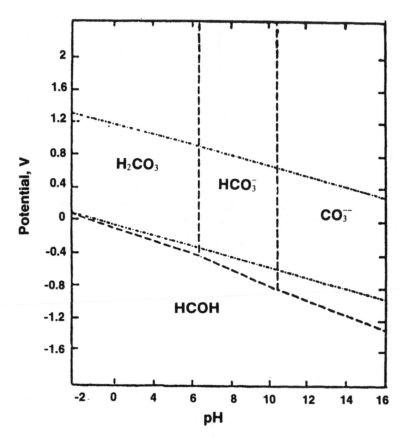

Fig. 12.2—Potential-pH equilibrium diagram for carbon-water system at 25° C, showing the relative predominance of carbon in the form of methanol and carbonates.

ions will be available for deposition. For instance, ethylenediaminetetraacetic acid (EDTA), which has a high stability constant, requires very careful control if plating is to occur.

Oxidation of the reducing agents employed in electroless copper plating invariably involves the formation of either hydrogen (H^+) or hydroxyl (OH^-) ions. Consequently, the pH of the plating solution changes during plating and thus affects the rate of deposition and the properties of the deposit. Therefore, buffers are added to stabilize the pH of the solution. These include carboxylic acids in acid media (which also act as complexants) and organic amines in alkaline solutions.

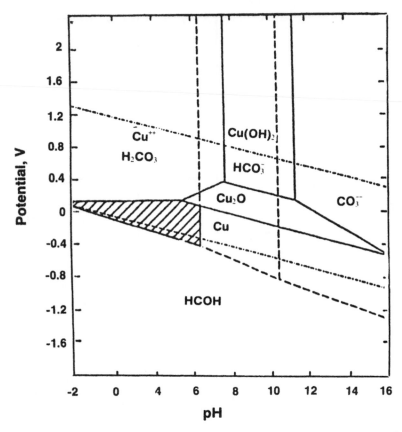

Fig. 12.3—Combined potential-pH diagram for copper-water and carbon-water systems, indicating the range of values within which metal deposition is possible.

Electroless plating formulations are inherently unstable and the presence of active nuclei such as dust or metallic particles can lead, over time, to homogeneous decomposition of the plating bath. The presence of the complexant in the correct concentration does not prevent this from occurring. To circumvent this problem, stabilizers such as 2-mercaptobenzothiazole are added to the bath in small concentrations. The stabilizers competitively adsorb on the active nuclei and shield them from the reducing agent in the plating solution. If, however, the stabilizers are used in excess, metal deposition may be completely prevented, even on the substrate itself.

The plating rate is sometimes inordinately lowered by the addition of complexants to the bath. Additives that increase the rate to an acceptable level without causing bath instability are termed *exaltants* or *accelerators*. These are generally anions, such as CN , which are thought to function by making the anodic oxidation process easier.

In summary, typical electroless plating formulations contain (a) a source of metal ions, (b) a reducing agent, (c) a complexant, (d) a buffer, (e) exaltants, and (f) stabilizers.

THE MIXED POTENTIAL THEORY

The Wagner and Traud (2) theory of mixed potentials has been verified for several corrosion systems (3,4). According to this theory, the rate of a faradaic process is independent of other faradaic processes occurring at the electrode and depends only on the electrode potential. Hence, the polarization curves for independent anodic and cathodic processes may be added to predict the overall rates and potentials that exist when more than one reaction occurs simultaneously at an electrode. Wagner and Traud (2) demonstrated the dissolution of zinc amalgam to be dependent on the amalgam potential but independent of simultaneous hydrogen evolution processes.

The concept of mixed potentials can be explained by first considering a redox reaction of the type:

$$Ox + ne \underset{k_a}{\overset{k_c}{\rightleftharpoons}} Red \qquad [6]$$

on an inert electrode. In Eq. 6, k_c and k_a are the rates for the cathodic and the anodic reactions, respectively. When equilibrium is established, the two opposing reactions occur at the same rate and no net current flows through the system. This condition may be expressed by the equation:

$$i_c = i_a = i^0 \qquad [7]$$

where i^0 is the exchange current density and i_c and i_a are the cathodic and anodic current densities, respectively. The potential at which this equilibrium occurs is described as the equilibrium potential, E^0, and may be determined in the thermodynamic sense by the Nernst equation.

We next consider the case where two or more reactions occur simultaneously at the electrode surface (5,6). A copper/formaldehyde electroless plating process is a perfect example of such a case. The anodic reaction is the oxidation of the reducing agent (formaldehyde, methylene glycol, or hydrated formaldehyde in alkaline media):

$$2H_2C(OH)_2 + 2OH^- \rightarrow 2HCOOH + H_2 + 2H_2O + 2e^- \tag{8}$$

$$(R^0 \rightarrow R^{z+} + Ze^-)$$

and the cathodic reaction is the reduction of the metal (copper) complex:

$$[CuL_6]^{(6n-2)-} \rightleftharpoons Cu^{2+} + 6L^{n-}$$

$$\downarrow 2e^- \tag{9}$$

$$Cu^0$$

$$(M^{z+} + Ze^- \rightarrow M^0)$$

R and M denote the reductant and the metal, respectively.

The current potential curves for the anodic and the cathodic partial reactions in an electroless plating process are shown schematically in Fig. 12.4. It is clear from Fig. 12.4 that a necessary condition for electroless plating to occur is that

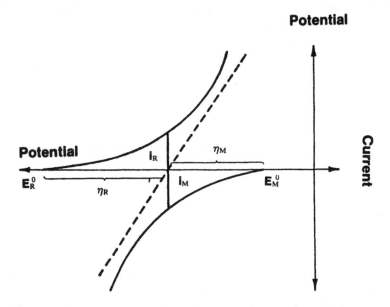

Fig. 12.4—Current-potential curves for the partial processes involved in electroless plating. Dotted line indicates the current-potential curve for the two simultaneous processes. E_R^0 and E_M^0 are equilibrium potentials.

the equilibrium potential for the reducing agent E_R is more cathodic than the corresponding potential E_M for the metal deposition reaction. At equilibrium, the Wagner-Traud postulate applies and the plating rate, $i_{plating}$, is given by:

$$i_{plating} = i_R = i_M \tag{10}$$

where i_R and i_M are the anodic and cathodic partial currents (with opposite signs). The potential associated with this dynamic equilibrium condition is referred to as the mixed potential E_{MP}. The value of the mixed potential lies between E_R^0 and E_M^0 and depends on parameters such as exchange current densities i_R^0 and i_M^0, Tafel slopes b_R and b_M, temperature, and others. The mixed potential corresponds to two different overpotentials:

$$\eta_R = E_{MP} - E_R^0 \tag{11}$$

and

$$\eta_M = E_{MP} - E_M^0 \tag{12}$$

If the anodic and cathodic half cells are coupled (short-circuited), the measured current-potential curve is the algebraic sum of the partial current-potential curves for each electrode reaction, as shown in Fig. 12.4 by the dotted curve.

APPLICATION OF THE MIXED POTENTIAL THEORY

In practice, the concept of mixed potentials is studied by constructing Evan's diagrams. In Fig. 12.5, a schematic Evan's diagram is constructed for the partial polarization curves shown in Fig. 12.4. The coordinates of intersection of the anodic and cathodic E vs. log i relationships give the value of the mixed potential E_{MP} and the plating rate in terms of a current density i_{dep}. This suggests that the dynamic relationship between the mixed potential and the copper plating rate can be obtained from the individual electrode processes if their current-potential relationships are known. Such polarization curves have been obtained by one or more of the following methods: (a) by applying the steady state galvanostatic or potentiostatic pulse method to each partial reaction separately; (b) by applying potential scanning techniques to a rotating disk electrode; or (c) by measuring the plating rate from the substrate weight-gain as a function of the concentration of the reductant or the oxidant (8,9). The plating rate is then plotted against the mixed potential to obtain the Tafel parameter (7,8). These methods suffer from the usual limitations associated with the theory of mixed potentials. For example, extrapolation of the polarization curve for the catalytic decomposition of the reducing agent to the plating potential is not valid if the catalytic properties of the surface change with potential over the range of interest. It is also not valid if the rate determining step, and hence the Tafel slope,

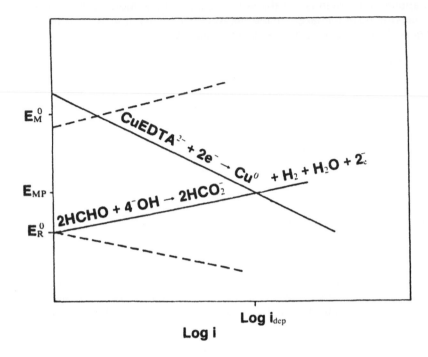

Fig. 12.5—A schematic Evan's diagram showing the two partial processes of the additive copper bath.

for any process changes in the potential range through which the polarization curve is extrapolated. Further, at least one of the two partial reactions involved in electroless metal plating is invariably diffusion-controlled. Therefore, the weight-gain method cannot be used to ascertain the plating mechanism unless the electrochemically-controlled partial reaction is first identified.

A further limitation to the extrapolation of polarization curves and to the application of the mixed potential concept to electroless plating, which is often not realized (10), is that the two partial processes are not independent of each other. For corrosion processes, such a limitation was first discovered by Andersen et al. (11). Bindra et al. (12) have found that same limitation to the application of the mixed potential theory to apply to electroless plating systems as well. These authors have developed a technique based on the mixed potential theory by which electroless plating processes may be classified according to their overall mechanisms. For example, electroless plating of metals can involve a reaction that proceeds at a rate limited by diffusion. Donahue (13) has shown that the plating rate of copper (in a certain concentration range) in a copper-formaldehyde bath is determined by the rate of diffusion of copper ions to the plating surface. The technique described by Bindra et al. (12) allows a clear distinction between those reactions whose rate is controlled by the rate of diffusion of reactants to the plating surface, and those reactions whose rate is limited by some slow electrochemical step. The theory of this technique is described below.

In order to achieve conditions of controlled mass transfer, a rotating disk electrode (RDE) may be employed. The current resulting from the diffusion of metal ions to such a geometrical surface is expressed in Eq. 13 (14):

$$| i_M | = B_M^i (C_M^x - C_M^d) \sqrt{\omega} \tag{13}$$

where C_M^x is the bulk concentration of the reducing agent; C_M^d the surface concentration, and B_M^i is a diffusion parameter (14,15).

$$B_M^i = 0.62\ n_M\ F\ D_M^{2/3}\ \nu^{-1/6}\ A \tag{14}$$

The symbols are defined in the appendix. For diffusion-controlled cathodic partial reactions, $C_M^d = 0$, and the limiting current i_M^D is independent of potential and takes the form:

$$| i_M^D | = B_M^i\ C_M^x\ \sqrt{\omega} \tag{15}$$

Similarly, the diffusion-limited current for the anodic partial reaction is:

$$i_R^D = B_R^i\ C_R^x\ \sqrt{\omega} \tag{16}$$

The concentration of metal ions at the surface may be expressed by the Nernst equation:

$$E_M = E_M^0 + \frac{RT}{n_M F}\ \ln C_M^d \tag{17}$$

which, by substituting Eqs. 13 and 15, becomes:

$$E_M = E_M^0 + \frac{RT}{n_M F}\ \ln |(i_M^D - i_M)| - \frac{RT}{n_M F}\ \ln B_M^i - \frac{RT}{n_M F}\ \ln \sqrt{\omega} \tag{18}$$

The corresponding equation for the reducing agent can also be worked out in a similar manner.

When the anodic partial reaction is under electrochemical control, the polarization curve is described by the equation:

$$E = E_R^0 - b_R \ln i_R^0 + b_R \ln i_R \qquad\qquad [19]$$

The anodic Tafel slope b_R is given by:

$$b_R = \frac{RT}{(1 - \alpha_R)\, n_R F} \qquad\qquad [20]$$

Similarly, when the metal deposition reaction is activation controlled, the kinetics are described by the cathodic Tafel equation:

$$E = E_M^0 + b_M \ln |i_M^0| - b_M \ln |i_M| \qquad\qquad [21]$$

where:

$$b_M = \frac{RT}{\alpha_M\, n_M F} \qquad\qquad [22]$$

Since each partial reaction is either under electrochemical control or under mass transfer control, the overall reaction scheme consists of four possible combinations. These will be considered next, and the dependence of E_{MP} on experimental parameters such as rotation rate ω, and C_R^∞ and C_M^∞ determined.

CASE 1: Cathodic partial reaction diffusion controlled—anodic partial reaction electrochemically controlled.

The diffusion-limited cathodic partial current depends on C_M^∞, D_M, and ω, and its magnitude is given by Eq. 15. Then, combining Eq. 15 with Eq. 19, which describes the anodic partial reaction, by means of Eq. 10, gives:

$$E_{MP} = E_R^0 - b_R \ln i_R^0 + \frac{b_R}{2} \ln B_M'^2 \, \omega + b_R \ln C_M^\infty \qquad\qquad [23]$$

Equation 23 shows that the mixed potential is a linear function of $\ln \omega$ and $\ln C_M^\infty$ and that the Tafel slope for the anodic partial reaction may be obtained by plotting E_{MP} against either of these experimental parameters. Similar functions have been obtained by Makrides (16) for corrosion processes. Case 1 is represented graphically in the symbolic diagram of Fig. 12.6.

Oxygen or air is frequently bubbled through electroless copper baths to oxidize any Cu(I) species formed, and thus avoid bath decomposition via Cu(I) disproportionation. Under these circumstances, there is another cathodic

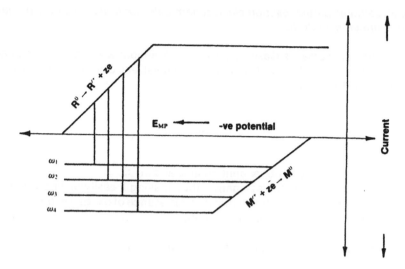

Fig. 12.6—Symbolic representation of the overall reaction scheme for electroless metal deposition in which the cathodic partial reaction is diffusion-controlled and the anodic partial reaction is kinetically-controlled.

current resulting from oxygen reduction, and the total cathodic current is the sum:

$$i_M' = i_M + i_{O_2} \qquad [24]$$

where i_{O_2} is the current for oxygen reduction. At the plating potential, i_{O_2} is diffusion limited, that is, it is independent of the electrode potential, and therefore, is equivalent to a cathodic current applied externally.

Under these circumstances, Eq. 15 becomes:

$$|i_M^{D'}| = (B_M' C_M^\infty + B_{O_2}' C_0^\infty) \sqrt{\omega} \qquad [25]$$

and Eq. 23 becomes:

$$E_{MP} = E_R^0 - b_R \ln i_R^0 + b_R \ln (B_M' C_M^\infty + B_{O_2}' C_{O_2}^\infty) + \frac{b_R}{2} \ln \omega \qquad [26]$$

Therefore, the diagnostic criteria for Case 1 remains unchanged in the presence of oxygen in the plating bath.

CASE 2: *Cathodic partial reaction electrochemically controlled—anodic partial reaction diffusion controlled.*

This is the converse of Case 1, so that the anodic and cathodic partial reactions are described by Eq. 16 and Eq. 21, respectively. Combining them with the help of Eq. 10 yields:

$$E_{MP} = E_M^0 + b_M \ln |i_M^0| - \frac{b_M}{2} \ln B_R^{i:} \omega - b_M \ln C_R^x \qquad [27]$$

Once again, E_{MP} is linearly dependent on $\ln \omega$ and $\ln C_R^x$.

The slope of E_{MP} vs. $\ln$ plot is negative, thus making it easily distinguishable from Case 1. It can easily be shown that in the presence of oxygen in the plating solution, the form of Eq. 27 does not change.

CASE 3: *Both partial reactions electrochemically controlled.*

CASE 4: *Both partial reactions diffusion controlled.*

Such cases are rarely encountered in electroless plating baths. Therefore, detailed relationships between the mixed potential and other parameters are not worked out. Suffice it to say that in both cases, the mixed potential is independent of rotation rate.

PARTIAL REACTIONS

A limitation to the application of the mixed potential theory to electroless plating is that the two partial reactions are not independent of each other. Such a limitation was alluded to as early as 1972 by Donahue (10), but *measurements* to clearly demonstrate the effect were not performed until much later (13). In an effort to demonstrate the interdependence of the partial reactions, Bindra et al. (12) performed kinetic and mechanistic measurements in the catholyte and the anolyte, as well as in the complete electroless copper bath separately. These measurements are discussed next.

Cathodic Partial Reaction

The copper deposition reaction was studied with the help of a rotating disk electrode. Some typical results for the copper deposition partial reactions in the catholyte are shown in Figs. 12.7 through 12.9. The polarization curves obtained by applying the potential scanning technique at various rotation rates are displayed in Fig. 12.7. If the kinetics are first order with respect to copper ions in solutions, then the experimental disk currents are related to the rotation rate by the Levich equation, where all quantities are considered as positive (15):

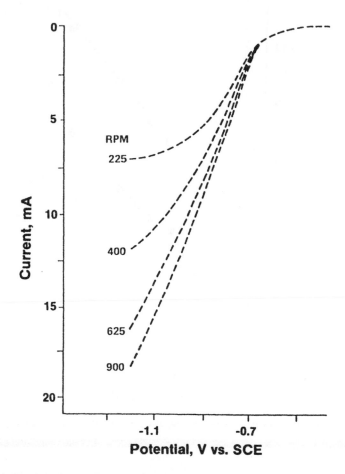

Fig. 12.7—Rotating disk data for copper deposition in the catholyte at 70° C. Electrode area = 0.458 cm².

$$\frac{1}{i_M} = \frac{1}{i_M^k} = \frac{1}{i_M^D} = \frac{1}{i_M^k} + \frac{1}{B_M \sqrt{\omega}}$$ [28]

It is clear from Eq. 28 and Eq. 15 that $B_M = B'_M C_M^x$.

Figure 12.8 depicts plots of $1/i_M$ vs. $1/\sqrt{\omega}$ for the data shown in Fig. 12.7. These plots are linear and parallel, indicating that the copper deposition reaction is first

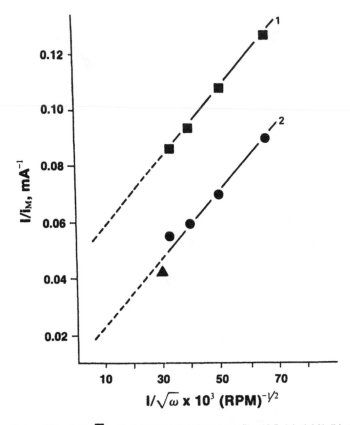

Fig. 12.8—Plots of I/i_M vs. $I/\sqrt{\omega}$ for the rotating disk data from Fig. 12.7. (a) -0.8 V; (b) -0.95 V.

order in copper ion concentration. Figure 12.9 shows a mass transfer corrected Tafel plot at 70° C. There is a large linear region that yields a Tafel slope of -165 mV/decade. The value of the transfer coefficient, α_m, calculated from this slope, comes close to 0.42. The polarization curve for copper deposition in the complete bath was also obtained and is shown in Fig. 12.10. This plot was obtained by the galvanostatic step method and yields a Tafel slope of -30 mV/decade.

A multistep, n-electron transfer metal deposition reaction may be written in the following form (17):

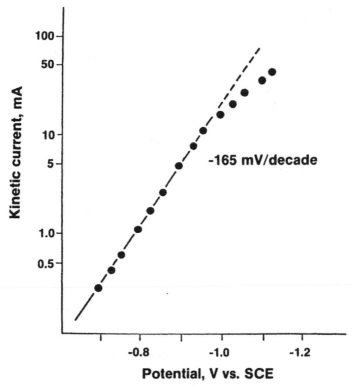

Fig. 12.9—Tafel plot for copper deposition in the catholyte at 400 rpm, 70° C, electrode area of 0.458 cm^2.

$$|E-E_M^{\mu}| = a - b_M \ln|i_M| \tag{29}$$

where b_M is given by Eq. 22 and α_M is expressed as:

$$\alpha_M = (\gamma/v + n_M \beta_M) \tag{30}$$

In Eq. 30, γ is the number of preceding steps prior to the rate determining step (rds), v the stoichiometric number, and β_M is the symmetry factor. The various pathways by which copper deposition in the catholyte may occur are as follows:

$$[CuEDTA]^{2-} + 2e^- \rightleftharpoons Cu^0 + EDTA^{4-} \tag{I}$$

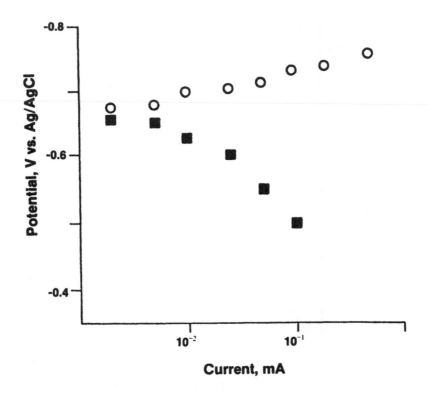

Fig. 12.10—Polarization curves for copper deposition (○) and formaldehyde oxidation (■) obtained in the complete bath by the galvanostatic pulse method at 70° C and pH 11.7.

$$[CuEDTA]^{2-} + e^- \xrightarrow{rds} [CuEDTA]^{1-} + e^- \rightleftharpoons Cu^0 + EDTA^{4-} \qquad [II]$$

$$[CuEDTA]^{2-} + e^- \rightleftharpoons [CuEDTA]^{1-} e^- \xrightarrow{rds} Cu^0 + EDTA^{4-} \qquad [III]$$

One need only identify the values of r, v, n_M, and β_M in order to determine the mechanism of copper deposition. Mechanism I is rejected on the grounds that a two electron transfer would require a very high activation energy. Besides, a β_M value of 0.21 obtained from Eq. 30 for mechanism I is also unlikely. For mechanism III, Eq. 30 gives a β_M value that is negative and therefore does not have a physical meaning. Hence, copper deposition in the catholyte occurs via mechanism II, i.e., in two steps with the cupric-cuprous step as the rds. In the presence of cyanide ions in the catholyte, it is possible that the cuprous ion is stabilized by complexation with the cyanide. Such a mechanism was not investigated in this study. It is, however, interesting to note that the mechanism of copper deposition in the catholyte is similar to the mechanism postulated for copper deposition from $CuSO_4$ solutions.

Anodic Partial Reaction

The oxidation of formaldehyde in the anolyte was investigated by the galvanostatic and the potentiostatic step methods (Figs. 12.10 and 12.11). Logarithmic analysis of the data gives linear plots of log i as a function of potential with Tafel slopes, which are considerably lower than those obtained in the complete electroless copper bath (Table 12.3). Such low Tafel slopes can only be interpreted in terms of a complex mechanism involving chemical and electrochemical steps. The proposed mechanism is shown in Reaction Scheme A.

Scheme A

$$HCHO + H_2O \rightleftharpoons H_2C(OH)_2 \tag{A1}$$

$$H_2C(OH)_2 + OH^-_{Ads} \rightleftharpoons H_2C(OH)O^-_{Ads} + H_2O \tag{A2}$$

$$H_2C(OH)O^-_{Ads} \rightleftharpoons HCOOH + \tfrac{1}{2}H_2 + e^- \tag{A3}$$

$$HCOOH + OH^- \rightleftharpoons HCOO^- + H_2O \tag{A4}$$

The observed Tafel slopes have a value around 110 mV/decade at 70° C (Table 12.3). Several reaction mechanisms could account for the experimentally observed kinetic parameters; three such mechanisms are described below:

Mechanism i
Step A3 is rate-controlling, and the electron uptake is an inner sphere electron transfer, i.e., $(1 - \alpha_R)n_R = 1$. The Tafel slope expected (under conditions of Langmuir adsorption) if steps A2 and A4 are in quasi-equilibrium is then 70 mV/decade at 70° C.

Mechanism ii
Step A2 is the rds, but the electron transfer is outer sphere. In such a case, the Tafel slope expected under conditions of Langmuir adsorption has a value of 140 mV/decade at 70° C.

Mechanism iii
Step A2 is rate determining and A3 is in quasi-equilibrium. This is quite possible, since the formation of methylene glycolate anion, which is the electroactive species, is base catalyzed and therefore depends on pH value. A Tafel slope of 70

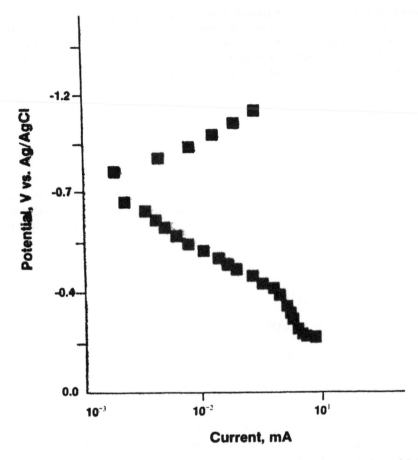

Fig. 12.11—Polarization curve for formaldehyde oxidation and reduction obtained in the anolyte by the potentiostatic method at 70° C and pH 11.7.

mV/decade at 70° C is expected in this case. However, the Tafel slope, and therefore the mechanism of formaldehyde oxidation would then be a function of formaldehyde concentration and pH. At pH equal to the pK value for step A2, this step is in quasi-equilibrium and mechanisms i or ii outlined above become operative.

It is clear that the observed Tafel slope of 110 mV/decade at 70° C cannot be accounted for by only one of the mechanisms outlined above. It is possible that

Table 12.3
Analysis of Formaldehyde Oxidation Data:
Current Density vs. Potential

Method	Electrolyte	Slope, mV	Transfer coefficient
$dE_{MP}/d\log \omega$	Complete bath	104	0.34
$dE_{MP}/d\log$ [HCHO]	Complete bath	210	0.33
$dE_{MP}/d\log$ [CuSO$_4$]	Complete bath	188	0.37
$dE_{MP}/d\log i_R$ (galvanostatic)	Complete bath	185	0.38
$dE_{MP}/d\log i_R$ (galvanostatic)	Anolyte	110	0.64
$dE_{MP}/d\log i_R$ (potentiostatic)	Anolyte	115	0.61
$dE_{MP}/d\log i_M$ (galvanostatic)	Complete bath	-30 ± 5	0.43
$dE/\log i_M$ (potential scanning)	Catholyte	-165	0.42

the components of the plating bath present in the anolyte (e.g., excess EDTA) affect formaldehyde decomposition and that the overall mechanism for this reaction is a combination of two or more mechanisms described above. In any case, it is clear that the mechanism of formaldehyde oxidation is more complex in the anolyte than in the complete plating bath, where it has been shown to be a catalytic process.

Measurements in the Complete Copper Bath

The overall mechanism of copper plating was also determined by the application of the mixed potential theory to a rotating disk electrode. The mixed potential was observed as a function of rotation rate, ω, of the RDE and the bulk concentrations C_M^x of metal ions. The data obtained is shown in Figs. 12.12 and 12.13. It is clear that each one of these plots is a straight line. Using the criteria developed earlier, it is relatively simple to assign a mechanism to the electroless plating process. We first note that the mixed potential increases with both rotation rate ω and the concentration, C_M^x, of metal ions in the plating bath; that is, the slopes for the plots in Figs. 12.12 and 12.13 are positive. This behavior is indicative of diffusion-controlled copper deposition partial reactions and activation-controlled formaldehyde decomposition reactions. The same mechanism with respect to the copper deposition partial reaction has been noted previously by Donahue (13).

Verification of the theory developed for this technique is obtained by comparing the measured slope of the rotation rate dependence, $dE_{MP}/d\ln \omega$, with

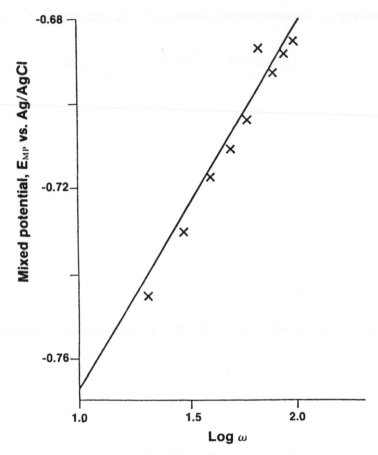

Fig. 12.12—Plot of the mixed potential of the plating solution as a function of rotation rate. Temperature = 70° C, pH = 11.7.

the concentration dependence $dE_{MP}/dlnC_M^x$. These slopes are reported in Table 12.3, which shows that:

$$\frac{dE_{MP}}{dlnC_M^x} = 2 \times \frac{dE_{MP}}{dln\omega} = b_r \qquad [31]$$

This result is predicted by Eq. 23. Clearly, this simple technique is capable of measuring *in situ* the Tafel slope for either the anodic partial reaction or the

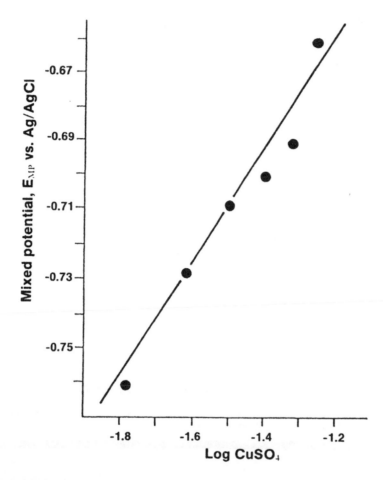

Fig. 12.13—Plot of the mixed potential of the plating solution vs. the logarithm of the CuSO₄ concentration. Temperature = 70° C, pH = 11.7.

cathodic partial reaction, depending on the overall mechanism of the plating process. In the electroless copper bath under study in this investigation, b_r is the Tafel slope for formaldehyde oxidation and has the value ~+210 mV/decade at 70° C. In order to further substantiate the validity of this approach, galvanostatic step measurements were also performed in the electroless copper bath. The Tafel plot obtained is shown in Fig. 12.14. The Tafel slope obtained has the value of +185 mV/decade at 70° C, which is in reasonable agreement with the value obtained by the technique based on observation of the behavior of the mixed

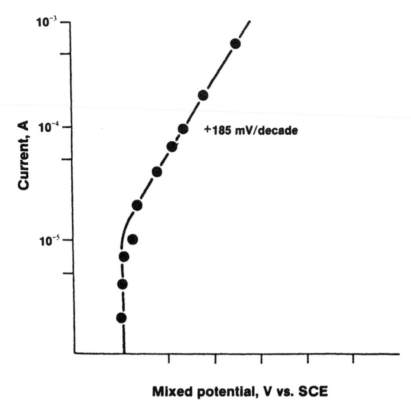

Fig. 12.14—Polarization curve for formaldehyde oxidation obtained in the complete bath by the galvanostatic pulse method at 70° C and pH 11.7.

potential. This latter technique is therefore important, not only from the point of view of ascertaining the overall mechanism of electroless metal plating, but also as a convenient method of determining the Tafel slope for the partial reaction under electrochemical control.

Tafel slopes in the range of 185 to 210 mV/decade correspond to values of the anodic transfer coefficient, $(1 - \alpha_R)n_R$, which are substantially lower than 0.5 (Table 12.3, column 4). The observed value of the transfer coefficient is known to deviate from 0.5 when the actual electron transfer occurs across only a fraction of the Helmholtz double layer, i.e., when the reacting species are specifically adsorbed and the electron transfer occurs in the inner Helmholtz plane. Such behavior is characteristic of catalytic reactions. The large value of the Tafel slope implies control by the first electron transfer under Temkin adsorption conditions.

Under the circumstances, the oxidation of formaldehyde in electroless copper plating may be described by scheme A.

Formaldehyde in aqueous solutions exists predominantly in the electro-inactive hydrated form. Reaction A1 above represents the quasi-equilibrium of the hydration reaction at high pH values. The anion of the hydrated form, which is the electroactive species, can then be created either in the bulk solution by a general base catalysis (18) or be generated on the substrate surface by interaction with adsorbed OH, as shown in Reaction A2. The high value of the Tafel slope indicates a catalytic reaction, and therefore supports the participation of the substrate surface, either as an antecedent step to the electron uptake or as a proceeding step to stabilize reaction intermediates (19,20). Measurements performed here do not allow a distinction as to whether the hydrogen abstraction in Reaction A4 precedes the electron uptake or follows it. Nonetheless, Reaction A3 is the rds in the overall mechanism. The overall stoichiometry of the reaction is in agreement with that proposed by Lukes (21).

Further support for the validity of the mixed potential technique can be found in the Tafel curve for the copper deposition partial reaction obtained in the complete bath (Fig. 12.10). The Tafel slope observed in this case has a value of -30 ± 5 mV/decade. For pure diffusion control by metal ions in the plating bath, the Tafel equation may be expressed as (19):

$$|E - E_M^0| = \frac{RT}{2F} \ln \left(1 - \left(\frac{i_M}{i_M^D} \right) \right) \tag{32}$$

At potential values where $i_M \ll i_M^D$, Eq. 32 reduces to the usual form of the Tafel equation:

$$|E - E_M^0| = \alpha' - \frac{RT}{2F} \ln |i_M| \tag{33}$$

where the term α' is different from the term α in the Tafel equation (cf. Eq. 29). Equation 33 exhibits a Tafel slope of -37 mV/decade at 70° C, suggesting that the cathodic partial reaction in the electroless copper process is diffusion controlled.

There are two possible mechanisms by which this could occur. These are (1) rate-controlling diffusion of [CuEDTA]²⁻ to the substrate, followed by dissociation of the complex prior to reduction; and (2) dissociation of the [CuEDTA]²⁻ complex in the bulk solution, followed by rate-controlling diffusion of aqueous copper ions to the substrate for reduction. The techniques used in this investigation do not allow a distinction between these two mechanisms. In either case, the rate of reduction of copper ions is much faster than the rate at which electrons are released by the reducing agent. The fact that the cathodic partial reaction in electroless copper plating is diffusion-controlled is in total agreement with the overall mechanism for electroless copper plating ascertained by observing the behavior of the mixed potential as a function of ω and C_M^x.

Interdependence of the Partial Reaction

To establish the interdependence or otherwise of the partial reactions in electroless copper plating, Tafel slopes for the partial reactions were obtained in the complete plating bath, as well as in the catholyte and anolyte separately. The Tafel slope for the copper deposition reaction (Table 12.3) in the catholyte indicates a stepwise reaction mechanism, with the cupric-cuprous step as the rds. In the electroless plating bath, the Tafel slope for the cathodic partial reaction has a value of -30 ±5 mV/decade, indicating diffusion control for this reaction. The difference in mechanisms in the catholyte and in the plating bath is attributed to the presence of the reducing agent and the anodic partial reaction in the bath. Electroless plating processes occur via two consecutive reactions: Electrons are released during the partial anodic reaction and are consumed by the cathodic partial reaction, which also happens to be the metal deposition reaction. The overall rate of the electroless plating process is therefore governed by the slower of the two partial reactions. In the bath formulation studied in this investigation, the exchange current density for the formaldehyde oxidation at E_{MP} is at least two orders of magnitude lower than the exchange current density for the copper deposition reaction. Therefore, whatever the mechanism of copper deposition in the catholyte, the electroless plating process is controlled by the kinetics of the formaldehyde oxidation reaction, i.e., the rate of the copper deposition partial reaction is totally dependent on the kinetics of the anodic partial reaction. Hence the two partial reactions in electroless copper plating are not independent of each other.

Further evidence in support of this theory is provided by the Tafel plots for formaldehyde oxidation in the plating bath and in the anolyte (Table 12.3). There is considerable difference in Tafel slopes in the two solutions, indicating a difference in the mechanism of the reaction in the two environments. Clearly, the anodic partial reaction in the electroless plating bath is affected by the presence of copper ions in the bath.

KINETICS OF ELECTROLESS COPPER DEPOSITION

A number of studies have emerged regarding the kinetics of electroless copper deposition (6,8,22-24). Most have analyzed results in terms of classical kinetic theory and utilize the following rate equation:

$$r = k[Cu^{2+}]^a [OH^-]^b [HCHO]^c [LIGAND]^d \qquad [34]$$

A summary of reaction orders, for various components, is shown in Table 12.4. The diversity of reaction orders, and in some cases negative reaction orders, is attributable to a number of factors. First, different substrates are employed that

**Table 12.4
Summary of Reaction Orders
For Electroless Copper Baths**

Cu	OH	HCHO	Ligand	Reference
0.47	0.18	0.07	- (tartrate)	56
0.37	0.25	0.08	0.19 (tartrate)	6
0.78	<0.02	0.13	<0.02 (tartrate)	8
0.43	-0.70	0.16	-0.04 (EDTA)	22
1.00	0.37	0.00	- (EDTA)	23[a]
0.00	1.00	0.68	- (EDTA)	23[b]
0.00	0.00	1.00	0 (EDTA)	5

[a]Final deposition rate.
[b]Initial deposition rate.

have varying degrees of catalytic activity. In some cases metal substrates are used, while catalyzed dielectrics are used in others. Second, the time frame of the measurement is critical since the rate has been observed to decrease with time (6,23). The initial rate (at copper coverages less than 25 μg Cu/cm^2) is generally larger than the subsequent steady state rate.

A third factor is associated with mass transfer effects. The importance of understanding how interfacial (surface) concentrations differ from bulk concentrations has been stressed (13). In general, the interfacial concentration of a species C is given by the following equation:

$$C = C_h + 10^1 \, N \, \delta/D \qquad [35]$$

where C_h is the bulk concentration, N is the flux of species C to the electrode surface, δ is diffusion layer thickness, and D is the diffusion coefficient for C.

For electroless plating baths, in the absence of forced convection, the primary means of mass transfer results from microconvection by hydrogen bubbles. Estimates for values of C, under this condition, have been given by Donahue (13).

Finally, most practical systems contain additional components to enhance metallurgy or stabilize the bath. These components certainly impact the kinetics of the system but render the system too complex for study. Thus, more studies are based only on the four primary constituents of the plating bath.

A variety of measurement techniques have been utilized to obtain kinetic data. Most investigators utilize some form of the mixed potential theory described in the previous section and correlate results to the weight gain observed at the cathode. A method based on a resistance probe has been described in which the cathode comprises one arm of a wheatstone bridge (25). The plating rate may be monitored by observing changes in resistance with time.

Dumesic et al. described an optical technique based on the absorption of monochromatic light at a sensitized transparent rotating cylinder (23). With this method they were able to clearly distinguish between changes in the initial rate and the final rate region. For example, the reaction order for formaldehyde changed from 0.68 during the initial stages of deposition to 0 during the final stage.

A method based on weight gain has been reported that utilizes a quartz microbalance (24). The advantage of this technique, compared to macroscopic weight measurements, is that the microbalance is an *in situ* technique.

The effect of cupric ion on the kinetics varies with concentration and is attributable to mass transfer effects (13). At low bulk cupric ion concentrations (i.e., less than 0.01M), the interfacial concentration (Eq. 35) is low and mass transfer plays a decisive role. For intermediate values of Cu_b^{2+} (i.e., $0.01M < Cu_b^{2+} < 0.2M$), the interfacial concentration falls in an intermediate range between 20 and 80 percent of the bulk concentration. Thus, the system is under mixed kinetic and mass transfer control. At higher concentrations the effect is primarily kinetic. Dumesic points out that during the initial stage of the reaction, the rate is independent of cupric ion concentration, assuming no mass transfer limitation (23).

The influence of formaldehyde concentration is generally small during steady state (final) deposition, i.e., reaction orders less than 0.1 have been reported. However, during initial stages the reaction was found to be first order with respect to formaldehyde.

The effect of hydroxide ion concentration is similar to that reported for formaldehyde. During steady state deposition the influence is minimal, but during initial stages the influence is greater. A maxima in the plating rate, as a function of pH, was reported by Schoenberg (26). This is attributable to competition between hydroxide ion and methylene glycolate for copper coordination.

Donahue examined a large number of plating bath compositions, based on mixed potential analysis, and obtained the following empirical rate equation (22):

$$r = 2.81 \frac{[Cu^{2+}]^{0.43} [HCHO]^{0.16}}{[OH^-]^{0.70} [EDTA]^{0.04}} \exp. \; 11.5 \frac{T-313}{T} \tag{36}$$

The correlation coefficients for formaldehyde and EDTA were low, although 90 percent of the measured plating rates were within 20 percent of the predicted values.

A more recent study is based on the following Arrhenius equation (24):

$$r = A \exp [-\Delta E^{\ddagger}/RT]C_{MG^-} \qquad [37]$$

The experimental conditions were such that the reaction was first order with respect to formaldehyde (actually methylene glycolate MG$^-$) and zero order for all other components. An activation energy of 60.9 kJ/mole was obtained. The proposed rds is the cleavage of the carbon hydrogen bond of methylene glycolate. Isotopic substitution of CD_2O for CH_2O resulted in a primary isotope effect of kH/kD = 5.

CATALYSTS FOR ELECTROLESS COPPER PLATING

The rate determining step for the electroless copper bath is most frequently associated with the anodic process, which is generally under kinetic control. In order to evaluate the catalytic activity of various metals, the oxidation of common reductants may be studied by electrochemical techniques. Ohno et al. have investigated the polarization behavior of a number of reductants on a variety of solid electrodes and determined that copper is most catalytically active for formaldehyde oxidation, as compared to other common reductants (27). Their results are shown in Fig. 12.15.* The potentials for each metal substrate are those observed when the electrodes were subject to a constant current of 10^{-4} A/cm^2. E_r is the standard reduction potential for each reductant. To achieve relatively thick electroless copper deposits, the plating reaction must be autocatalytic with respect to copper, and therefore, most commercially available electroless copper baths are formaldehyde-based.

The electrolcatalytic activity of various metals for formaldehyde oxidation has been evaluated by a number of workers, primarily by cyclic voltammetry (Fig. 12.16). Bindra et al. determined that palladium and platinum were among the best catalysts in alkaline media (28). Results were expressed in terms of a volcano plot (Fig. 12.17), which compares the normalized peak current for formaldehyde oxidation to the enthalpy of formation of the metal formate. The metal formate is a proposed intermediate in the formaldehyde oxidation pathway. Enyo evaluated several metals at more negative potentials (i.e., closer to the onset of the electrocatalytic oxidation) and concluded that Au was a better catalyst than Pd or Pt in alkaline media (Fig. 12.18) (29).

In acid media, Au has very low catalytic activity compared to Pt and Pd. This is attributed to the importance of the surface oxide structure in catalyzing the electrooxidation reaction. For Pt and Pd, the onset of formaldehyde oxidation occurs near the onset of surface oxide formation (Fig. 12.16). Since oxide formation occurs at more positive potentials on Au, compared to Pd or Pt,

---

*Reprinted by permission of the publisher, The Electrochemical Society, Inc.

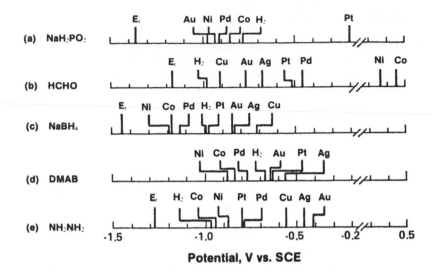

Fig. 12.15—Catalytic activities of metals (the potentials at 10^{-4} A/cm^2) for anodic oxidation of different reductants. E_R = oxidation-reduction potentials of reductants. H_2 = reversible hydrogen potentials. Conditions: (a) 0.2M NaH$_2$PO$_2$ + 0.2M Na-citrate + 0.5M H$_3$BO$_3$, pH 9.0, 343 K; (b) 0.1M HCHO + 0.175M EDTA·2Na, pH 12.5, 298 K; (c) 0.03M NaBH$_4$ + 0.175M EDTA·2Na, pH 12.5, 298 K; (d) 2.0 g/dm^3 DMAB + 0.2M Na-citrate = 0.5M H$_3$BO$_3$, pH 7.0, 298 K; (e) 1.0M N$_2$H$_4$ + 0.175M EDTA·2Na, pH 12.0, 298 K.

formaldehyde oxidation is correspondingly shifted to more positive potentials. However, in alkaline media the role of the surface oxides is less predominant since oxygen atoms for the reaction may be supplied by solution species, e.g., OH$^-$. Therefore, Au is a far better catalyst in alkaline media than in acid media.

Alloys have been examined for their catalytic activity for formaldehyde oxidation and in many cases show enhanced activity when compared to constituent metals. Enyo showed that Pd and Au alloys were more active catalytically than Pd (Fig. 12.18) (29). The activity increased montonically with increasing Au content. Similarly, Au and Pt alloys were determined to be more active than Pt (30). It has also been demonstrated that Cu and Ni alloys are more active for formaldehyde oxidation than Cu (31). The composition Cu$_{89}$Ni$_{11}$ had the highest activity and decreased with increasing Ni content. This observation has a further consequence, since the inclusion of small quantities of Ni^{2+} in a copper plating bath results in alteration of the physical properties of the deposit, attributable to inclusion of nickel in the deposit. Therefore, it is possible to

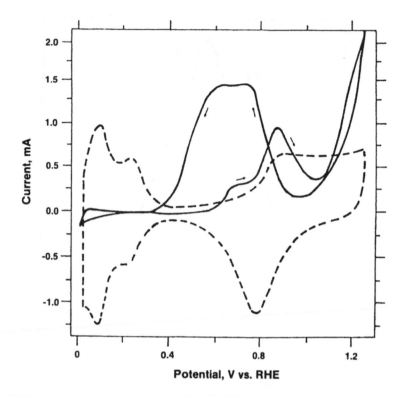

Fig. 12.16—Cyclic voltammetry curves for Pt in 1M HClO₄ (dashed curve, scale on right) and 1M HClO₄ + 0.1M HCHO (solid curve). Ar saturated, electrode area = 0.458 cm², scan rate = 0.1 V/sec, temperature = 25° C.

enhance both deposit metallurgy and catalytic activity by judicious inclusion of foreign metals in the bath.

The mechanism for formaldehyde oxidation is dependent on the nature of the catalytic substrate. In some cases (e.g., group VIII metals), the reaction is not accompanied by hydrogen generation and proceeds as follows (28):

$$H_2C(OH)O^- + 2OH \rightarrow 2HCO_2^- + 2H_2O + 2e^- \qquad [38]$$

However, for group 1B metals, the reaction is accompanied by hydrogen production, as shown in Eq. 39:

$$2H_2C(OH)O^- + 2OH^- \rightarrow 2HCO_2^- + 2H_2O + H_2 \uparrow + 2e^- \qquad [39]$$

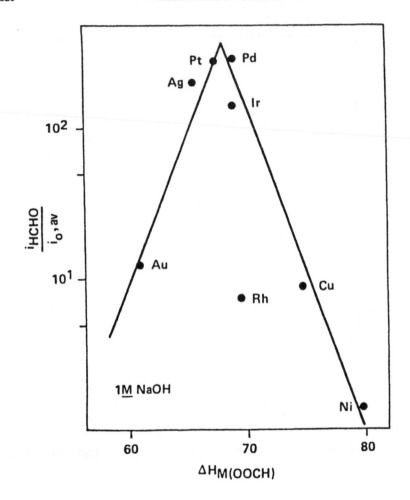

Fig. 12.17—Plot of HCHO oxidation currents in 1M NaOH vs. enthalpy of formation of metal formate.

This may be explained by the disposition of the intermediate metal hydride (adsorbed hydrogen radical). The hydrogen radical, when adsorbed on Pt, is oxidized at potentials where formaldehyde oxidation occurs. Therefore, radical recombination to form hydrogen gas is unfavorable. In contrast, on copper, recombination of hydrogen radicals does occur and hydrogen gas evolution is

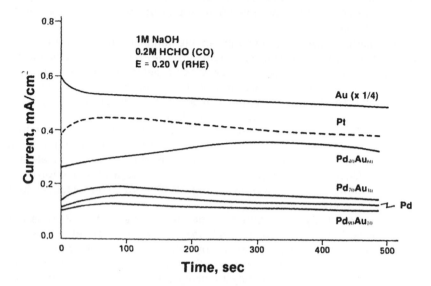

Fig. 12.18—Time dependence of the currents of electrooxidation of 0.2M HCHO in 1M NaOH solution at 30° C. Curve for Au is shown in ¼ times reduced scale.

observed. Oxidation of the hydrogen radical is kinetically unfavorable at the potentials where formaldehyde oxidation is observed.

In practice, the most commonly utilized catalysts are comprised of Pd and Sn. A two step catalysis (seeding) process has been described where a non-catalytic surface is first treated by immersion in an acidic solution of $SnCl_2$, followed by immersion in an acidic solution of $PdCl_2$ (32). Stannous is thermodynamically capable of reducing Pd. The resultant reduced Pd is an active catalyst for formaldehyde oxidation.

Single step Pd and Sn catalysts have also been described whereby a colloid consisting of Pd and Sn is formed by digestion of an acidic $PdCl_2$ and $SnCl_2$ mixture (33-35). Such catalysts are the most widely used in practice. In the early stages of digestion, a complex is formed with a Pd:Sn ratio of 1:3 (36,37). As digestion proceeds, a colloid is formed and is comprised of a reduced Pd metal nucleus. The reaction proceeds according to Eq. 3, where $(Pd-Sn)_x$ denotes the complex.

$$(Pd-Sn)_x \rightarrow Pd^{0} + Sn^{4+} + 2Sn^{2+} \tag{40}$$

The colloid is stabilized by a shroud of $SnCl^{1-}$ ions imparting a negative charge to the particle. At least three species of tin have been identified by Mossbauer spectroscopy and include stannic and two stannous components (36,37). The two stannous components are identified as part of the Pd-rich core and the protective shroud.

Colloid particle sizes are generally in the range of 10 to 20 angstroms, depending on the method of measurement (38). The Pd:Sn ratio is in the range of 0.5 to 1.0, depending on the particle size. When a dielectric surface is immersed in a solution of colloidal catalyst, the colloid is adsorbed on the surface. The coverages of Pd and Sn, as a function of exposure time, are shown in Fig. 12.19a (38).* Maximum coverage is usually obtained within several minutes.

It is usually necessary to treat the catalyzed dielectric surface with an accelerator whose function is to selectively remove the stabilizing stannous ions, thus exposing the catalytically active Pd nucleus. Typical accelerating agents are HCl, NaOH, and HBF_4 (39). Pd and Sn coverage as a function of time in an accelerating solution are shown in Fig. 12.19b (38).* The selectivity for tin removal is apparent.

The catalytic activity of the colloid is dependent on the method of preparation and may be verified by electrochemical measurements. Osaka et al. examined several colloidal catalyst preparations by dipping a gold electrode into the various catalyst mixtures and observing the electrochemical anodic stripping peaks in 1.0M HCl (40). The catalyst activity was proportional to the charge passed for the Pd stripping peak, although the results were only qualitative. This behavior was later quantified by employing a glassy carbon electrode, which has significantly lower background currents in this potential region (41). Coulometry data obtained from stripping curves correlated well to data obtained by microprobe.

Alternate electrochemical methods include observation of the hydrogen adsorption/desorption region on catalyzed gold electrodes and comparing them to the response obtained at bulk Pd (42). This region has been well characterized by Pd by cyclic voltammetry. The response for active catalysts resembles bulk Pd, while less active catalysts show little or no structure in the hydrogen region.

Electrochemical measurements have also demonstrated that colloidal Pd/Sn catalysts are intrinsically more active for formaldehyde oxidation than either bulk Pd or Au (42,43). This is attributable to Pd/Sn alloy formation and is in agreement with the previous discussion regarding the enhanced activity of certain alloys.

PROPERTIES OF ELECTROLESS COPPER DEPOSITS

The properties and grain structure of electroless copper deposits are dependent on the substrate upon which deposition occurs. The grain structure associated with a conducting catalytic substrate differs markedly from that obtained on an

---

*Reprinted by permission of the publisher, The Electrochemical Society, Inc.

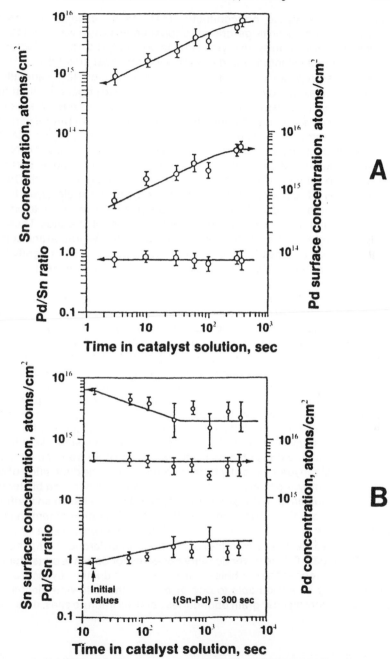

Fig. 12.19—Pd and Sn surface concentration vs. (a) immersion time in S catalyst solution, and (b) accelerator time for an S solution with a catalysis time of 300 sec.

activated or catalyzed nonconductor (44). As an example of the former, electroless deposits on large-grained copper sheets show well defined epitaxial growth. This is in contrast to crystal growth on catalyzed nonconductors, which is of more practical interest and has received the most attention.

Electroless deposits obtained on catalyzed nonconductors generally exhibit fine grain equiaxial growth initially with conversion to larger columnar grains as film thickness increases (Fig. 12.20). Fine grain structure is obtained initially because the catalyst (most frequently colloidal Pd/Sn) is highly dispersed on the nonconducting surface. The catalytic particles, on the dielectric surface, are 10 to 50 nm in diameter and function as discrete centers of nucleation. Notice that there is agglomeration of the particles on the surface since particle sizes measured in solution are approximately an order of magnitude less. During very early stages of exposure in the electroless bath, the grain sizes for copper crystallites at the nonconductor interface are generally observed to be nearly an order of magnitude larger than the catalyst particle agglomerates. Thus, it seems likely that agglomeration of crystallites occurs (45,46). Migration and subsequent agglomeration is possible, since the catalyst particles are not tightly bound to the surface, and probably continues to occur until a complete, uniform film is obtained.

Observation of crystal growth at a catalytic amorphous substrate (e.g., Pd-Cu-Si alloy) exhibits very finely-grained deposits (44). In this case, agglomeration is prevented because the catalytic sites are not mobile and cannot migrate.

In the very early stages of copper deposition on Pd/Sn catalysts, EDAX measurements demonstrate a surface composed primarily of Cu and Pd rather than Cu, Pd, and Sn (47). This may be attributable to the exchange of Cu for Sn, resulting in a solid solution of Cu and Pd. This is supported by the observation that Pd/Sn catalysts are intrinsically more active than Pd. It appears that the galvanic exchange of Sn for Cu is an essential aspect of the catalytic mechanism. Further, the lattice constant observed for the Cu/Pd particles is intermediate between that of Cu and Pd. It therefore seems likely that a solid solution of Cu and Pd is obtained with exclusion of Sn from the lattice.

For formaldehyde-based electroless copper baths, the deposition of copper is accompanied by the liberation of hydrogen, which may cause blistering of the deposit. Inclusion of hydrogen within copper deposits has been verified by the observation of microscopic gas bubbles (44). This phenomena has a direct impact on the physical properties of the deposit. Shown in Table 12.5 (44)** are data from two identical electroless baths, the only difference being the presence of NaCN in bath B. The lower ductility observed for bath A is consistent with the higher occurrence of gas bubbles. The reduction in ductility by incorporated gas bubbles may be attributable to one or both of the following mechanisms: (1) bubbles occurring near grain boundaries serve as fracture sites, thus increasing

---

**Table 12.5 has been reprinted with permission from *Acta Metallurgica*, Vol. 31, S. Nakahara and Y. Okinaka, 1983, Pergamon Journals, Inc.

Table 12.5
Properties of Electroless Copper Deposits*

Bath	Ductility, %		Grain size, μm	Gas bubble density/cm³
	As-deposited	After 6 months		
A	1.2	4.8	0.5 to 1.0	9×10^{15}
B	3.6	4.8	0.5 to 1.0	9×10^{14}

*Plexiglass substrate.

the possibility of fracturing; (2) finely dispersed bubbles may act as obstacles to the passage of slip dislocations, resulting in increased hardness for the deposit.

It is interesting to note that the ductility for each bath composition improved with time. This may be attributed to the diffusion of hydrogen out of the copper. The diffusivity of hydrogen through copper occurs via an interstitial mechanism and is quite high, although it is impeded to some extent by the presence of voids left by the gas bubbles. This hypothesis is supported by the fact that annealing improves deposit ductility, presumably by accelerating the hydrogen diffusion process (48).

In general, ductility in the range of 4 to 7 percent elongation and tensile strength in the range of 30,000 to 80,000 psi are characteristic of crack-free copper deposits (49). The ductility of electroless copper deposits improves with increasing plating temperatures and decreases with increased plating rate at a given temperature. The tensile strength increases with decreasing plating rate (48). It is clear that lower plating rates tend to yield more favorable metallurgy. However, there continues to be a substantial effort to increase plating rates without sacrificing deposit metallurgy. This may be accomplished, to some extent, by increasing the bath temperature, although bath stability is an issue. One novel approach is to operate the bath at elevated temperatures in a closed system, which results in an increase in pressure (50). The result is an increased plating rate, compared to ambient pressure, with no sacrifice in physical properties.

There are numerous reports, primarily in the patent literature, on the influence of additives on the physical properties of electroless copper deposits (51-52). Additives include cyanide or related nitrile compounds, organo-sulfur compounds, silanes, and surfactants such as polyethylene glycol. Cyanide, as

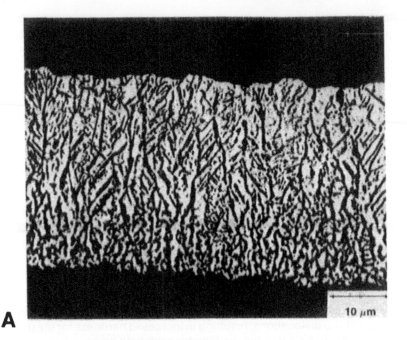

A

10 μm

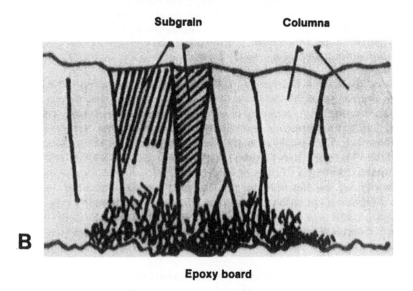

Subgrain   **Columna**

B

Epoxy board

Fig. 12.20—Optical micrograph showing cross-section of thick deposit.

illustrated above, not only stabilizes the plating bath but enhances deposit ductility. The mechanism is likely to be that of competitive adsorption of the strongly adsorbed CN^- ion and the intermediate adsorbed hydrogen radical, thus precluding hydrogen entrapment and embrittlement. Addition of polyethylene glycol to electroless copper baths generally improves ductility, particularly at higher molecular weights (53). The mechanism for this and other surfactants is lowering of the overall plating rate compared to an additive-free bath.

Glycine has been observed to decrease the plating rate and increase ductility and tensile strength for copper deposits, although the proposed mechanism is somewhat different than described above (54). Glycine and formaldehyde form a condensation product according to the following reversible equilibrium:

$$NH_2CH_2COOH + HCHO \rightleftharpoons H_2C = NCH_2COOH + H_2O \qquad [41]$$

Therefore, increasing the glycine concentration tends to lower the free formaldehyde concentration, which in turn reduces the plating rate and improves the deposit metallurgy. In addition to the above homogeneous reaction, glycine is also a complexant for cuprous ion and may function as a stabilizer in much the same fashion as CN^-, although the stability constant is lower for the glycine complex.

Small quantities of metal ions present in the plating solution may also alter the deposit properties. For example, Ni^{2+} ions present in the plating solution have been reported to increase tensile strength and resistivity, but did not alter the ductility (55). Analyzed copper films showed the presence of 0.1 to 1.2 percent nickel incorporated into the deposit. Incorporation of foreign metals in the deposit may not only alter the physical properties, but may also result in a surface with enhanced catalytic activity.

REFERENCES

1. M. Pourbaix, *Atlas of Electrochemical Equilibria*, Pergamon Press, London, 1966.
2. C. Wagner and W. Traud, *Z. Electrochim.*, **44,** 391 (1938).
3. J.V. Petrocelli, *J. Electrochem. Soc.*, **97,** 10 (1950).
4. E.J. Kelly, ibid., **112,** 124 (1965).
5. M. Paunovic, *Plating,* **51,** 1161 (1968).
6. S.M. El-Raghy and A.A. Aho-Solama, *J. Electrochem. Soc.*, **126,** 171 (1979).
7. F.M. Donahue and C.U. Yu, *Electrochem. Acta,* **15,** 237 (1970).
8. A. Molenaar, M.F.E. Holsrinet and L.K.H. van Beek, *Plating,* **61,** 238 (1974).

9. F.M. Donahue and F.L. Shippey, *Plating*, **60,** 135 (1973).
10. F.M. Donahue, *J. Electrochem. Soc.*, **119,** 72 (1972).
11. T.N. Andersen, M.H. Ghandehari and M. Ejuning, ibid., **122,** 1580 (1975).
12. P. Bindra and J. Tweedie, ibid., **130,** 1112 (1983).
13. F.M. Donahue, ibid., **127,** 51 (1980).
14. V.G. Levich, *Physicochemical Hydrodynamics*, Prentice-Hall, Englewood Cliffs, NJ, 1962.
15. A.C. Riddiford, *Advances in Electrochemistry and Electrochemical Engineering*, Vol. 4, Interscience Press, New York, 1966.
16. A.C. Makrides, *J. Electrochem. Soc.*, **107,** 869 (1960).
17. K.J. Vetter, *Electrochemical Kinetics*, Academic Press, New York, 1967.
18. D. Barnes and P. Zuman, *J. Electroanal. Chem.*, **46,** 323 (1973).
19. D. Manousek and J. Volke, ibid., **43,** 365 (1973).
20. W. Vielstich, *Fuel Cells*, D.J.G. Ives translator, Wiley-Interscience Publishers, London, 1970.
21. R.M. Lukes, *Plating*, **51,** 1066 (1964).
22. F.M. Donahue, K.L.M. Wong and R. Bhalla, *J. Electrochem. Soc.*, **127,** 2340 (1980).
23. J. Dumesic, J.A. Koutsky and T.W. Chapman, ibid., **121,** 1405 (1974).
24. R. Schumacher, O.R. Melroy and J.J. Pesek, *J. Phys. Chem.*, **89,** 4338 (1985).
25. D. Vitkavage and M. Paunovic, *Plating*, **70,** 48 (1983).
26. L.N. Schoenberg, *J. Electrochem. Soc.*, **118,** 1521 (1971).
27. J. Ohno, O. Wakabayashi and S. Haruyama, ibid., **132,** 2323 (1985).
28. P. Bindra and J. Roldan, ibid., **132,** 2581 (1985).
29. M. Enyo, *J. Electroanal. Chem.*, **186,** 155 (1985).
30. M. Beltowska-Brzezinksa and J. Heitbaum, ibid., **183,** 167 (1985).
31. M. Enyo, ibid., **201,** 47 (1986).
32. See for example N. Feldstein and J.A. Weiner, *J. Electrochem. Soc.*, **120,** 475 (1973), and references therein.
33. R.J. Zeblisky, U.S. patent 3,672,938.
34. C.R. Shipley, U.S. patent 3,011,920.
35. E.D. D'Ottavio, U.S. patent 3,532,518.
36. R.L. Cohen and K.W. West, *J. Electrochem. Soc.*, **120,** 502 (1973).
37. R.L. Cohen and K.W. West, *Chem. Phys. Lett.*, **16,** 128 (1972).
38. R.L. Meek, *J. Electrochem. Soc.*, **122,** 1177 (1975).
39. R.L. Cohen and R.L. Meek, *Plating and Surf. Fin.*, **63,** 47 (Jun. 1976).
40. T. Osaka, H. Takematsu and K. Nihei, *J. Electrochem. Soc.*, **127,** 1021 (1980).
41. E.J.M. O'Sullivan, J. Horkans, J. White and J. Roldan, *IBM Journal of Research and Dev.*, **32,** 591 (1988).
42. J. Horkans, *J. Electrochem. Soc.*, **130,** 311 (1983).
43. J. Horkans, ibid., **131,** 1853 (1984).
44. S. Nakahara and Y. Okinaka, *Acta Metall.*, **31,** 713 (1983).
45. R. Sard, *J. Electrochem. Soc.*, **117,** 864 (1970).
46. A. Rantell, *Trans. Inst. Metal Finish.*, **48,** 191 (1970).
47. J. Kim et al., *IBM J. of Res. Develop.*, **28,** 697 (1984).

48. J.J. Grunwald, H. Rhodenizer and L. Slominski, *Plating*, **58**, 1004 (1971).
49. M. Paunovic and R. Zeblisky, ibid., **72**, 52 (1985).
50. S.K. Doss and P. Phipps, U.S. patent 4,594,273.
51. O.B. Dutkewych, U.S. patent 3,475,186.
52. F.W. Schneble et al., U.S. patents 3,310,430 and 3,257,215.
53. C.J.G.F. Janssen, H. Jonker and A. Molenaar, *Plating*, **58**, 42 (1971).
54. S. Mizumoto et al., ibid., **73**, 48 (1986).
55. M.L. Khasin and G.A. Kiteau, *J. Appl. Chem. USSR*, **57**, 1971 (1984).
56. F.L. Shippey and F.M. Donahue, *Plating*, **60**, 43 (1973).

GLOSSARY

A	Electrode area	i_M^D	Diffusion-limited current density for total cathodic reactions
a	Tafel slope intercept		
B_M	Diffusion parameter for [CuEDTA] complex	i_{O_2}	Current density for oxygen reduction
		$i_{plating}$	Plating current density
B_{O_2}	Diffusion parameter for dissolved oxygen	i_R	Current density for HCHO oxidation
		i_R^0	Exchange current density for HCHO oxidation
B_R	Diffusion parameter for HCHO		
b_M	Tafel slope for cathodic partial reaction	i_R^D	Diffusion-limited current density for HCHO oxidation
b_R	Tafel slope for anodic partial reaction	n_M	Number of electrons transferred in metal deposition reaction
C_M	Bulk concentration of copper ions		
C_{O_2}	Bulk concentration of dissolved oxygen	n_R	Number of electrons transferred in the HCHO oxidation reaction
C_R^s	Surface concentration of HCHO	R	Gas constant
C_R	Bulk concentration of HCHO	T	Absolute temperature
D_R	Diffusion coefficient of HCHO	v	Stoichiometric number
E	Electrode potential	α_M	Transfer coefficient for metal deposition
E_M	Thermodynamic reversible potential for the metal deposition reaction		
		α_R	Transfer coefficient for HCHO oxidation reaction
E_M^0	Standard electropotential for copper deposition		
		β_M	Symmetry factor
E_{MP}	Mixed potential	γ	Number of preceding steps prior to rds
E_R	Thermodynamic reversible potential for reducing agent reaction		
		η_M	Overpotential for metal deposition
E_R^0	Standard electropotential for HCHO	η_R	Overpotential for HCHO oxidation reaction
F	Faraday constant		
i_M	Current density for metal deposition	ν	Kinematic viscosity
i_M^t	Total cathodic current density	ω	Rotation rate of rotating disk electrode
i_M^k	Kinetic controlled current density for metal deposition		
i_M^0	Exchange current density for metal deposition		
i_M^D	Diffusion-limited current density for metal deposition		

Chapter 13
Electroless Copper
In Printed Wiring Board Fabrication

Frank E. Stone

Electroless copper is employed in the manufacture of printed circuit boards that will ultimately have plated-through holes. Its purpose is to make this hole path conductive enough to permit further build-up of this path with copper metal deposited electrolytically to the thickness specified by the board designer, usually 1 mil (0.001 in.) or slightly more, or to build the entire circuitry on its own. The electroless copper process is a series of steps required to accomplish electroless copper deposition. Each step is critical to the overall process.

The purpose of this chapter is not to detail the fabrication of printed wiring boards, but rather to detail the important aspects of electroless copper deposition as they relate to PWB fabrication. Those readers who want to know how to make a PWB are advised to consult other texts, some of which are listed in the bibliography section of this chapter.

The concept of "plated-through-hole" centers around one or both of the following purposes:

- To hold a component lead wire.
- To interconnect circuitry or "printed wires".

A printed circuit board is simply a non-conductive composite substrate material (epoxy-glass, phenolic-paper, polyester-glass, etc.) upon which circuitry is either etched (if the composite was copper-clad to begin with) or plated (if the composite was not copper-clad).

Some typical composites used in circuitry manufacturing are:

Polyimides—For flexible circuitry or for high-temperature applications.
Paper/phenolic—Can be punched; NEMA grades FR-2, XXX-PC.
Paper/epoxy—Better mechanical properties than paper/phenolic; NEMA grades CEM-1, FR-3.
Glass/epoxy—Woven glass fabric—good mechanical properties; NEMA grades FR-4, FR-5, G-10, G-11.
Random glass/polyester—Suitable for some applications; NEMA grade FR-6.

Table 13.1
Electroless Copper
Foil Thicknesses

Oz/ft²	Foil thickness
0.5	0.7 mils (17.5 μm)
0.25	0.35 mils (8.75 μm)
1	1.4 mils (35 μm)
2	2.8 mils (70 μm)

Holes through the nonconductive substrate (dielectric) are made conductive to either interconnect both sides, better solder in the joining operation, or for *both* reasons. Internal to the dielectric there may exist internal layers of circuitry—etched circuits that were prepared prior to the lamination of the dielectric composite. In this case, the result is what is termed a "multilayer" board. In the case of a multilayer composite, the plated-through hole serves not only to connect the two external sides, but also to interconnect with the internal layer(s) if this is desirable in a particular hole through the dielectric.

Most printed circuit boards presently produced are *subtractive* in process characteristic. That is, the external layers of the dielectric material are supplied clad with a certain thickness of electrolytically produced copper foil. The foil thickness is designated by ounces of copper metal per square foot or fractions or multiples thereof. These designations translate into actual foil thickness as shown in Table 13.1.

The dielectric thickness may also vary, so that we may have a very rigid substrate or even one that is so thin that it is flexible.

If the substrate is non-copper clad to begin with, as for *additive* processing, then the function of the electroless copper is not only to make the holes conductive, but to make the surface conductive as well, for either further electroplating or for full electroless deposition to the full thickness desired, and to the full surface circuit pattern desired.

Since most boards presently produced are of the subtractive processing variety, whether double-sided or multilayer, the discussion will begin with attention to subtractive-type processing; most of the principles involved in pretreatment for electroless deposition as well as some of the principles discussed relative to the electroless copper bath itself, will pertain, whether subtractive processing or additive processing is being considered.

SUBTRACTIVE PROCESSING

Since the through-hole processing of multilayer boards with electroless copper brings some extra factors into consideration, a sketch of multilayer processing

Drilling or Punching

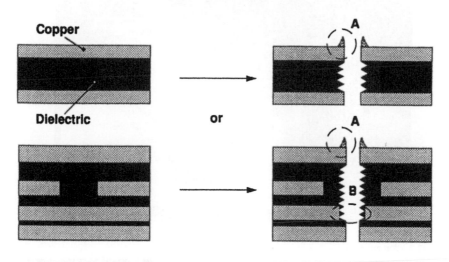

Fig. 13.1—Burrs and smear.

will be carried through this presentation where pertinent. The complete, complex aspects of multilayer board production will not be covered in this chapter.

So, the process then starts with a sheet of laminate, copper-clad both sides, which may or may not contain an internal layer or layers of circuitry (previously prepared).

The first step is to produce the holes, in designated positions, by either drilling or punching through the copper cladding and through the dielectric material. Whether to drill or to punch is determined by the thickness of the sandwich, the ease of punching or drilling the particular dielectric material, the quality obtained by either choice, and in some cases, the quantities of material to be perforated (amortization of die costs, etc.).

Once the holes have been produced, a few things should be taken care of before proceeding to the wet processing involved in plated- through-hole electroless copper deposition (see Fig. 13.1).

First of all, after drilling or punching, burrs of copper (A) will exist on the copper on the external copper layers of the sandwich. These should be removed prior to further processing to avoid plating nodulation at the burr site, tearing problems in dry-film lamination, or screen damaging if screen pattern generation is used. The burrs are indicated by Area A in Fig. 13.1. The burrs are generally removed by wet sanding using a polymeric brush with abrasive tips. Area B will be referred to later.

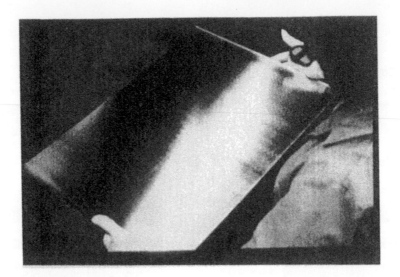

Fig. 13.2—Wet sanding.

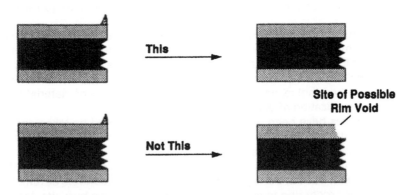

Fig. 13.3—Rim voids.

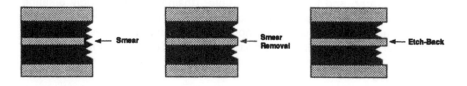

Fig. 13.4—Smear removal.

In wet sanding (Fig. 13.2), care must be taken not to apply too much pressure at this step, to avoid rounding the corner of the hole and exposing bare dielectric material, since the adhesion of the electroless copper subsequently deposited will not be very good to bare surface dielectric substrate and later rim-voids may be apparent (see Fig. 13.3).

Secondly, the purpose in plating through the hole produced is to make this path conductive to the external copper layers, and, in the case of multilayers, to those internal layers of copper that abut the pierced hole. When drilling or punching through any copper-polymer-copper sandwich, particularly in the case of multilayer boards where there are copper layers internal to the sandwich, there is a high degree of probability that polymer will be smeared over the copper edges internal to the hole (Fig. 13.1, Area B). The "smear" must be removed, particularly from the internal copper abutting the hole, to insure electrical continuity between the laminated internal layers and the copper that will ultimately be plated through the hole.

Sometimes it is desirable to "etch-back" a small amount of resin also, to produce a tiny protrusion of the inner-layer copper *into* the hole, thereby creating a better bond of plated-through copper to the inner layer copper.

Smear removal (Fig. 13.4) may be accomplished by the same techniques as used for etch-back. Smear removal alone may be accomplished mechanically by "wet-blasting" or vapor-honing; these techniques employ fine abrasives, such as glass beads or alumina, in a wet slurry that is nozzle-blasted through the holes. The chemical materials employed for both etch-back and/or smear removal dissolve polymer resin. Commonly (for epoxy resin systems) either concentrated sulfuric acid or concentrated water solutions of chromic acid are used. Both chemical methods require post-attention so that problems with through-hole electroless deposition are not *created* by wet processing for smear removal.

Sulfuric Acid

It is important to follow this solution with a very good water rinse, preferably using warm water, and to avoid, if possible, rinsing in any *strongly* alkaline solution. The formation of sodium salt of any residual sulfonated epoxy resin in the hole is undesirable, since this compound is very hard to remove by rinsing, once formed. Its presence in the hole as a residual will cause plating difficulties.

Chromic Acid

The presence of residual Cr^{+6} in the pierced hole will cause coverage problems, since it will tend to destroy the tin-palladium activation by an oxidation mechanism, and interfere with electroless copper reduction. Voids are a typical result of this interference. A second pass through the activation step will usually correct this problem, but second passing is cumbersome, particularly in automatic processing, and at best—expensive.

It is typical to follow chromic acid with a neutralization step, typically using $NaHSO_3$ (sodium bisulfite) to reduce any Cr^{+6} to Cr^{+3}. This is best accomplished by using warm water (100° F) solutions of sodium bisulfite, and following the neutralization with a hot water rinse (120 to 150° F) to prevent interference with activation later on by any *bisulfite* that may otherwise tend to be dragged into other process solutions.

Other Systems

Several other chemical methods are in common usage for etch-back/desmear. Among these are systems that employ a combination of organic solvents (to swell the polymer) and permanganate, used either after treatment with concentrated sulfuric acid, or even instead of either sulfuric acid or chromic acid.

THE ELECTROLESS COPPER PROCESS

Prior to the actual *plating* of electroless copper, some pretreatment of the parts to be plated is necessary to insure:

- Adhesion of electroless copper (in holes) to the dielectric material.
- Continuity of the electroless copper deposit.
- Adhesion of electroless copper to the copper cladding external to the laminate sandwich.
- Adhesion of electroless copper to any internal layers of copper.

The classical steps of pretreatment will now be taken in sequence.

Cleaning

The first step in any pretreatment line is to clean the entire composite—the laminated copper and the pierced dielectric material. The cleaner is typically alkaline in nature, although neutral or acidic materials may also suffice if not much classical "cleaning" is required. The purpose of the cleaning step or steps are:

- Remove soil from the copper foil and the holes (the dielectric).
- Remove light oils from the copper foil and the holes.
- Help remove stains and heat-treatment oxides from the copper foil.
- Roughen slight smear from drilling (existent on the polymer resin).
- Remove impacted drilling debris from the holes.

The cleaning solution is one of the key solutions in the pretreatment line. Inadequate cleaning can give rise to a number of problems, as illustrated in Fig. 13.5 and the following two figures.

The contaminated area will not adequately hold the needed activator to allow coverage with electroless copper—and a "void" or non-plated area will result. Tiny voids after electroless copper may bridge-over with electroplate, but in that case there is no adhesion of the electroplate to anything underneath it. In this case, a pull-away or blow-holing may be the ultimate result.

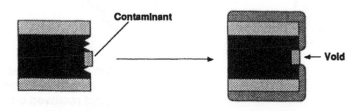

Fig. 13.5—Inadequate cleaning—voids.

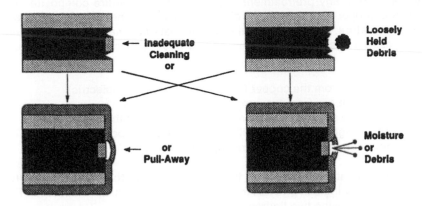

Inadequate
Cleaning
or

Loosely
Held
Debris

or
Pull-Away

Moisture
or
Debris

Fig. 13.6—Inadequate cleaning—pull-away or voiding.

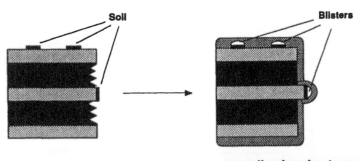

Soil

Blisters

poor adhesion due to residual
soil or lack of proper etch due to
residual soil not removed until
partway through etching

Fig. 13.7—Inadequate cleaning of copper.

The mechanical stresses within the electroplated deposit put on after electroless copper or the force of any moisture entrapped in the laminate as a result of plating that will tend to stress the plated deposit from behind during any operation involving *heat* (baking of resists, solder fusion, etc.) tend to pull the plating away from the dielectric, resulting in what would be termed "hole-wall pull-away". By the same measure, loosely-held debris that is not removed in the cleaning step will be plated over, meaning that the plate has adhered to a structure that itself is not adhering too well to the side of the hole. Pull-away may again be the ultimate result in this case. For both cases in Fig. 13.6, whether pull-away is evident or not, the adhesion may be so poor and the stresses on this area so great during any thermal shock, that the plating continuity may be destroyed, particularly in a solder fusing or wave-solder operation, and the result will be what is called a blow-hole, due to out-gassing from the dielectric behind this weakened area.

If the electroless copper is deposited over soil existent on either the copper laminate or copper protruding to the hole, as in the case of multilayers, then the bond between the electroless plate and the underlying copper will not be as good as if the underlying copper were clean. Bond failure may result, evidenced as blisters if the soil is spotty, or peeling electroless deposition, if the soil is more widespread (see Fig. 13.7).

Important factors to be considered about the cleaning solution are:

- Proper cleaner choice to do the job.
- Temperature of cleaner.
- Concentration of cleaner.
- Time in the cleaner.
- Work agitation in the cleaner.
- Cleanliness of the cleaner—when to dump.
- Rinsing after cleaning.

Temperature is a key factor that is often overlooked in cleaner operation. Most cleaners have a lower limit of temperature operation below which they rapidly lose effectiveness.

Rinsing of the cleaner from the work in process is actually as important a step as the cleaner-solution step itself. Residual cleaner left on the printed circuit board can act as a contaminant to the board itself, as well as a contaminant to other key process solutions—namely, the etch solution and the activator solution. The best type of rinse at this stage would be a rinse that is (a) not ice cold (preferably 60° F or higher), (b) aerated, and (c) equipped with spray nozzles to run fresh water over the boards as they exit the rinse. Condition C is not often used, but Conditions A and B should be easily achievable (see Fig. 13.8).

An adequate flow of fresh water must enter the rinse. The fresh water flow to the rinses (all rinses in general) will be predicated upon the following factors:

Fig. 13.8—Boards emerging from cleaner: (left) normally adequate cleaning and rinsing; (right) probably inadequate cleaning and rinsing.

- The drag-out from the process tank (mL/rack).
- The work load to be rinsed (ft^2/rack).
- Whether the rinse is to be followed by another rinse, which will be cleaner than the rinse preceding it (cascading rinse system).

Conditioning

Following classical cleaning, some processes, and indeed some particular laminate types (notably multilayers, because of the resin-altering nature of the etch-back chemistries), require the use of a step called *conditioning*. The function of the conditioners is to "super-wet" the dielectric material, and in some cases, to provide this surface with a uniformity of polarity such that the later adsorption of activator will be more easily facilitated (see Fig. 13.9).

Sometimes conditioning materials are built into the cleaner formulation, resulting in what is termed a cleaner/conditioner, although the separation of the two functions into two separate process steps, with water rinsing in-between, is generally more effective for both cleaning and conditioning. Conditioning materials typically contain surfactants to do the job intended. Rinsing after this step is an *extremely* important consideration, since too little rinsing can allow the surfactant to remain on the laminated copper surfaces throughout the rest of the process and act as an interference to etching, activation, and the ultimate copper-to-copper bond—resulting in poor electroless copper to laminate-copper bond. Care must be taken to avoid too cold a rinse water temperature and to allow for *sufficient* rinsing quantities of water. The concentration of the

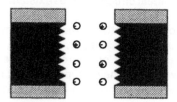 

Fig. 13.9—Hole conditioning.

conditioning solution has to be watched carefully—too strong a conditioner solution should be avoided—a little bit usually goes a *long* way.

Etching

The next step in the electroless deposition line is the micro-roughening or micro-etch stage. The purpose of this step is primarily to give a micro-toothed surface structure to all copper surfaces that will ultimately be plated-over with electroless copper. Without proper etching, the bonds obtained between electroless copper and the laminated copper are usually very poor.

The etched surface serves a number of purposes:

• The surface area of the copper laminate is substantially increased, providing more opportunity for intimate contact between the electroless copper and the laminated copper.

• Some residual surface contamination may be removed in the etch, being undermined by the action of the etch on the underlying copper, if the contamination was not totally removed in the soak cleaner. Relying on the etch to do a substantial amount of cleaning is *not* good practice, however, since uneven etching is generally the result. This is because the residual contaminant does not allow the etch to start working on the underlying copper at the same time it starts to work on uncontaminated copper areas.

• A properly etched surface will provide anchoring sites for the activator (catalyst) material, which can now mechanically "bond" to the laminated copper surface.

• The deposited electroless copper can mechanically "bond" to the laminated copper by "keying" into a satisfactorily etched surface (see Fig. 13.10).

To accomplish what is desired, the depth of etch as well as the type of etch must be correct. Usually a minimum depth of 0.1 mil is recommended, although lesser depths can give satisfactory results under ideal circumstances existent in the rest of the total process line.

Copper removal *alone* from the laminate is not really all that is required. Some etchant chemicals may remove the proper amount of copper, but leave a polished surface. This is contrary to our goal. A *matte* or micro-roughened surface is needed. Some etchants that work well when their copper content is low tend to become micropolishers when their copper content becomes too high.

Temperature is an important parameter with etch solutions. Too low a temperature can result in little or no etching, or a micropolishing effect. Too high a temperature can result in runaway etch activity or etching solution decomposition, with a correspondingly difficult-to-control exotherm.

Too vigorous an etch is undesirable from two standpoints, as shown in the next two figures.

In Fig. 13.11, the adhesion of electroless copper to dielectric at Point A will most likely be less than adequate, and will fail if tested. The dielectric at Point A has not been roughened as it has in the drilled hole.

In Fig. 13.12, the inner-layer copper has been severely etched so that it is receded into the dielectric. This phenomenon is termed *reverse etch-back* and is undesirable from the standpoint that less sure-fire bonding of electroless copper to inner-layer copper can result.

Poor electroless copper adhesion to the copper surfaces of the laminate is generally the result of inadequate etching. Tape testing for adhesion after electroless copper deposition, as a once or twice a day routine quality control test, can help ward off potential problems. Unfortunately, most shops tape test only when they are fairly sure that a problem exists—then the tape test merely confirms everyone's fears.

The tape test best suited for most shops uses a six inch strip of ½- to 1-inch-wide tape, sticky enough for the purpose and strong enough to be rapidly ripped away from the copper surface without tearing or breaking. The tape should be applied to the copper surface (after electroless deposition, rinsing, and drying) across flat areas and some holes, by pressing and rubbing with the edge of one's thumbnail or a coin. The tape is removed with a quick snap motion, and with enough force so that the board has to be held down tightly with the hand that is not pulling the tape. If any copper is removed, it will be observed on the tape.

Copper removal around a hole may indicate an over-etching condition. Copper removal elsewhere may indicate under-etching or *other* problems, such as residual wetting agents on the board, overactivation, or a poorly cohesive electroless copper deposition.

**No
Mechanical
Keying**   **Adequate
Keying
Sites**

Fig. 13.10—Micro-roughened switches.

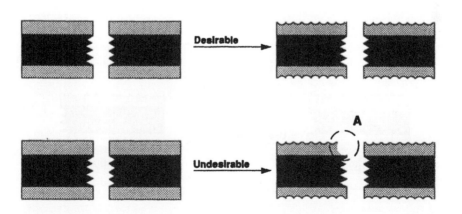

Fig. 13.11—Bare dielectric exposure.

Rinsing prior to etching is important to insure that residual cleaning agents from the cleaner are not dragged into the etch nor are residual to the copper surface. With all etchant materials, the introduction of alkaline components will dramatically reduce their effectiveness.

Other precautionary steps are:

- The electroless plating racks should be stripped in a separate step because all etchant materials are sensitive to the introduction of metals like palladium, which promote etchant decomposition.

- A good rinse after etching is also desirable so as not to drag copper down the electroless processing line.

Typical chemicals employed as etchants in the electroless line are: ammonium persulfate, sodium persulfate, other sulfuro-oxide type-oxidants (all typically 1-½ lb/gal with 1% sulfuric acid added), and peroxide-sulfuric-type etchants. Cupric chloride is not used due to the fact that it leaves a very difficult to remove residue on copper surfaces.

Depending on the type of etch used, the processing steps prior to and after the etch step may vary slightly, as in the following:

Undesirable

Fig. 13.12—Inner-layer copper attack.

Persulfate Type Etch
* 1. Cleaner
  2. Rinse
  3. Conditioner (optional)
  4. Rinse (optional)
  5. Etch
  6. Rinse
  7. Dilute sulfuric acid
  8. Rinse

Peroxide-Sulfuric-Type Etch
1. Cleaner
2. Rinse
3. Conditioner (optional)
4. Rinse (optional)
5. Dilute sulfuric acid
6. Rinse (optional)
7. Etch
8. Rinse

With the peroxide-sulfuric-type etchants, all of which operate at elevated temperatures, it is usually recommended to insert a sulfuric acid immersion step after cleaning and prior to the etch step, to keep the etch solution free of dragged-in alkaline materials.

With other types of etchants, the etch will be followed by a sulfuric acid immersion step, following etch rinsing, to remove etch residues. This is generally not necessary with the peroxide-sulfuric acid-type etchants.

If a persulfate-type etch is used, a dilute sulfuric acid (10 to 15 percent by vol.) dip is recommended to remove etch residues, followed by a satisfactory rinse. As with all acids to be used in the electroless line, care should be taken that the acid is free of heavy metals, to prevent an adhesion-disrupting immersion deposit onto the copper surface.

Activator Pre-Dip Solutions
Prior to the activation step, it is most common to immerse the work into a solution containing ions common to those of the activator itself. After this *pre-dip*, the work is then brought directly to the activating bath *without* an intervening water rinse. Since most activating solutions are chloride-ion based, most pre-dips are based upon chloride ions. In this fashion, the work that is brought to the activating solution is residual with an ion and has a pH common to the activator.

Every time a copper-clad laminate enters rinse water, a slight surface oxide will develop, particularly if the work is being rinsed of an acid. Immersion *into* any mildly acidic pre-dip solution will remove slight amounts of copper oxide and therefore put copper ions into this solution. Activators are usually acidic to some degree. Therefore, a mildly acidic pre-dip solution takes the copper ion into itself, saving the activator to some degree. Most activators are sensitive to copper contamination.

The pre-dip is dumped frequently enough (it is orders of magnitude less expensive than activating solution), based upon copper content, to "save" the activator from copper buildup. Usually a pre-dip will be dumped before it reaches 1000 ppm copper content.

Due to the fact that a strongly acidic pre-dip can attack the black-oxide layer on the internal copper layers of multilayer boards and promote "reverse etchback" as well as residual hydrochloric acid along the internal copper layer, the trend in recent years has been to go to reduced acid strength pre-dips (1 to 2 percent by vol. as hydrochloric acid), as well as low acidity activator solutions.

Activation
Proper activation is perhaps the *key* step in the electroless copper process, other than the electroless deposition itself. However, in order to properly activate, the preceding steps must all be in good working order.

The activating species is held primarily mechanically to both the copper surface of the laminate (in sites created by etching) and the surface of the dielectric (in sites created by piercing).

Activating solutions (or catalysts) typically contain palladium held in a reduced state by stannous ions. These solutions are typically high in chloride ion and possess some degree of acidity, typically lower than had been the case five or ten years ago (to prevent reverse etchback). Figure 13.13 depicts the type

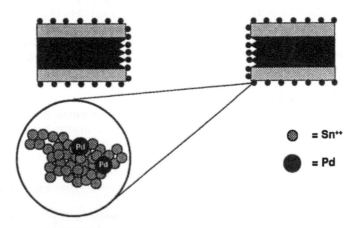

Fig. 13.13—Activation (catalyzation).

of species that is deposited all over the work, copper surfaces and dielectric alike. Typically, dielectric surfaces will take on a brownish color.

Typically, the number of tin atoms is 50 to 100 times greater than the number of palladium ions and the species is manufactured to be as small as possible. *Without* the presence of *stannous* tin, the palladium will readily oxidize to a non-activating form of palladium. The catalytic ability of the species is not only dependent on the palladium content, but also on the structure of the compound. Activating solutions are typically deep black-brown in color. By chemical analysis, maintenance of the solution requires monitoring the stannous tin content to insure that the activating solution has enough stannous tin to prevent the solution from becoming useless.

Activator solutions should *never* be aerated, because this will oxidize the stannous tin and make the palladium itself susceptible to oxidation—rendering it useless.

Since activator (catalyst) is a layer of material *between* the copper cladding and the subsequent electroless copper, it can prevent intimate copper-to-copper contact and interfere with adhesion. It is desirable *not* to put too much activator on the work being processed. Too *long* an immersion time is to be avoided. Palladium activator materials are such that a little goes a long way.

With improper activation, poor copper-copper bonding will probably result. Also, palladium consumption, and hence cost, will be very high (Fig. 13.14).

When withdrawing work from activating solutions, care should be taken to drain parts adequately. In most shops, more activator is dragged out of the

Electroless Copper

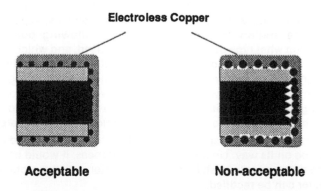

Acceptable   **Non-acceptable**

Fig. 13.14—Improper activation.

activating solution than is normally consumed by having activator mechanically adhere to the work.

Tin-palladium activating solutions are sensitive to several ions that act as contaminants, notably copper and antimony. Relatively small amounts of antimony (10 to 20 ppm) can lead to tiny voids in the electroless copper deposit. Activators tend to have different tolerances for copper contamination, based upon their formulation. The tolerance is also somewhat dependent upon what type of laminate is being processed. A difficult-to-cover multilayer laminate (due to changes effected on the dielectric by smear removal technique) may exhibit voiding problems when the copper content in the activator is around 800 ppm, while the same effect may not be observed on a standard double-sided laminate until copper concentrations of 1500 to 2000 ppm or *higher* exist in the activator.

Copper may be kept to a minimum in the activation solution by observing these precautions:

- Keep the activation time nominal, about 4-6 minutes.
- Dump the activator pre-dip frequently.
- Avoid lifting work out of the activator and re-immersing it.

Rinsing After Activation

Too long a rinse time can undermine the activator and render it useless. Long rinse times can lead to copper oxidation of surfaces.

The rinse directly following the activator usually gets cloudy due to the formation of a tin precipitate. Poor water flow usually will result in this rinse becoming very concentrated with this precipitate, which may ultimately contaminate the activated surfaces.

Aeration, while promoting beneficial rinsing, may serve to oxidize the catalytic species, making it less active or totally inactive, depending upon the severity or duration of aeration. Avoid aeration if possible, or use very low aeration.

The use of recycled water sometimes creates problems, especially in this particular rinse station and the rinse station following post-activation. Depending upon what type of recycling system is used, and which rinse waters are recycled, there arises the occasional probability that the recycled water will contain wetting agents and/or chelating agents, and/or oxidizing agents. These materials can render the activating species inactive, resulting in voiding. Once the activating species has been deposited onto the work in process, every precaution should be taken to insure its being held in its active state.

A yellow color in these rinses, or a foam development, is an indication that trouble may be on its way. Under these circumstances, it would be prudent to switch these rinses over to fresh incoming city water until the problem with the recycled water can be rectified.

Many useful rinsing habits may be obtained from an AESF Illustrated Lecture, *Rinsing, Recycle and Recovery of Plating Effluents*, by D.A. Swalheim. Suggestions made in this chapter will help in *all* rinsing considerations.

Post-Activation

The purpose of the post-activation step (or acceleration step, as it is sometimes called) is to render the activating species deposited in the activation step as "active" as possible prior to immersing the work in process into the electroless copper bath.

If the activating species is thought of as a closed flower bud, with palladium in the center of the bud, surrounded by many layers of tin, then the post-activator can be thought of as functioning to open the bud, removing some tin and perhaps some palladium in the process of doing so, and letting the inner palladium be more readily seen by the electroless copper solution.

If the post-activation step were skipped, the initiation of deposition in the electroless copper bath would be prolonged and the risk would be run of contaminating the electroless copper bath with any marginally adherent palladium-containing activating species.

Hydroxides of tin are formed in the rinses after activation, as can be seen by the cloudy nature of this rinse, and can coat the activator species on the board and mask its functionality. The hydroxides of tin are gelatinous in nature and coat the palladium metal particles. The purpose of the post-activator, then, is also to solubilize these compounds, leaving the activating species as free as possible from residual tin compounds.

Some tin is removed and small amounts of palladium are cut loose to the post-activator solution. The new activating species is "more active" and able to quickly initiate electroless deposition (Fig. 13.15).

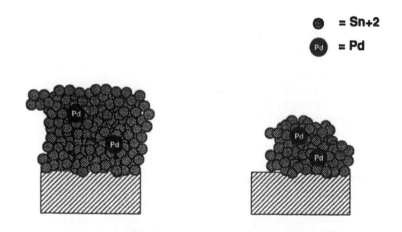

$$Sn_b Pd_a \longrightarrow Pd_x + Sn_y^{+4}$$
$$\text{plus } Sn_{b-y} Pd_{a-x} \text{ (New Activating Species)}$$

Fig. 13.15—Post-activation (acceleration).

Most post-activating solutions are acidic, and will also remove copper oxide that has formed on the copper surfaces due to the rinsing steps between activator and post-activator. Post-activators are generally considered spent when they reach a copper content of 1000 ppm.

Post-activator solutions usually contain materials that will dissolve stannous tin from the species. For this reason, care must be taken not to stay too long in these solutions. In the extreme case, enough tin can be removed to undercut the tin that is attached to the surfaces coated with activating species, and these surfaces will then be devoid of *any* activating species.

Also, the most common post-activators are based upon fluoride compounds and *some*, although minimum, amounts of fluoride ion will be ever present. The fluoride ion can attack glass fiber, which may be at the surface in a pierced hole, and thereby undercut the activator species that attached mechanically to the glass.

Some post-activators may contain reducing agents whose function it is to be dragged *into* the electroless copper solution with the activating species, to promote rapid initiation of the electroless copper deposit.

So far, a lot of time has been spent discussing everything *but* electroless copper. The reason for this is that in order to deposit electroless copper, many steps have to be carefully executed. Electroless copper is the *end* product in the electroless copper process, and as such, it takes a great deal of undeserved blame when the end result is not as anticipated.

In summary, the typical process steps are as follows:

1. Cleaning or cleaning/conditioning—3 to 5 min
2. Water rinse—30 sec to 2 min
3. Conditioning (optional)—3 to 5 min
4. Water rinse—30 sec to 2 min
5. Etch—2 to 3 min
6. Water rinse—30 sec to 1 min
7. Sulfuric acid, 10 to 15 percent by vol.—1 to 3 min
8. Water rinse—30 sec to 1 min
9. Activator pre-dip—1 to 2 min
10. Activator—4 to 7 min
11. Water rinse*—30 sec to 1 min
12. Post-activator—3 to 6 min
13. Water rinse*—20 sec to 1 min
14. Electroless copper—10 to 30 min

---

*Preferably a double rinse.

Steps 5 and 7 may be reversed, depending upon the type of etch used. These times are typical only, and may vary depending upon the proprietary process used.

It must be mentioned at this point that other activation chemistries have been commercially successful, although none enjoy the present widespread popularity of the tin-palladium chemistry presented in this discussion.

Basically, two other types of systems now exist:

- A neutral to alkaline pH palladium-containing activation system. This system is intended to obviate the problem of oxide attack along inner-layers in a multilayer package, thereby eliminating entrapped acid-chloride compounds and reducing the "pink ring" possibility. Also, reverse etchback, an outgrowth of inner-layer attack, should be greatly reduced with systems of this type.
- A copper-based activation system. This system provides a very cost effective activator, and utilizes a strong reducing agent type of post-activator to put the copper into a proper form to be receptive to electroless copper deposition.

Electroless Copper

Typical electroless copper bath formulations will contain the following ingredients:

- Copper salts
- Reducing agent (formaldehyde)
- Alkaline hydroxide
- Chelating agents (quadrol, EDTA, Rochelle salts, etc.)
- Stabilizers, brighteners, etc.
- Wetting agents (optional)

The formaldehyde and the hydroxide ions provide the reducing force necessary for the deposition of metallic copper. The deposition reaction must be initiated by a catalytic species on the surface of the work to be plated. This is provided through the activator step.

The main plating reaction can be postulated to be:

$$Cu(II)(chel) + 2HCHO + 4OH^- \rightarrow Cu^0 + 2HCOO^- + (chel)^= + 2H_2O + H_2$$

The reaction as written is only autocatalytic with respect to copper metal in the presence of an activated surface or when hydrogen is being generated. That is to say, if one were to place a clean piece of copper in the electroless bath, no plating would occur. Plating would only occur on this copper if the copper were catalyzed, or if another catalyzed substrate was plating near this copper and liberating hydrogen. Sitting in the bath and not plating, the copper would most likely become oxidized. Catalysis (activation) starts the reaction. A broader scope of electroless copper chemistry is provided in the Appendix to this chapter.

The choice of which bath to use is predicated upon several factors. Typical electroless copper baths are:

Slow speed baths
   Room temperature
   1-2 millionths in./min (plating speed)

High speed baths
   Room temperature—21 to 32° C (70 to 90° F)
   2-6 millionths in./min

High speed baths
   Elevated temperature—38 to 60° C (100 to 140° F)
   2-6 millionths in./min

If the electroless copper is to be overplated to full thickness for full panel plating, or the electroless is to be overplated with some minimal thickness (0.1 to 0.3 mil) of electrolytic copper to insure through-hole coverage before proceeding with imaging and pattern plating—as is often the case in very expensive multilayer board manufacture, then the thickness of electroless copper need only be minimal, such as 10 to 20 millionths of an inch deposit. A slow speed bath, which affords the easiest control, would be the most likely choice.

The trend in the printed circuit industry has been to get away from full panel plating and into pattern plating for economical and functional reasons, and also to get away from the additional racking and unracking steps necessary to overplate the electroless copper with electrolytic copper (flash plating) prior to imaging and pattern plating. With the proper choice of cleaning techniques, after imaging and prior to pattern plating, electroless copper alone may be satisfactory, and usually is, provided that the thickness and integrity of the deposit is suitable to withstand the cleaning technique.

If not too much cleaning is required, and cleaning can be implemented insuring that the attack of electroless copper will be minimal, then low thicknesses (20 to 40 millionths in.) of electroless copper may be adequate to the purpose, or a high speed bath may be chosen if speed of production is critical (see Fig. 13.16).

If more vigorous cleaning techniques or chemicals are required after imaging and prior to electrolytic pattern plating, then higher thicknesses of electroless copper will be needed to insure that this deposit remains intact during the cleaning. In these cases, thicknesses of 60 to 100 millionths in. are usually needed, and the use of the high speed baths is generally favored, for time considerations.

With respect to high speed baths, the elevated temperature variety seems to be increasing in popularity, since solution growth (because of needed chemical additions) is minimized due to water evaporation at the elevated temperature. Also, it is economically easier to maintain a constant temperature, which is

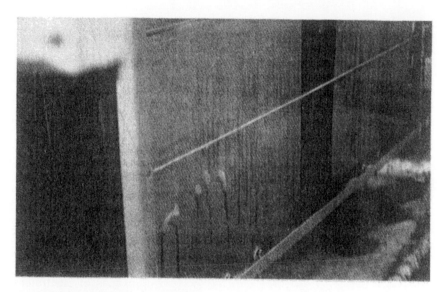

Fig. 13.16—High-speed electroless copper.

important to the rate of deposition, with an elevated temperature bath, since only heating is required. Room temperature high speed baths will require both heating and cooling facilities to keep the temperature fairly constant throughout the year in the more seasonal parts of the country.

The physical structure of the electroless copper also plays a very important part in the choice of which commercial bath should be chosen, if flash electroplating is to be eliminated. It is generally acknowledged that electroless copper is of a different physical structure than electrolytic copper. This is supported by the following facts:

• The density of electroless copper is 2 to 5 percent less than that of electrolytic copper.
• Electrolytic copper on an equal thickness basis will be less rapidly dissolving than electroless copper in any etching medium—indicating a more open structure for electroless copper.

So, if one electroless copper of 60 millionths in. thickness will withstand immersion in the cleaning process solutions for the same time as another electroless copper of 100 millionths in. thickness, the thinner copper is obviously the more economical. Rate *alone* is not the determining factor of best choice.

Another factor that can enter the picture on choice of electroless coppers is

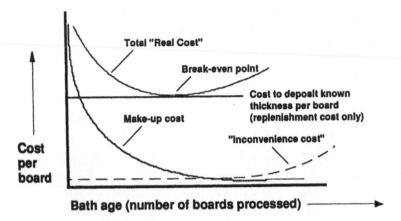

Fig. 13.17—Real electroless copper plating costs.

that of "useful economic life". All electroless copper baths build up by-products during use. All electroless copper baths require continual replenishment for copper ion, formaldehyde, sodium hydroxide and other proprietary agents during use. Depending upon the type employed, these additions will be made with concentrates, either relatively concentrated in constituents, or not so concentrated. Typically, copper ion and formaldehyde are in one component and sodium hydroxide is in the other.

Usually solution must be decanted from the bath and thrown away to make room for additions. Depending upon the concentration of the additive solutions and the bath's operating temperature, this throw away will be negligible to sizeable. If throw away is negligible, by-product accumulation in the bath will be rapid, minimizing the life of the bath. If the throw away is sizable, by-product accumulation will be less significant and the life of the bath will be longer. Whether we throw away the whole bath frequently, or throw away *portions* of the bath frequently and the entire bath infrequently, is a choice that needs to be made, depending upon *overall* throw away over a long period of time and the nature of bath operation and deposit characteristics over a long period of time.

Every bath has a "break-even" point based upon its cost to deposit a known thickness, its make-up cost amortized over its useful life, and its "inconvenience cost" with age (more plate-out with age, more chances of voiding, more erratic operation, etc.). This overall operation is depicted in Fig. 13.17.

Keeping a bath in operation beyond the break-even point is basically not good economics.

Temperature control is perhaps one of the most important physical parameters affecting rate control and quality of deposition, and is probably the most widely abused parameter to be mentioned. Every electroless copper bath is *designed* chemically to be operated in a certain temperature range. Low temperature baths usually have the widest window of temperature operation (see Fig. 13.18).

Below a minimum temperature, all baths will fail to initiate properly, and voiding will be the most likely result. Above a certain temperature, the rate of deposition will increase beyond the point where a physically sound deposit will result, and several of the side reactions mentioned later will predominate, leading to a lessening of overall bath stability. It is therefore very important to pay close attention to electroless copper bath operating temperature.

Physical Parameters of Electroless Copper
Aeration

Some electroless copper baths are somewhat dependent upon aeration for bath stability and integrity of deposit. It is thought that the mechanism involved here is the aid in removing hydrogen gas (evolved in the plating reaction), which can act as a secondary reducing medium causing localized hyperactivity.

Other electroless copper formulations are such that the introduction of air will destroy the ability of the solution to deposit upon activated surfaces.

The *amount* of aeration required depends upon the particular proprietary formulation being used.

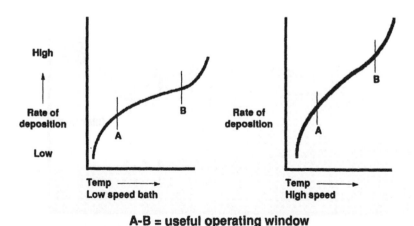

A-B = useful operating window

Fig. 13.18—Temperature vs. rate of deposition.

Filtration

For slow speed baths, periodic batch filtration through 1-micron polypropylene filter cartridges is usually recommended on a once-a-day to once-a-week schedule, to rid the bath of extraneous copper particles. Continuous overflow filtration is better, but not critical.

For high speed baths, continuous overflow filtration through a 10- to 25-micron polypropylene filter bag is generally recommended (see Fig. 13.19).

The solution is allowed to overflow the main plating chamber, falling through a filter bag to a sump, from which it is then pumped back into the main plating chamber. High speed baths may generate a more vigorous evolution of hydrogen, and side reactions leading to the formulation of extraneous copper are more predominant than in slow speed baths, making this continuous bath filtration more desirable.

Bath loading

The concentration of replenishment concentrates is usually based upon an *average* bath loading (surface square feet of work processed per tank volume) at a certain operating temperature. Side reactions, notably the reaction of *formaldehyde* with *sodium hydroxide* to form methanol and formate, proceed at a definite rate at a given temperature and are a function of time and temperature. The formaldehyde and sodium hydroxide ratio in the concentrates to replenished copper is based upon a certain loading prediction.

If the electroless copper bath were loaded with workloads below what the manufacturer of the electroless copper concentrates recommends, the result

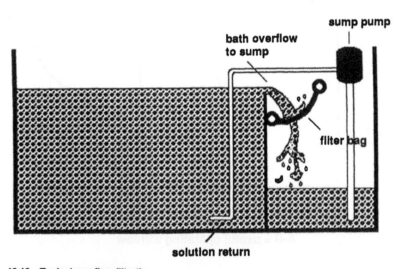

Fig. 13.19—Typical overflow filtration.

would be that the formaldehyde and sodium hydroxide content of the bath would fall off and extra additions of these components would be required. If one were to load the bath with workloads greater than those recommended, the formaldehyde and sodium hydroxide content would climb with time.

Also, with very low workloads in an elevated temperature bath, evaporation losses would exceed replenishment volumes, so that bath volume would decrease with time, making the addition of water possibly necessary.

Typically, electroless copper baths are designed to accept 1 to 1.5 ft^2/gal as the *optimum* loading for the considerations described above.

Work agitation

It is generally desirable to agitate the work in the electroless copper bath with through-hole agitation, to bring fresh solution to the areas being plated, and to help remove hydrogen gas, which is generated during plating, from any recesses.

Also, care should be taken that the boards do not touch one another and that they do not come into contact with the walls of the tank, nor any solution or aeration spargers. When the boards enter the tank, they are coated with activating species, and the activator material can smear onto other surfaces— resulting in these surfaces becoming catalytic and thereby receptive to electroless copper plating.

Rack materials

Electroless copper plating racks should be constructed from 316 stainless steel. The electroless copper will plate *any* rack material. What we want is for the electroless copper plated on the rack to adhere and not flake off into the plating solution, where it will continue to plate. "Inert" racks such as plastic (PVC) or Teflon will eventually start to plate because they will hold some activator, but the plate will be loosely adherent, fostering flake-off.

Tank cleaning

Eventually, every tank will exhibit plate out on horizontal surfaces (tank bottom, overflow weir lips, etc.) due to extraneous copper particles failing to overflow and falling down through the solution. For this reason, tanks should be thoroughly cleaned and rinsed periodically by removing the electroless copper bath, preferably through a filter, doing the necessary cleaning, and returning the copper bath to the clean tank.

Tank design

Polypropylene and polyethylene are good tank construction materials since they are relatively inert, fairly resistant to impact cracking, and will not soften appreciably at the temperatures that most hot electroless copper baths are run. One should choose a thickness of material for construction that will not allow the plating tank to bow at the sides when filled with solution at intended operating temperature. PVC-lined steel tanks should be avoided, since the insides of the tanks will invariably get scratched over time, and linings tend to become cut, allowing the electroless solution to get behind the lining. Electroless baths can become contaminated in this fashion. Also, the use of disposable PVC liner bags should not be recommended because folds will exist where solution and evolved gas will become entrapped, promoting extraneous plating in these pockets, leading to premature bath decomposition.

Tanks must be large enough in volume to handle the intended workload per immersion (1 to 1.5 ft^2/gal), including the rack surface area, and also to accommodate air spargers (if used), solution return spargers if overflow filtration is employed, and a satisfactorily large external sump—again, if overflow filtration is used.

The rack immersion volume should be large enough to allow racks to be immersed and withdrawn without bumping or scraping the tank walls. Scraping will obviously put scratches in the otherwise smooth tank walls, which become sites for extraneous plating. If the scratch is created from the rack at entry, activator from the rack will be imbedded in the scratch, promoting electroless plating in the scratch. Some allowance should be made for rack (through-hole) agitation, again allowing no bumping of the walls of the tank with the rack.

A bottom drain on the plating tank is both a convenience and a nemesis. Since valves have a way of not being fully closed at times, it is practical to employ double valving for bottom drains. If an overflow sump is employed in the electroless plating tank, it too should have a bottom drain, and this drain should be the one used to control bath volume adjustments during operation, to correct for additional volume growth.

The overflow sump must be large enough so that it won't flood over when the plating rack is immersed into the plating section of the tank.

Heating, if required, is best done with hot water flowing through Teflon spaghetti coils. It is highly recommended that the coil or coils be placed in an external sump so that the rack of parts is not allowed to hit and possibly rupture a coil strand. A puncture in the coil will result in the tank of electroless copper becoming rapidly diluted with water. Coils should be periodically checked for leaks and leaking strands cut and tied off if necessary. Electric immersion heaters are difficult to use in overflow sump arrangements due to the inherent fluctuations of the liquid level in the sumps during normal operation.

The tank should also have an absolute minimum of horizontal shelf areas. Obviously a tank and sump bottom are a necessity, but other internal struts or boxes are to be avoided, since extraneous copper will settle and plate on these areas.

Racking

In addition to preferred construction materials, several other considerations are important:

• Enough space must be provided between circuit panels to avoid their touching one another and thereby interfering with the plating, and to allow for the efficient release of gas bubbles that are generated during plating. With the release of gas from holes being plated, if panels are too close to one another, a shadow image of an actual hole on one board can develop on the surface of an adjacent board. If panels are very long, they may have to be spot supported at a few points over their length, in addition to top and bottom, to prevent them from bowing in the plating solution and touching an adjacent board. In general, spacings of 0.3 to 0.5 inches, board center to board center, are usually adequate.

• The rack should be free of sharp edges that can either scratch parts being racked or unracked, or scratch the smooth surfaces of the plating tank itself.

• The rack should be of thin construction with respect to support rods, so as to minimize the contact area to the panel being plated—otherwise the rack can serve to mask portions of the panel being plated.

• Compromises must be made to minimize the overall surface area of the rack in general and yet have a rack that is structurally supportive, strong and sound. Rack surface area must be added to panel surface area when figuring bath loading and copper consumption during plating. The higher the rack surface area, the less efficient the plating operation.

• Racks should have no shelf or "cup" recesses that will allow them to act like ladles to carry solution from one tank to another, thereby increasing drag-out and increasing chances for contamination of process solutions or lowering the effectiveness of rinsing. Racks should be open enough in structure to allow efficient solution flow, or in the case of drying, air flow.

• The racks should be easy to load and unload with panels, so that a minimum of *handling* is required.

Handling
Panels should be handled by their edges, and gloves should be worn during racking and unracking after electroless copper. The reason for this is two-fold:

• Gloves will help protect the handler from getting glass slivers in his hand.

• Gloves will help prevent the handler leaving finger oils and skin-acid on the surface of the panels.

Panels should always be *dry* after electroless copper plating before they are handled at all. In any acid-fume environment, such as a plating area, boards will stain very easily by sitting in the environment. This staining will be accelerated if the panel is not dry. Electroless copper plated surfaces are very active chemically and *very* prone to staining.

Chemical Parameters of Electroless Copper
The chemical balance in any electroless copper formulation is very important. The bath is designed to function most optimally when the ratios of the ingredients to one another are close to the ratios existent when the bath is initially made up.

The main constituents to be monitored in any electroless copper solution are copper metal, formaldehyde and hydroxide (as sodium hydroxide, usually). Secondary to these main ingredients, but maybe requiring periodic attention would be the chelating agent (or complexor) and stabilizers.

Copper metal

Copper is consumed as the work being processed is plated, and requires frequent replenishment to keep the bath in chemical balance. The frequency and amount of replenishment required will depend upon the bath's deposition rate, as well as the workload (ft^2/gal) being processed. If the deposition rate is very slow, moderate additions will only be required from every 2 to 3 hours to once or twice in an 8-hour work shift. With high speed baths, replenishments will have to be made every 30 minutes or so, since the copper content is being rapidly depleted. With high speed baths where the rate consistency is more sensitive to chemical ratios, it is generally advisable to monitor in small additions of constituents on a continual or frequent intermittent-continual basis.

Formaldehyde

As additions to the electroless copper bath are made to replenish copper that is being depleted, it is common that formaldehyde is brought into the bath along with the copper ion, and usually the two ingredients are placed in the same replenisher concentrate. If the bath were to be worked *continually* at *optimum* workload for the particular formulation, no further additions of formaldehyde would normally be required. However, electroless copper baths are rarely operated 24 hours a day, day after day, and workloads do vary.

The Cannizzaro reaction, or reaction of sodium hydroxide and formaldehyde to form methanol and sodium formate, goes on continually in any bath at a somewhat constant rate for that particular bath, as long as the temperature is constant. The reaction is much faster in warm solutions than in room temperature solutions. The effects of the Cannizzaro reaction (formaldehyde and caustic loss) can be corrected for by designing replenisher concentrates that will add back enough formaldehyde to account for losses due to the plating reaction *and* the Cannizzaro reaction, but once workload is lower than expected and bath downtimes enter the picture, the formaldehyde loss will be higher than can be corrected for, based upon copper consumption alone.

Therefore, corrections will have to be made for the formaldehyde level independently. The frequency of correction may be as much as 2 to 3 times per day, where elevated temperature high speed baths are concerned. A formaldehyde correction will almost always be necessary at morning bath start-up with high speed baths, particularly elevated temperature baths.

When the formaldehyde content of an electroless copper bath is too low, plating initiation may be very slow or faulty, leading to voiding. When the formaldehyde content is too high, bath hyperactivity may result, leading to the predominance of side reactions and bath instability, along with a poor physical structure to the electroless copper deposit.

Hydroxide

Due to the fact that hydroxide is consumed not only in the main plating reaction, but also in the Cannizzaro reaction previously discussed, the same replenishment comments pertain to hydroxide as pertain to formaldehyde—that is, independent analyses will have to be made and the hydroxide levels corrected independently of corrections that are made through replenisher concentrates added to the bath at some ratio to the copper/formaldehyde replenisher concentrate(s). Carbon dioxide in the air also reacts with the sodium hydroxide in solution to form carbonate.

Hydroxide consumption increases when baths are run at elevated temperatures through the Cannizzaro reaction, and is higher for aerated baths due to the reaction with carbon dioxide in the air.

Sodium hydroxide additions to most baths are made with proprietary replenisher concentrates, which may or may not contain other ingredients that are rate-altering. Therefore, in some systems high hydroxide may produce high plating rates, while in other systems, higher hydroxide content will lessen the plating rate. No matter what system is used, it suffices to say that if the hydroxide content is not correct, problems with either bath hyperactivity and stability, or problems with plate initiation can be the result.

Chelating agents

Some electroless copper baths may require a periodic analysis and subsequent correction for chelating agent to insure keeping the copper ion soluble in the electroless copper bath, which is highly alkaline. Without enough chelator, copper ion will precipitate as copper hydroxide in alkaline solution and the bath will become extremely cloudy in appearance.

Stabilizers

Most present day electroless copper formulations contain so-called "stabilizer" materials to keep the bath from depositing copper randomly in solution. The types of materials used are such that they are thought to interfere with the side reactions promoting the formation of the cuprous ion. Cuprous ion can react to put extraneous copper into the plating bath.

Stabilizers are generally used in the 1 to 100 ppm concentration range. Too high a concentration usually will lower the plating rate and also may stop the bath from initiating deposition entirely. Typically, the types of materials used are strong chelators for palladium also, and it is thought that this is how they can interfere so dramatically with initiation.

Replenishment concentrates usually contain stabilizers. Periodic (usually once a day) additions of stabilizer are normally required, independent of other additives, to control the plating rate and provide bath stability during shutdown periods. This is particularly true with high speed plating baths.

Automatic controllers

Several very good automatic controllers (analyzers) are commercially available that will monitor electroless copper baths for copper, formaldehyde, and hydroxide content independently of one another and that will activate chemical metering addition pumps to replenish the bath for these components when additions are necessary, based upon preset low concentration points.

These devices provide for close chemical control and, consequently, optimum operation with respect to the attainment of consistent plating rates. They are particularly useful when high speed plating baths are being used (Fig. 13.20).

Some models contain strip-chart recorders that leave the user with a record of the parameters involved.

Post-Electroless Copper Considerations

A good deal of thought should be given to how the just-electroless-plated panels are processed directly after electroless deposition. Much of what is done here will have a major impact on how easy the subsequent imaging operation will be, or how the electroplated (either flash-plated or panel-plated) deposit will look.

When panels exit the electroless copper solution, they will be coated with a highly alkaline film of solution, which must be rinsed well. Unless completely and adequately rinsed before drying, an alkaline surface on the panel will result in rapid oxidation of the copper in moist air. This heavy oxide coating will run a spectrum of colors—typically purple, blue, red, and brown. Later in processing, when the oxide is removed, the imprint of the oxide will be visible, even after electroplating, since an oxided surface area is one that has been attacked chemically and has a different surface structure, even after oxide removal, than surrounding areas that have not developed an oxide coating.

For this reason, it is usual after the electroless copper rinse step to immerse the plated panels into a mild acid solution, typically a 1 to 5 percent sulfuric acid solution, or a mild organic acid solution such as citric acid or malic acid, or a proprietary mild organic acid solution designed to be very free rinsing.

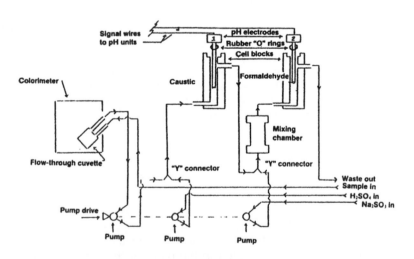

Fig. 13.20—Typical electroless copper controller plumbing.

The important consideration, if an acid dip is employed, is to *rinse* very well, so that panels do not end up with an active acid surface prior to drying. Sulfuric acid solutions are difficult to rinse satisfactorily, and therefore the popularity of the milder acid dips has gained in recent years.

If an acidic surface film is allowed to remain on the panel, the panel will oxidize over time to give a brownish oxide surface.

Once the surface has been properly neutralized, the panels should be dried quickly so that the wet or moist surface doesn't become exposed to the acid or alkaline fume-laden shop air, which defeats the purpose of the neutralization process. A dryer tank in the electroless copper line is highly recommended, where within 5 to 10 minutes, panels can be forced-hot-air dried.

A recommended practice is to immerse panels into an anti-tarnish solution prior to drying. Typically, these solutions leave a mono-molecular film on the copper surface that is somewhat water-repellent or aids in water shedding, thus allowing the panels to dry quickly and without oxidation. The residual anti-tarnish film allows panels to sit in shop air for longer periods without tarnishing, and has been shown to actually increase the adhesion of subsequently-applied inks and dry- film photoresists—in some cases eliminating the need for panel scrubbing prior to primary image plating resist application. Care must be taken to use a *very dilute* anti-tarnish solution. It is usually preferable to operate these solutions at 120 to 140° F to aid in the drying process.

Summary
Suffice it to say that there are many critical parameters to be aware of in electroless copper processing. Once routine monitoring procedures are established, however, operation of the electroless copper line becomes relatively simple, and work can be routinely processed, yielding consistent high quality.

ADDITIVE PROCESSING

The distinction between additive and subtractive processing centers about the fact that in additive processing, no *outer-layer* copper foil exists on the laminate outer surfaces prior to processing. There may be inner-layer copper, as in the case of a multilayer laminated package, but no copper on the outer surface.

One of the differences in the through-hole plating process is, then, to prepare the non-copper, or polymer, surfaces to be receptive to electroless copper deposition and adhesion. In the case of a simple double-sided package, the panel may be treated as a plating-on-plastic exercise, but in the case of multilayers, one must also contend with the hole-interfacing inner-layer copper for desmearing operations and plating pretreatment to insure good copper-to-copper bonding between this inner-layer copper and the electroless copper deposition.

The basic substrate may consist of one of the following types:

• A treatable polymer—cast, laminated, or molded—that can be prepared to accept activation and, ultimately, electroless copper.

• A polymer composite material, such as epoxy/glass, epoxy/paper, polyester/glass, etc., covered with a polymeric coating that can be pretreated prior to the through-hole plating process in order to get acceptable electroless copper coverage and adhesion.

• An epoxy/glass substrate, or epoxy/paper substrate that is "resin-rich" at the surfaces, with a layer or "butter-coat" of epoxy to insure some minimal thickness of epoxy resin on the surface before one encounters glass or paper. In this way, the epoxy itself may be treated to get adhesion promotion for complete covering and adherent electroless copper.

The concept used in preparing non-copper-clad surfaces for electroless copper deposition is *site* development. A pocketed surface must be produced on which catalyst can be anchored, or, if pre- catalyzed polymer is used, to expose catalyst. The ultimate goal in both cases is to finally anchor electroless copper to the polymer surface.

Several options are available:

• The molded polymer, the adhesive used to coat the substrate, or the entire laminated package may be pre-impregnated with electroless copper-catalytic material. The job, then, would be to open the surfaces to expose catalyst and produce adhesion pockets, and to "develop" or "activate" the catalyst prior to electroless copper immersion. Several commercial techniques are available to accomplish these prerequisites—the choice is dependent on the polymers in question—but, typically, strong oxidizers such as chromic/sulfuric acid mixtures, chromic acid alone, or potassium permanganate solutions are employed.

• In the case of laminate substrate coated with an "adhesive" or plateable coating that is non-catalyzed, the panel is simply pre-treated to promote receptive pockets for catalyst and electroless copper deposition, and then treated with activator, post-activator, and electroless copper. Here again, strong oxidizing solutions are generally used to etch the polymer coating. In most instances, these solutions will also serve to desmear the drilled hole in the case of a multilayer composite.

• In the case of epoxy-butter-coated or epoxy-rich-surface epoxy/glass laminate composites, it is usual to first "swell" the polymer with an organic solvent to make it receptive to oxidative "etching" before entering the oxidizing solution. Other polymers may be "conditioned" in this manner as well. This technique is often referred to as "swell and etch".

• A metal foil that has been anodized or oxidized on one side may be laminated to the polymer surface prior to the start of board manufacture. The pre-treatment process prior to catalyzation, then, would involve etching away the metal from the outer surfaces, thereby leaving pockets in the underlying polymer. Aluminum has been successfully employed for this purpose.

Where oxidants are employed, a step generally follows in the production sequence *prior* to activation, to neutralize, reduce, or otherwise remove all oxidizers from the panel.

In all of these techniques, one *very* important factor is that the polymer coating where electroless deposition is to occur and adhesion is to develop, must be of *consistent* satisfactory thickness to ensure that catalyst and electroless copper come into contact with this polymer coating *only*, and not with other materials, such as glass-fiber, paper, or underlying nontreatable polymer—otherwise, adhesion will be lost at these points of contact, eventually resulting in blistered plating. The coated polymer integrity must be such to ensure that the polymer pretreatment does not take away too much of it, either. These concerns relate primarily to surfaces, and not usually to holes in the panel, since adhesion in the holes is governed by the same mechanism that is involved in subtractive processes. Usually the holes are rough already, and the plating is somewhat "locked" into the hole because it is completely plated through, with no edge to be undercut, as in the case of circuitry lines on board surfaces.

In almost every method employed, at some step in the manufacturing process *after* electroless copper deposition, the panel will be heated or baked to lock in the electroless copper to the polymer. This is generally how "ultimate" adhesion is obtained.

Additive processing may be grouped into two general categories: semi-additive and fully additive.

Semi-Additive Processing

In this mode of processing, the panel is prepared to create surface polymer adhesion sites, neutralized or washed free of any oxidizing agents used, then activated (if the composite is not pre-activated), and immersed in electroless copper solution, so that the entire panel becomes covered with electroless copper. Usual thicknesses in semi-additive processing are 20 to 100 millionths of an inch of electroless copper. Panels may be flash-electroplated at this point if desired. Otherwise, the panel may now be treated as though it were copper-clad to begin with, and a plated-through-hole operation has simply been performed. Very little change in the overall mode of operation is required for a production facility to change from standard subtractive processing to semi-additive processing.

The benefit of processing in this fashion is that after final pattern plating, very little final etching is required to produce final circuitry. In fact, a solder overplate to circuitry as a final etch resist may not be necessary, since only 20 to 100 millionths in. of electroless has to be etched away as background copper, and this thickness can usually be sacrificed on the circuitry traces. This feature becomes most attractive when one is producing all copper circuits for solder-mask-over-bare-copper applications, whether or not hot air leveling or hot-oil leveling will be used.

Another feature of not having to use tin or tin-lead as a final etch resist overplate is that one is free to choose a wider range of final etchant chemistries, notably, cupric chloride or ferric chloride—these choices not possible when tin or tin-lead is used as a final etch resist.

An obvious advantage to eliminating tin-lead plating in particular is the elimination of the tin-lead stripping operation for SMOBC or for finger plating with nickel/gold. Also eliminated is a hazardous waste (lead) that would otherwise have to be dealt with from spent tin-lead stripping solutions.

The types of electroless copper plating solutions used in semi-additive processing are very similar to those used for standard subtractive processing, with one exception—the initiation of plating speed is generally inhibited to prevent too *rapid* a start of plating. This is done to ensure a low-porosity deposit. If the initiation is too rapid, spongy, high-porosity deposits will be obtained. A standard subtractive-processing electroless copper bath will usually initiate much faster on a polymeric coating because the polymer usually has a much higher density of palladium present, due to its surface structure being more highly "pocketed" than an etched copper-clad surface.

Fully-Additive Processing

Fully-additive processing differs from semi-additive processing in the fact that only circuit traces and holes will receive electroless copper—not the entire panel. Thus, all the plating must be done electrolessly, since there will be no underlying conductive path to allow electrolytic plating.

The process steps through activation are the same as for semi-additive processing. *Prior* to electroless copper immersion, however, those areas of the surface where copper is not wanted are masked with a plating resist.

The plating resist must be of a special type, since in fully-additive electroless copper plating, immersion times are very long (8 to 14 hours for 1 mil deposition) to achieve desired circuit thicknesses, and panels are immersed in a highly alkaline medium (the electroless copper) at elevated temperatures (usually 120 to 140° F). The resist must be very tenacious, and thus, will be hard to strip off after use. As a matter of fact, there is no reason to strip it off the panel at all, since no later processing step requires its removal. In most cases, then, the resist becomes an integral part of the final board. Because of this, however, the resist should possess electrical insulation characteristics and flammability requirements that the final product is intended to meet.

Once the resist is applied, it is usual to post-activate the panel to be plated, prior to electroless copper deposition.

The electroless copper formulations used for fully-additive processing are usually quite different from those formulations used in standard subtractive processing. The entire physical properties of the copper circuitry are now dependent on those of the electroless copper alone. Much research has been done to develop electroless coppers that will have physical properties adequate to meet the standards set by the printed circuit industry. Electroless coppers are available commercially that are adequate to this task.

Table 13.2 shows the minimum physical property values of electroless copper deposition that an electroless copper bath of this type is required to produce.

Generally, electroless copper densities are slightly lower than those of electrolytically deposited copper. This is probably a result of the slightly more

Table 13.2
Minimum Physical Properties
Of Fully-Additive Electroless Copper Deposits

Property	Units	Electroless copper*	Electrolytic copper**
Purity	%	99.2% minimum	99.9%
Density	g/cm³	8.8 ±0.1	8.92
Resistivity (volume)	μohm-cm at 20° C	1.90 maximum	1.72
Tensile strength	lb/in.²	30,000 maximum	44,000
Percent elongation	%	3.0 minimum	10*

*According to IPC specification IPC-AM-372.
**Typical high-throw copper sulfate with brighteners.

open structure of the electroless deposit. In general, electroless copper deposits will etch somewhat faster than electrolytically deposited copper on an equal thickness basis.

Fully-additive processing is satisfactory to produce boards of many types, but there are some drawbacks when certain auxiliary electroplating operations are desired.

Table 13.3 serves to illustrate some of the differences between what is achievable through standard subtractive processing, semi-additive processing, and fully-additive processing.

A big advantage of using fully-additive processing is that *no* final etching is required, making this process ideally suited for fine line and close spacing board production. Also, since no electroplating is required, one need not worry about the skewness of a circuit pattern on the panel to be plated; therefore, circuitry of uniform height will be produced with fully-additive processing. This uniformity is not easily attainable through *pattern* electroplating if there is any degree of skewness in the board. One must electrolytically panel-plate to full thickness if one is to have any chance of obtaining uniform-height circuitry when skewed patterns are involved, usually as a result of uneven current density distribution across the panel.

The more important processing step differences between standard subtractive, semi-additive and fully-additive processes in the manufacture of a bare copper board (for solder mask over bare copper) are illustrated in Table 13.4.

Finally, control of the electroless copper bath is critically important in fully-additive processing. It is important to monitor the consumption of all the critical bath ingredients and to keep the plating rate steady throughout the long immersion times. This control is needed to insure that the correct physical properties are obtained. While physical properties are important with the thin electroless deposits used for standard subtractive and semi-additive processing, these deposits contribute only to a minor extent to the *overall* physical properties of the circuitry produced, since more than 90 percent of the circuitry thickness in these instances is produced with electrolytic copper, which is the major physical property contributor. In the case of fully-additive processing, if the electroless copper goes out of control in any way, the results can be disastrous. Automatic controllers are *highly* recommended in any additive circuitry processing for electroless copper maintenance.

Table 13.3
Comparison of Subtractive, Semi-Additive, And Fully-Additive Electroless Copper Processes

Desired operation, achievement or feature	Subtractive	Fully-additive	Semi-additive
Solder/tin circuitry	Standard	Thin only (immersion)	Standard
Nickel/gold on tabs	Standard	Extremely difficult (bussing required)	Standard
Physical properties (thermal shock)	Standard	High degree of electroless copper control necessary	Close to standard
Plating resist	Standard	Special	Standard
Laminate	Standard	Special	Special
Process control	Standard	Critical	Close to standard
All copper circuits	Much etching and stripping required	Easy	Relatively easy
Etching costs	25 to 100% depending on foil thickness	0%	5 to 10% depending on electroless thickness
Drill life	Standard	Increased	Increased

Table 13.4
Comparison of Process Steps

Step	Subtractive	Semi-Additive	Fully-Additive*
1.	Cut panels	Cut panels	Cut panels
2.	Drill holes	Drill holes	Drill holes
3.	Deburr	Adhesion promote	Adhesion promote
4.	Clean	Water rinse	Water rinse
5.	Water rinse	Neutralize	Neutralize
6.	Microetch	Water rinse	Water rinse
7.	Water rinse	Pre-activate	Pre-activate
8.	Acid dip	Activate	Activate
9.	Water rinse	Water rinse	Water rinse
10.	Pre-activate	Post-activate	Dry
11.	Activate	Water rinse	Apply electroless resist
12.	Water rinse	Electroless copper	Post-activate
13.	Post-activate	Rinse, neutralize	Water rinse
14.	Water rinse	Dry	Electroless copper
15.	Electroless copper	Apply electroplating resist	Rinse, neutralize
16.	Rinse, neutralize	Clean	Dry
17.	Dry	Water rinse	Apply solder mask
18.	Apply electroplating resist	Microetch	
19.	Clean	Water rinse	
20.	Water rinse	Acid dip	
21.	Microetch	Electrolytic copper	
22.	Water rinse	Water rinse	
23.	Acid dip	Dry	
24.	Electrolytic copper	Strip resist	
25.	Water rinse	Water rinse	
26.	Acid dip	Final etch (FAST)	
27.	Water rinse	Rinse	
28.	Tin-lead or tin plate	Dry	
29.	Water rinse	Apply solder mask	
30.	Dry		
31.	Strip resist		
32.	Water rinse		
33.	Final etch		
34.	Rinse		
35.	Dry		
36.	Solder or tin strip		
37.	Apply solder mask		

ALTERNATIVE PROCESSES

Electroless copper is used in many other processes that have commercial viability in the production of printed wiring boards, using substrates in some instances that deviate from the typical laminates discussed so far in this chapter.

Processes exist for bonding electroless copper to ceramic substrates, thermoplastic materials (permitting the molding and circuitizing of boards *molded* into any three-dimensional shape), and for metallizing holes only, using print-and-etch or other schemes to produce discrete wiring patterns on board surfaces.

Generally speaking, the alternative processes use electroless copper solutions that operate similarly to the types of electroless copper discussed in this chapter. The differences in processing relate to surface preparation of the substrate involved and the imaging/activation techniques employed. With the trend toward finer lines and spacings on printed wiring boards, several processes have been developed that employ no classical *resist* materials to define circuitry for plating, but rather utilize *printing* (photoimaging) of an activated substrate to either activate or deactivate selected areas prior to electroless plating. In this fashion, the line definitions obtainable in the process are not subject to any limitations inherent to the plating resist itself.

Many interesting alternative processes, either using resist or photoactivation (resistless) techniques, have been developed, and references are given in the bibliography of this chapter for further investigation by the reader.

Much research has been done in recent years, and is continuing, on the use of activation systems for electroless copper that obviate the use of classical palladium activation, or for that matter, the use of any metallic-type activation at all.

SUMMARY

The purpose of this chapter has been to acquaint the reader with general considerations about the use of electroless copper deposition in the manufacture of printed wiring boards. For further reading, the Bibliography references many texts and articles that have appeared in the literature, which the author feels will help to provide the reader with further depth into topics that may be of interest and could only be briefly mentioned in this chapter.

APPENDIX

The reactions pertinent to electroless copper deposition are as follows:

$$Cu(II)(chel) + HCHO + 3OH^- \rightarrow Cu^0 + HCOO^- + 2H_2O + (chel)^= \qquad [1]$$

$HCHO + OH^- \xrightarrow{(act)} HCOO^- + H_2$ [2]

$Cu(II)(chel) + 2HCHO + 4OH^- \rightarrow Cu^0 + 2HCOO^- + (chel)^= + 2H_2O + H_2$ [3]

$2HCHO + OH^- \rightarrow CH_3OH + HCOO^-$ [4]
(The Cannizzaro Reaction)

$2Cu(II)(chel) + HCHO + 5OH^- \rightarrow Cu_2O + HCOO^- + 3H_2O + 2(chel)^=$ [5]

$Cu_2O + H_2O + (chel)^= \rightarrow Cu^0 + Cu(II)(chel) + 2OH^-$ [6]

$Cu(II)(chel) + HCOO^- + 3OH^- \rightarrow Cu^0 + (chel)^= + CO_3^= + 2H_2O$ [7]

$2OH^- + CO_2 \rightarrow CO_3^= + H_2O$ [8]
(Reaction with carbon dioxide in air)

Combining Eqs. 1 and 2, the overall reaction is expressed in Eq. 3:

$Cu(II)(chel) + 2HCHO + 4OH^- \rightarrow Cu^0 + 2HCOO^- + (chel)^= + 2H_2O + H_2$ [3]

After the reaction is initiated with catalyst, the reaction becomes autocatalytic as a result of the presence of hydrogen gas in solution. As long as H_2 gas is being produced, reduction of more copper on freshly deposited copper surfaces continues. Reaction 3 can be made more efficient with respect to Reaction 1 by the inclusion of suitable additives to the electroless bath. This will result in less hydrogen evolution per unit of copper metal deposited and, in general, leads to better deposit characteristics.

Some notable side reactions that occur are as follows:

$2HCHO + OH^- \rightarrow CH_3OH + HCOO^-$ [4]
(The Cannizzaro Reaction)

This reaction tendency increases with the alkalinity of the bath. While it can be retarded to some degree, it is difficult to stop entirely.

$2Cu(II)(chel) + HCHO + 5OH^- \rightarrow Cu_2O + HCOO^- + 3H_2O^2 + 2(chel)^=$ [5]

Particular attention has been paid to this reaction, since Cu_2O is only sparingly soluble, and early formulations contained no specific complexing agent (or "stabilizer") for Cu in the +1 valance state. Any Cu^{+1} in the bath will not be as readily reduced to Cu^0 via the catalytic reduction mechanism (Reaction 3) as will Cu^{+2}, but has a marked tendency to undergo spontaneous disproportionation to Cu^0 and Cu^{+2}, to produce random copper particle formulation throughout the solution. This tendency is shown in Reaction 6:

$Cu_2O + H_2O + (chel)^= \rightarrow Cu^0 + Cu(II)(chel) + 2OH^-$ [6]

A sizeable volume of research has been done to find ways of mitigating the effects of Cu(I) in electroless copper baths by either prevention of Cu_2O formation, inactivation of the Cu^{+1} so formed, or retarding the disproportionation reaction.

The formate ion formed in many of the reactions presented thus far is, in itself, a good reducing agent in alkaline media, because of its aldehyde "character" and may react as such in the electroless copper bath.

$$Cu(II)(chel) + HCOO^- + 3OH^- \rightarrow Cu^0 + (chel)^= + CO_3^= + 2H_2O \qquad [7]$$

Formate will not, however, catalytically or spontaneously reduce Cu(II) to Cu^0, unless some formaldehyde is also present. When formate is present in considerable amount, the tendency is to promote the occurrence of Reaction 5, with resultant loss in solution stability.

$$2OH^- + CO_2 \rightarrow CO_3^= + H_2O \qquad [8]$$

This reaction occurs as a result of bath aeration and is most noticeable when baths are not utilized for long periods of time, yet continue to be aerated and replenished for sodium hydroxide. High carbonate levels will ultimately yield the electroless copper bath totally inactive. As the carbonate level rises, the alkalinity level of the bath by analysis will be progressively erroneous, because some contribution to the alkalinity reading will be made by the Na_2CO_3. The value obtained, then, is not all NaOH.

BIBLIOGRAPHY

General Texts
1. Raymond H. Clark, *Handbook of Printed Circuit Manufacturing,* Van Nostrand Reinhold, New York, 1984.
2. Clyde F. Coombs Jr., *Printed Circuits Handbook,* Second Edition, McGraw Hill, New York, 1979.
3. Norman S. Einarson, *Printed Circuit Technology,* Printed Circuit Technology, Burlington, MA, 1977.
4. Theresa Kiko, *Printed Circuit Board Basics,* PMS Industries, Alpharetta, GA, 1984.
5. H.R. Shemilt, *Printed Circuit Troubleshooting,* Electrochemical Publications Ltd., Ayr, Scotland.
6. J.F. Walker, *Formaldehyde,* Third Edition, Reinhold Publishing Corporation, New York, 1964.

Specifications
1. IPC-A-600C, *Acceptability of Printed Boards,* Institute for Interconnecting and Packaging Electronic Circuits, Evanston, IL, 1979.
2. IPC-AM-372, *Electroless Copper Film for Additive Printed Boards,* 1978.

Selected Papers and Articles

A. Cleaning and Pretreatment

1. B. Arden, "Chromic Acid Desmear and Etchback Guidelines," *Printed Circuit Fabrication,* **7** (Mar. 1984).
2. P.E. Bell, "Surface Preparation for Copper," ibid., **6** (Jun. 1983).
3. R.L. Cohen and R.L. Meek, "Role of Rinsing in Palladium—Tin Colloid Sensitizing Processes, II, An Improved Processing Sequence," *Plat. and Surf. Fin.,* **63** (Jun. 1976).
4. C.I. Courduvelis, "Method for Selective Removal of Copper Contaminants from Activator Solutions Containing Palladium and Tin," U.S. patent 4,304,646, 1981; assigned to Enthone, U.S.A.
5. H.J. Ehrich, "The Application of Alkaline Activators to Complicated PCBs," *Printed Circuit Fabrication,* **7** (Jul. 1984).
6. W.T. Eveleth and L. Mayer, "Plated Through Hole Technology," ibid., **8** (Aug. 1985).
7. B. Jackson et al., "Through Hole Plating Technology," ibid., **8** (Mar. 1985).
8. T.D. Loy, "Chemical Cleaning: Processing and Practices, Part I," ibid., **9** (Nov. 1986).
9. T.D. Loy, "Chemical Cleaning: Processing and Practices, Part II," ibid., **9** (Dec. 1986).
10. G. Malthaes, "The Techniques of Surface Cleaning Professional PCBs and Multilayers," ibid., **4** (Dec. 1981).
11. C.H. de Minjer and P.F. v.d. Boom, "The Nucleation with $SnCl_2$-$PdCl_2$ Solutions of Glass Before Electroless Plating," *J. Electrochem. Soc.,* **120** (Dec. 1973).
12. F.E. Stone, "Preplating Steps are Critical when Working with Electroless Copper," *Insulation Circuits,* **22** (Feb. 1976).
13. F.E. Stone, "Preplating Steps are Critical in Printed Circuit Board Processing," *Institute of Printed Circuits Technical Paper TP 100,* April 1976.
14. F.E. Stone, "Peelers—An Old Gremlin," *Products Finishing* Mar. 1981).
15. D. Swalheim, *Rinsing, Recycle and Recovery of Plating Effluents,* AESF Illustrated Lecture.
16. Y.J. Wang and C.C. Wen, "The Kinetics of $PdCl_2$/$SnCl_2$ Activating Solutions for Electroless Plating," *Plat. and Surf. Fin.,* **69** (Aug. 1982).

B. Electroless Copper

1. W.M. Beckenbaugh and K.L. Morton, "Method of Depositing a Stress-Free Electroless Copper Deposit," U.S. patent 4,228,213, 1980; assigned to Western Electric, U.S.A.
2. K.F. Blurton, "High Quality Copper Deposited from Electroless Baths," *Plat. and Surf. Fin.,* **73** (Jan. 1986).
3. C.B. Castner, "Controller for Electroless Copper," *Products Finishing,* **46** (Jul. 1982).
4. C.I. Courduvelis and G. Sutcliffe, "Accumulation of Byproducts in Electroless Copper Plating Solutions," *Plat. and Surf. Fin.,* **67** (Sep. 1980).

5. A.T. El-Mallah et al., "Some Aspects of the Structure of the Electroless Copper Deposit," ibid., **67** (Oct. 1980).
6. J.J. Grunwald et al., "Some Physical Properties of Electroless Copper (II)," *Plating,* **60** (Oct. 1973).
7. G. Herrmann, "Measurement of the Deposition Rate in Electroless and Galvanic Plating Baths," *Galvanotechnik,* **67** (Apr. 1977).
8. R.E. Horn, "Continuous Regeneration of an Electroless Copper Bath," *Plat. and Surf. Fin.,* **68** (Oct. 1981).
9. T.S. Krishnaram et al., "The Role of Stabilizers in Electroless Copper Baths," *Printed Circuit Fabrication,* **5** (Feb. 1982).
10. R.M. Lukes, "The Chemistry of the Autocatalytic Reduction of Copper by Alkaline Formaldehyde," *Plating,* **51** (Nov. 1964).
11. S. Mizumota et al., "Mechanical Properties of Copper Deposits from Electroless Plating Baths Containing Glycine," *Plat. and Surf. Fin.,* **73** (Dec. 1986).
12. K.J. Murski and D.F. DiMargo, "A Method of Measuring Electroless Copper Plating Thickness," *Printed Circuit Fabrication,* **6** (Mar. 1983).
13. R.A. Nesbitt and C. Courduvelis, "Chelator Effect on Deposition Rate of Electroless Copper on Various Surfaces," *Proc. AES 8th Symp. on Plating in the Electronics Industry,* 1981.
14. R.A. Nesbitt, "Selecting the Electroless Copper Bath for Your Printed Circuit Shop," *Proc. 1st AES Electroless Plating Symp.,* 1982.
15. F.J. Nuzzi, "Accelerating the Rate of Electroless Copper Plating," *Plat. and Surf. Fin.,* **70** (Jan. 1983).
16. P&SF Report, "Developments in Electroless Plating," ibid., **71** (Jul. 1984).
17. M. Paunovic, "Electrochemical Aspects of Electroless Deposition of Metals," *Plating,* **65** (Nov. 1968).
18. M. Paunovic and R. Zeblisky, "Properties and Structure of Electroless Copper," *Plat. and Surf. Fin.,* **72** (Feb. 1985).
19. F. Polakovic, "Contaminants and Their Effect on the Electroless Copper Process," *Printed Circuit Fabrication,* **8** (Apr. 1985).
20. E.B. Saubestre, "Electroless Copper Plating," *Proc. AES Annual Technical Conf.,* 1959.
21. E.B. Saubestre, "Stabilizing Electroless Copper Solutions," *Plating,* **59** (Jun. 1972).
22. C.R. Shipley Jr., "Historical Highlights of Electroless Plating," *Plat. and Surf. Fin.,* **71** (Jun. 1984).
23. F.E. Stone, *Electroless Copper,* AESF Illustrated Lecture.
24. D. Vitkavage and M. Paunovic, "Maximum Rate of the Cathodic Reaction in Electroless Copper Deposition," *Plat. and Surf. Fin.,* **70** (Apr. 1983).

C. Additive Circuitry

1. B. Barclay, "Semi-Additive Processing of Multilayer Circuits," *Circuits Manufacturing,* (May 1985).

2. O.R. Cundall and J. Quintana, "Fully Additive Processes for Fabrication of Printed Circuit Boards," U.S. Army Missile Command, Redstone Arsenal (Mar. 1982).
3. D.J. Esposito, "New Ideas in Electroless Copper," *Printed Circuit Fabrication,* **9** (Aug. 1986).
4. G. Herrmann, "Fully Additive Technique for the Production of Printed Circuit Boards," *Metalloberflaeche* (Nov. 1977).
5. J. Mettler, "Additive Printed Circuits," AESF Merrimack Valley Branch 3rd Annual Printed Circuit Workshop, 1974.
6. R.E. Smith Jr., "Process Options for Fine Line Circuitry," *Printed Circuit Fabrication,* **6** (Aug. 1983).
7. F.E. Stone, "Advances in Electroless Copper Technology: Semi-Additive Processing—The Right Time for an Old Idea," *Proc. Printed Circuit World Expo,* Lake Publishing Corp., 1980.

D. Alternative Techniques

1. J.F. D'Amico et al., "Selective Electroless Metal Deposition Using Patterned Photo-Oxidation of Sn(II) Sensitized Substrates," *J. Electrochem. Soc.,* **118** (Oct. 1971).
2. J.F. Dennis-Browne, "Circuit-Board Technology," *Printed Circuit Fabrication,* **8** (Jul. 1985).
3. D.A. Luke, "Through-Hole Plating Without a Palladium Catalyst," *Trans. Institute of Metal Finishing,* **60** (Autumn 1982).
4. K. Murakami et al., "Process for Forming Printed Wiring by Electroless Deposition," U.S. patent 4,239,813, 1980; assigned to Hitachi, Japan.
5. "Philips Says Yes to Resistless Etchless Printed Circuit Boards," *Electronics,* **51**(9) (Sep. 1977).
6. M.E. Pole-Baker, "Printed Circuits: Origins and Development, Part II," *Printed Circuit Fabrication,* **8** (Jan. 1985).

E. General

1. G.L. Fisher et al., "Measuring PTH Copper Ductility, Part I," *Printed Circuit Fabrication,* **9** (Oct. 1986).
2. G.L. Fisher et al., "Measuring PTH Copper Ductility, Part II," ibid., **9** (Nov. 1986).
3. M.E. Pole-Baker, "Printed Circuits: Origins and Development, Part III," ibid., **8** (Feb. 1985).
4. A. Porter, "In Plating Printed Circuits, Automata Lives Up to Its Name," *Products Finishing,* **71** (Jun. 1984).
5. F.E. Stone, "How Much Space Does a Hole Take Up?" *Printed Circuit Fabrication,* **5** (Jun. 1982).

Chapter 14
Plating on Plastics

John J. Kuzmik

The plating of non-conductors has been achieved for many years. Articles plated were mainly decorative, and adhesion of the plate to the substrate was minimal. In the early 1960s, due to technological advances in chemical processing techniques, plating on plastics began on a commercial level. Industries that utilize plated plastics include the automotive, plumbing, appliance and electronics.

One of the early plastics to be plated on a large scale was polypropylene, which gave adhesion values of over 20 pounds per linear inch. Even though adhesion was excellent, other problems occurred that ultimately led to the demise of plating polypropylene. These problems included (1) failure to pass thermal cycling due to its high coefficient of linear thermal expansion (i.e., the amount in inches per inch the part shrinks in the mold—68 to 95 x 10^{-6} in./in./° C for polypropylene); (2) brittleness after plating due to notch sensitivity (a phenomenon whereby a crack in the plate will propagate through a normally flexible plastic, causing it to break easily); and (3) "sink" marks caused by the plastic shrinking in the mold, especially over bosses and ribs and showing up as dimples or "sinks" after molding.

Some of the reasons various industries are interested in plating plastics include (1):

- Lower cost
- No secondary operations (i.e., no deflashing or buffing)
- Design freedom (i.e., the ability to mold large and complex parts)
- Weight reduction

Weight reduction became a very real benefit to the automotive industry when the gasoline crunch occurred in the early 1970s.

Many plastics are plated today, including:

- ABS
- Polypropylene
- Polysulfone
- Polyethersulfone
- Polyetherimide
- Teflon

- Polyarylether
- Polycarbonate
- Polyphenylene oxide (modified)
- Polyacetal
- Urea formaldehyde
- Diallyl phthalate
- Mineral-reinforced nylon (MRN)
- Phenolic

Some typical applications of plated plastics are shown in Figs. 14.1 and 14.2.

Several of the above-mentioned resins have little or no adhesion and/or unacceptable appearance. Of the resins listed, ABS, acrylonitrile-butadiene-styrene, has found the widest acceptance in the plating industry. ABS is a terpolymer thermoplastic that has an acrylonitrile-styrene matrix with butadiene rubber uniformly distributed in it. This quality makes it unique for plating, as the butadiene can be selectively etched out of the matrix (Fig. 14.3), leaving microscopic holes that are used as bonding sites by the electroless plate. Other factors influencing the choice are:

- Low cost
- Low coefficient of thermal expansion
- Ease of molding
- Good metal adhesion to the substrate
- Good appearance after plate

Because ABS is the most widely plated plastic, it will be the material mainly referred to in this chapter. Scanning electron micrographs of ABS surfaces are shown in Fig. 14.4.

Quality plating on plastics involves the following three basic steps:

- *Molding*—converting plastic pellets into the desired part.
- *Preplate*—processing the molded part through an electroless bath in order to render it conductive.
- *Electroplating*—building additional metal thickness using current.

Because of the importance of molding, which must be done properly to insure a part with the high quality necessary for plating, it will be advantageous to mention some general guidelines that should be followed. Molding consists of transforming the plastic pellets into the shape (part) to be plated. A mold of the part must be built and in doing so, certain design features must be taken into account for the finished part. These include (2):

- Gates (fill areas) should be put in a non-appearance area.
- Integral parts should be used to avoid welded joints.
- Ribs and bosses should be designed to eliminate "sink" marks.
- Texturing can be used to break up large flat surfaces and hide any defects, such as scratches.

Fig. 14.1—Typical applications of plated plastic parts.

Fig. 14.2—More applications of plated plastic parts.

- Draft angles should be at least one degree (1°) for easy removal of the part from the mold.
- Parting lines (where the mold opens) should be put in a non-significant area if possible.
- Close tolerance fits must include the final plate in the initial part design.
- Wall thickness should be sufficient to insure rigidity.
- Plate uniformity, which results from current density distribution, must be considered in the initial design. Use no 90° angles, no V grooves, keep letters close to the surface, make angles as large as possible, and crown large flat surfaces (Figs. 14.5 and 14.6).

Butadiene

Styrene acrylonitrile

Fig. 14.3—Etching of ABS.

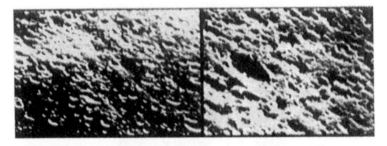

Fig. 14.4—SEM photomicrographs of ABS: (left) as molded; (right) after etching.

Some other design considerations are: location of drainage holes, no blind holes, and rack contact areas on the part.

If the above features are taken into account when designing a mold, a part suitable for plating should be produced. These are not all the considerations, but some that must be looked at.

Certain molding parameters should also be followed to insure good parts. Some of these are (2):

• *Highly polished mold*—Poor mold surfaces can cause defects in the molded part such as pits, which show up in the final plate and cause rejects.

• *Proper drying of resin*—Moisture in the resin can cause "splay" or delamination on the part, which may result in a blister.

• *Proper melt temperature*—If the melt temperature is too low, stress can be incorporated in the part, which could cause uneven etching and possible thermal cycle failure. Consult the resin manufacturer for the proper parameters.

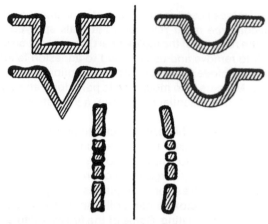

Fig. 14.5—Design considerations for plating thickness uniformity: (left) 90° angles and V-grooves result in poor current distribution; (right) suggested design.

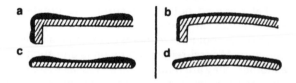

Fig. 14.6—Sharp corners (a) result in poor plating uniformity; (b) rounded corners, both internal and external, improve coverage. Expansive surfaces (c) should be crowned for a convex surface as in (d).

- *Proper mold temperature*—If the mold is too cold, "skinning" can occur, that is, the first material to hit the mold hardens instantly and the hot material under it flows, causing a surface skin that may cause delamination.
- *Proper fill speed*—A fast fill speed can overpack a mold, thus making it harder to etch, resulting in adhesion loss. Best results are obtained using a slow fill speed.

These are some of the more obvious parameters to be considered when molding a part to be plated. The mold design and molding parameters are extremely important to insure good parts that are free of stress and other imperfections. Most plastics can be evaluated for stress. In polycarbonate and other clear plastics, stress can be detected using polarized light. ABS is immersed in glacial acetic acid for one minute, rinsed and dried. Any peeling indicates delamination, white areas depict stressed areas, which usually etch differently than the rest of the part, causing adhesion and thermal cycle problems. Cracks or actual breakage of parts indicate severe stress and should not be plated at all.

The type of mold release used can have a profound effect on the molded part. For example, silicon-type mold releases (materials sprayed on the molded surface to help in the removal of the molded part) should not be used, as they are extremely difficult to remove and usually lead to adhesion failures (blisters) or misplate because they hindered the processing solutions from performing their functions properly. Molders who mold plastic parts for plating generally do not use a mold release. If it is essential to use one, a stearate or soap type may be used sparingly.

The addition of fillers (4) to the resin can create problems when plating of the parts is involved. In some cases fillers are added to the resins to increase strength (i.e., glass fibers), to impart color (i.e., carbon black or titanium dioxide), or for fire retardancy (organic phosphates, antimony oxide). In any case, problems can arise. Glass, for example, usually gives a rough surface to the finished part. Titanium dioxide can build up in the processing tanks and cause a rough surface after plating. Some of the fire retardants and other fillers can leave a non-adherent film on the surface that results in no adhesion after plating. If the film can be removed prior to catalyst, good results are usually obtained.

Other fillers such as calcium carbonate are added to facilitate etching. These fillers are preferentially etched out of the surface to create bonding sites required for adequate adhesion. If these particles are large, a poor surface results. Generally, a happy medium is reached where adhesion is adequate and the surface is acceptable. Mineral-reinforced nylon is one such material.

When parts are molded, they are ready to be plated, as there is no secondary operation required. Parts are racked and run through the processing solutions, which include cleaners and/or predips, etching, neutralization, preactivation, activation, acceleration, and finally electroless plating (see Fig. 14.7). These steps are known as the preplate cycle (5).

CLEANERS

Cleaners are used for removing light soils such as fingerprints, dirt, and other debris from the parts. They are usually mild alkaline cleaners. In some cases, a highly wetted chromic acid solution may be used to insure complete wetting of a part in knurled or lettered areas prior to etching. This will minimize pinpoint skips in these areas. Cleaning prior to etching is optional and generally not used if parts are reasonably clean.

Troubleshooting these materials usually only involves chemical analysis to insure they are up to concentration.

PREDIPS

Predips, which are generally solvents, such as dimethylformamide for polysulfone, are used prior to etchants for two types of plastic problems. The

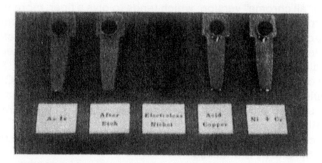

Fig. 14.7—Process steps in plating plastics.

first use is to help improve the surface of a normally plated but poorly molded, highly stressed part. This is done on ABS and its alloys by slightly swelling the surface, allowing the etchant to more uniformly attack the surface. By using a predip, non-uniform etch conditions are reduced on a stressed part, giving better overall adhesion. As stated earlier, proper molding can eliminate stress as well as the need for a predip in the above case.

The second use of predips is to attack or swell the surface of normally hard to etch plastics. These materials include polypropylene, polycarbonate, polysulfone, and others. This swelling action of the predip enables the chemical etchant to readily attack the surface of the normally resistant resin. Generally, a different solvent is needed for each resin in order to be successful.

Troubleshooting the predips involves analysis to insure proper balance.

ETCHANTS

Etchants are usually strong oxidizing solutions that eat away the plastic surface to varying degrees, accomplishing two purposes. First, the surface area is greatly increased, making the part turn from a hydrophobic (water-hating) to a hydrophilic (water-loving) material. Second, the microscopic holes left in the surface of the plastic by the etchant provide the bonding sites for the deposited metal. These sites are needed for adhesion between the plastic and the metal.

When ABS is exposed to an etchant, the butadiene is selectively removed, thus leaving small holes or bonding sites. Commonly used etchants for ABS are:

Chromium trioxide—375 to 450 g/L
Sulfuric acid—335 to 360 g/L

This is the so-called "chrome-sulfuric" version, or,

Chromium trioxide—900+ g/L

This is the "all chrome" version, which tends to give a finer, more uniform etch.

Both types are used in production. All-chrome etchants require several milliliters of sulfuric acid per liter of etchant to retard tank deterioration if chemical lead is used. It is also desirable to have about 40 g/L trivalent chromium in a new all-chrome bath to avoid over-etching the ABS surface. Etchants are generally operated at 140 to 160° F for 4 to 10 minutes on ABS. Other plastics may require higher or lower temperatures and times. Polycarbonate, for example, uses an etchant at 170° F for 10 minutes, while mineral-reinforced nylon (MRN) uses one at 105° F for 2 minutes.

As the chrome-based etchants are used, the following reduction reaction occurs as the plastic is attacked:

$$Cr^{+6} + 3e^- \rightarrow Cr^{+3} \qquad\qquad\qquad [1]$$
$$\text{reduction}$$

After a certain build-up of Cr^{+3} (about 40 g/L in the "chrome-sulfuric" type etchant), the etchant starts to lose its ability to perform properly. Adding more Cr^{+6} in the form of chromium oxide extends the life of the etchant for a while, but it is better to eliminate the excess Cr^{+3}. This can be done in two ways:

• Decant part or all of the etchant and rebuild it. This method may cause a waste disposal problem if provisions for the spent material are not made.

• The excess trivalent chrome can be electrolytically regenerated. This can be accomplished by using porous pots containing an electrolyte of 10 to 50 percent by vol H_2SO_4. Copper or stainless steel cathodes are then inserted in the pots. The tank or lead straps may be used as the anode. The usual anode to cathode ratio is 2:1. The voltage used is 9 to 12 V at a temperature of 100 to 155° F. The current is 100 to 150 amps for a 15-cm-diameter by 45-cm-long porous pot. Efficiency of this regeneration process is best when more than 20 g/L of trivalent chromium are present in the bath. The reaction involved is:

$$Cr^{+3} - 3e^- \rightarrow Cr^{+6} \qquad\qquad\qquad [2]$$
$$\text{oxidation}$$

at the anode.

Using this method and a flash evaporator for the rinses results in a tight closed loop system. One problem with this system is that all impurities in the etch such as dissolved metals and titanium dioxide filler are returned to it. This causes a build-up of impurities that eventually results in problems such as "stardusting," a fine roughness usually occurring on the shelf of a part.

The etchant is the most critical step in obtaining an acceptable finished part. An underetched part can result in poor adhesion and possible skip plate. Overetching a part can degrade the surface causing poor adhesion and cosmetics.

Some problems that may be encountered are shown in Table 14.1.

Table 14.1
Troubleshooting Etchants for Electroless Nickel

Problem	Cause	Solution
Shiny parts after electroless nickel	Underetching	Increase time
		Increase temperature
Poor adhesion	High Cr^{+3}	Check chemistry
Dull or almost black parts after electroless nickel	Overetching	Decrease time
Poor surface—low adhesion with black layer on plastic and metal		Decrease temperature Check chemistry
Stardusting or roughness on part	TiO_2 build-up	Install spray rinse after etch
Chemistry good, but parts have low adhesion and look shiny out of electroless nickel	Poor molding Mold release on part	Check parts for stress
Parts warp out of etch	Temperature too high Too much tension on rack	Lower temperature Check rack tension and adjust or change rack
CrO_3 additions do not show up in analysis	H_2SO_4 too high CrO_3 is salting out	Check chemistry and rebalance bath
Poor adheion at gate area only	Part stressed at gate	Check with glacial acetic acid to confirm. Reduce etch time and/or temperature. Use a predip

NEUTRALIZERS

After etching, the parts are thoroughly rinsed in water and then put into a neutralizer. These are materials such as sodium bisulfite or any of the proprietary products available that are designed to eliminate excess etchant from the parts and racks, usually by chemical reduction. Hexavalent chromium (Cr^{+6}) is harmful in the ensuing steps. Even with excellent rinsing, etchant may be trapped in blind holes. If this is not rinsed out or reduced to trivalent chromium (Cr^{+3}), "skip plate" can occur as a result of "bleedout" of etchant in subsequent steps.

Neutralizers are generally cheap to make up and are usually dumped and remade on a regular basis. They are usually run at 70 to 110° F for 1 to 3 minutes with air agitation. Troubleshooting is relatively simple, as seen in Table 14.2.

Table 14.2
Troubleshooting Neutralizers for Electroless Nickel

Problem	Cause	Solution
Cr^{+6} seen dripping off parts coming out of neutralizer	Bleed-out of blind holes on rack tips Bath spent	Check chemistry Check bath color—if orangish, dump Increase air and/or temperature Check rack tips
Drip skip noted after electroless nickel	Bleed-out of hexavalent chromium	As above

PREACTIVATORS

After neutralization and rinsing, a preactivator may be employed for certain resins, such as polypropylene or polyphenylene oxide. These proprietary materials are designed primarily to enhance activator absorption. Generally they do not reduce hexavalent chromium. Preactivators help make otherwise unplateable resins plateable by conditioning the resin surface by forming a film or changing the surface charge. One must be careful in selecting a preactivator, as excessive conditioning can lead to resist overplate and rack plating. Typical parameters for these baths are 70 to 120° F for 1 to 3 minutes.

If the parts are fully covered out of the electroless plating bath, the preactivator is doing its job. Troubleshooting guidelines for preactivators are given in Table 14.3.

ACTIVATORS

Activators or catalysts, as they are sometimes called, are materials that in most cases contain some precious metal such as palladium, platinum or gold.

The early version of activation was a two-step procedure. Step 1 was a stannous chloride/hydrochloric acid solution in which the stannous ion (Sn^{+2}) was adsorbed onto the surface. The part was then rinsed well. Step 2 was a palladium chloride/hydrochloric acid solution which, when the part with the stannous ion was immersed in it, caused the Pd^{+2}2 ion to be reduced to Pd^0 according to the following reaction:

$$Sn^{+2} + Pd^{+2} \rightarrow Sn^{+4} + Pd^0 \qquad\qquad [3]$$

Table 14.3
Troubleshooting Preactivators for Electroless Nickel

Problem	Cause	Solution
Skip or no plate out of electroless bath	Bleed-out of Cr^{+6} Preactivator out of spec	Use better rinsing Increase air Check chemistry of bath—if spent, dump
Rack or resist overplate	Temperature too high	Lower temperature and/or time
	Concentration too high	Check chemistry

The Pd sites formed the catalytic surface needed to deposit the chemical nickel. Present day catalysts are essentially the earlier two step version combined. In other words, the palladium chloride, stannous chloride and hydrochloric acid are in one solution. What we get is a palladium-tin hydrosol, which is a solution of complex ions and colloidal particles whose activity and stability depend on the chloride and stannous ion concentrations (6,7).

A typical working activator bath today would consist of the following:

Stannous chloride—6 g/L
Palladium (metal)—20 to 100 ppm
Chloride ion—2.5 to 3.5 N

The chloride ion may be maintained with hydrochloric acid or sodium chloride.

When processing a part, the first noticeable change will be evident after the activator. The part acquires a tan to brownish color. This coloration, unless the part is black, can easily be seen and is an indication the activator is working. The absence of a coloring on the part usually indicates a problem that can result in skip plate and possibly low adhesion.

The primary purpose of activators is to provide catalytic sites on the plastic surface. The usual operating conditions for the activator bath are 80 to 95° F for 2 to 5 minutes. Because this is the most expensive bath in the system, care is taken to analyze regularly. Even when the bath is not being used, analysis for tin is required, as tin is continually oxidized to the stannic ion (Sn^{+4}). When almost all the stannous is oxidized, the bath "falls out".

Troubleshooting tips for the bath are given in Table 14.4.

Table 14.4
Troubleshooting Activators for Electroless Nickel

Problem	Cause	Solution
Silver mirror seen on bath surface	Low stannous ion	Replenish stannous ion
Roughness on parts	Particulates in the bath	Batch filter to the bath. (Expect to lose 2-3% due to absorption on the filter
	Stannous may be low	Raise stannous level Install spray rinse after activator
Solution clear or tea colored	Bath fell out due to low stannous, impure HCl, or an addition of HNO₁	Make up a new bath
Rack plate and/or stop-off overplate	High temperature High concentration	Lower temperature Check chemistry
Skip plate after electroless bath	Low temperature Low concentration	Raise temperature Check chemistry
Blotchy surface (light and dark areas	Stress in part	Confirm with glacial acetic acid
Parts very dark	High temperature	Lower temperature Check chemistry
Parts very light	Low temperature	Raise temperature Check chemistry Agitate parts

ACCELERATORS

After rinsing following the activator, metallic palladium is present on the surface of the part surrounded by hydrolyzed stannous hydroxide. The excess stannous hydroxide must be removed from the part before the palladium can act as a catalyst.

The role of an accelerator is just that. It is to remove the excess tin from the part while leaving the palladium sites intact for the deposition of the electroless bath. Tin will inhibit the action of the electroless bath, resulting in skip plate.

This step is important, as too short a time in bath, as well as too long a time, can cause skip plate. These baths are usually organics or mineral acids.

The biggest problem with accelerators is the effect of metallic contamination. Most proprietary baths contain a built-in reducer to eliminate any hexavalent chromium. Chromium may be dragged through the line in blind holes or bad rack tips. These contaminants, such as hexavalent chromium, iron, and other metals, can cause the accelerator to become over-aggressive. This causes the accelerator to remove not only the stannous tin, but also palladium. When this occurs, skip plate can be the result. Some materials are designed to minimize the effect of these contaminants.

Accelerators are usually run at 110 to 140° F for 2 to 5 minutes. Some problems that may occur are as shown in Table 14.5.

Table 14.5
Troubleshooting Accelerators for Electroless Nickel

Problem	Cause	Solution
No plate on edges of part	Over-acceleration High temperature	Check chemistry Lower temperature Lower time
Parts have no plate in major areas or slow takeoff in the electroless that goes from the edges inward	Under-acceleration Low temperature	Check chemistry Raise temperature Increase time Agitate parts in solution
No plate at all out of electroless bath	Contamination that causes aggressiveness	When parts come out of the activator brown, but are clean after the accelerator, contamination and/or high temperature may be the problem. In this case, lower the temperature or make up a new bath. If parts are brown out of the accelerator, under-acceleration is probably the cause. In this case, increase the time, temperature, and/or agitation
	High temperature	Check chemistry

ELECTROLESS PLATING

After rinsing, the parts are put into the final step in the preplate cycle. This is the electroless bath, which deposits a thin, adherent metallic film, usually copper or nickel, on the plastic surface by chemical reduction. This is accomplished by using a semi-stable solution containing a metal salt, a reducer, a complexor for the metal, a stabilizer and a buffer system. When idle, the baths are stable, but when a palladium-bearing surface is introduced into the solution, a chemical reduction of metal occurs on the palladium sites, and through autocatalysis, continues until the part is removed. The basic reactions for copper and nickel are:

$$Cu^{+2} + 2HCHO + 4OH^- \xrightarrow{Pd} 2HCOO^- + 2H_2O + Cu^0 + H_2 \qquad [4]$$

$$Ni^{+2} + H_2PO_2^- + 3OH^- \xrightarrow{Pd} HPO_3^{2-} + 2H_2O + Ni^0 \qquad [5]$$

The electroless nickel used is an alkaline bath, usually partially complexed with ammonium hydroxide. It is operated at 80 to 100° F for 5 to 10 minutes at a pH of about 9.0. Nickel thickness can range from 8 to 12 μin. and should have a resistance of 50 to 20 ohms/in. The pH is controlled with ammonium hydroxide.

If an electroless copper is used, it is usually reduced by formaldehyde and complexed by various materials, depending on the supplier. Copper baths are operated at 100 to 140° F at a pH of 12 to 13. The thickness ranges from 20 to 50 μin.

For most applications, electroless nickel is adequate, as the primary function of the electroless bath is to render the surface conductive. The automotive industry has done studies that show that electroless copper tends to exhibit less of a tendency to blister in a humid, corrosive environment (7,8). Because of this, most exterior automotive parts are processed through electroless copper.

Nickel baths are relatively easy to control, while copper baths generally require automatic analysis for control. Copper baths are more susceptible to problems than nickel. Some troubleshooting hints are given in Tables 14.6 and 14.7.

In all cases, if an automatic controller is being used to analyze and add the necessary chemicals to the bath, the first thing to check is the controller if something is out of spec. Be sure the pumps are working and the reagents and/or replenishment solutions are not used up. When the controller is operating properly, over-the-side additions of chemicals are rare.

Following the electroless deposition, the parts are generally run through an electrolytic copper or Watts nickel strike (9). The purpose of the strike is to build up the thin electroless deposit prior to subsequent electroplating. This build-up, usually to 0.0001 in., prevents "burn-off", the loss of contact between the part and the rack tips.

The electrolytic strike is usually followed by an electrolytic bright acid copper plating solution, which is normally a proprietary product. This step will build a copper thickness to about 0.0005 to 0.001 in. and give a bright surface. Copper, being ductile, will act as a buffer layer between the plastic substrate and the final plate as the part sees large thermal changes. This buffer layer helps prevent blisters and/or cracking of the plated deposit.

If good corrosion and abrasion resistance is desired, as for an exterior automotive part, the next step will be an electrolytic semi-bright nickel plating

Table 14.6
Troubleshooting Electroless Nickel Plating Baths

Problem	Cause	Solution
Skip plate	Low reducer	Check chemistry
	Low temperature	Raise temperature
	Contamination	Check for contamination
Roughness	Particulates in bath	Filter bath
		Install spray rinse after bath
		Agitate work
Burn-off in strike	Electroless coating too thin	Increase time
		Increase temperature
		Check bath rate
		Check chemistry
	Bad racking, not enough contacts	Check racks
Bath overactive	High temperature	Lower temperature
	High reducer	Check chemistry
	Palladium contamination	
	Low complexor	
	Tank plate-out	Clean and strip tank
	Low stabilizer	Check stabilizer
Bath sluggish	pH out of range	Check chemistry
	High stabilizer	
	Low reducer	
	Low temperature	Raise temperature
	Contamination	Check for contamination

solution. Normally 0.0003 to 0.0008 in. of semi-bright nickel are plated, followed by an electrolytic bright nickel plating solution, which will deposit about 0.0002 to 0.0004 in. of metal. The deposit from the latter bath is generally very specular. For the best corrosion resistance, a layer of about 0.0001 in. of microporous nickel can be electrolytically deposited on the bright nickel, followed by an electrolytic chromium deposit from either a hexavalent or trivalent bath. This layer, only about 5 to 10 millionths of an inch thick, is the final step in producing a finished part. Be aware that if the surface of the substrate is pitted, scratched, dented, or marred in any way, electroplating will enhance these blemishes. Care must be taken when handling and racking plastic parts to avoid damaging them (1). Parts are usually plated to meet various service conditions, which are:

- *SC1—Mild.* Includes indoor exposure to normally dry, warm atmosphere and subject to minimal wear and abrasion.
- *SC2—Moderate.* Indoor exposure where moisture condensation occurs. Examples are kitchen and bathroom fixtures.

Table 14.7
Troubleshooting Electroless Copper Plating Baths

Problem	Cause	Solution
Bath slow—dark deposits	Low copper	Check chemistry, plating rate
	Low caustic	
	Low formaldehyde	
	High stabilizer	
	Low temperature	Increase temperature
	Too much air	Decrease air
Bath overactive—dark, grainy deposits	High caustic	Check chemistry, plating rate
	High formaldehyde	
	Low stabilizer	Add stabilizer
	High temperature	Lower temperature
	High loading factor	Decrease work in bath
	Tank plate-out	Clean and strip tank
		Increase air
Roughness on parts—precipitate forms when additions are made	Low complexor	Check chemistry

- *SC3—Severe.* Includes frequent wetting by rain or dew, and in some cases strong cleaners and salt solutions. Examples are outdoor furniture, bicycle parts, and hospital furniture.
- *SC4—Very severe.* Includes likely damage from dents, scratches, and abrasive wear in addition to being in a corrosive environment. Examples are boat parts and exterior automotive parts.

The various plate thicknesses required for each type of service are given in Table 14.8.

Chromium is not the only finish that can be applied to plated plastics. Once the electroless is put down, any plate combination can be used. Final finishes can be brass, gold, silver, or any of the other finishes put on plated metal articles.

After plating to the desired specification, most platers perform quality control testing on plated parts. These tests include:

Adhesion (Jacquet Test)
The part, usually one with a flat surface, is plated with 1 mil of bright acid copper. A 1-in.-wide strip is cut and the plate is separated from the plastic at one end. The plate is then pulled at 1 in./min at 90° to the surface. Values obtained in pounds per linear inch are recorded. ABS, for example, has from 5 to 15 lb/linear in. pull.

Table 14.8
Minimum Plating Thickness Requirements
Under Various Service Conditions

Service	Copper	Total nickel	Chromium
SC1	0.6 mils	-0.3 mils	10 μin.
SC2	0.6 mils	0.6 mils	10 μin.
SC3	0.6 mils	0.9 mils	10 μin.
SC4	0.6 mils	1.2 mils	10 μin.

Thermal Cycling

Because various plastics have various coefficients of thermal expansion, some do not readily pass this test, even though adhesion is good. To counteract this, more copper may be plated on the part to act as a cushion.

In thermal cycling, the part is first put into an air-circulating oven for 1 hr at 180° F. The part is then removed and held at room temperature for 15 min. The part is then placed into a cold chamber at -20° F for 1 hr. This is repeated a minimum of three times. Loss of adhesion, blisters, or cracking indicate a failure. Part design and molding can also affect thermal cycling. This test was designed for ABS and does not necessarily apply to other substrates.

C.A.S.S.

The C.A.S.S. (10), or Copper Accelerated Acetic Acid Salt Spray Test, checks the corrosion resistance of the final plate. This test is generally performed in a cabinet, according to ASTM B-368. Requirements are as follows:

- SC1—None
- SC2—1 8-hr cycle
- SC3—2 16-hr cycles
- SC4—3 16-hr cycles

Failure is indicated by white or green corrosion.

The above are the most widely used tests. Others include:

- N.S.S. (10)—5 percent neutral salt spray, which is also used in another corrosion test—ASTM B-117.
- Plate thickness—Checking plate thicknesses by either a destructive or non-destructive method.
- Outdoor exposure—Setting industrial samples in environments they will be normally exposed to.

- C.S.A. B-125, LRP-15 (11)—A Canadian test for plumbing goods that includes 450 cycles of dipping the parts in alternating hot (175° F) and ambient (70° F) water baths for 40 seconds. One cycle is 40 seconds in the hot water, and 40 seconds in the ambient water.

New technologies in plating on plastics are leaning toward plating engineering resins for printed circuits. Resins with high strength and heat distortion are also being looked at to replace metals in functional as well as decorative roles.

ELECTROMAGNETIC INTERFERENCE SHIELDING

EMI shielding is another area where plating-on-plastics technology can be utilized. Electromagnetic interference is electrical "noise" generated by a piece of electrical or electronic equipment that causes a problem of interference with the operation of another piece of electrical or electronic equipment. An example is the annoying lines on a TV set caused when an electric razor or other small appliance is in use. Sources of EMI are shown in Fig. 14.8.

Some of the more common sources of EMI are:

- Airplanes
- Computers
- Electric motors
- Ignition systems
- Power lines
- Radio and television transmitters
- Video games
- Lightning

Some equipment affected by EMI "noise" includes:

- Radio and television receivers
- Computers
- Telephones

In the early days of electronic packaging, EMI radiation emitted from equipment was not a major concern. Today, these emissions are subject to controls, it has become necessary to redesign and effectively shield electronic equipment. The electromagnetic radiation spectra, which covers a wide range of frequencies, is shown in Fig. 14.9.

Metal enclosures were used for many years as the main shielding material. This provided structural integrity as well as a shielding function. Some drawbacks of metal were weight and design limitations. As housings for electronic equipment changed from metal to plastic, it became necessary to apply shielding to protect the equipment. Several types of technical approaches have been used in an attempt to effectively solve the shielding problems of plastic housings. One such method could be the use of electroless-plated

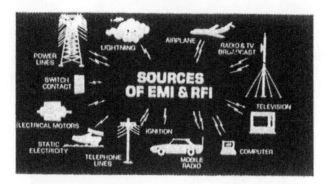

Fig. 14.8—Sources of EMI and RFI.

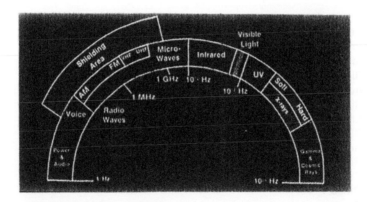

Fig. 14.9—Electromagnetic radiation spectrum.

copper and/or nickel. The process used would be similar to that presently used to plate on plastics. Steps would include:

1. Pretreatment (if necessary)
2. Rinse
3. Etch
4. Rinse
5. Neutralize

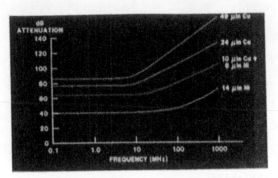

Fig. 14.10—Attenuation as a function of thickness of metal deposits.

6. Rinse
7. Preactivate (if necessary)
8. Rinse
9. Activate
10. Rinse
11. Accelerate
12. Rinse
13. Electroless copper
14. Rinse
15. Activate
16. Rinse
17. Electroless nickel
18. Rinse
19. Dry

The electroless copper (Step 13) generally would be deposited to a thickness of 40 to 60 millionths of an inch. The copper is a very effective shield by itself, but because of its corrosive nature, an electroless nickel (Step 17) topcoat of about 10 to 20 millionths of an inch is applied to retard the corrosion. Nickel can be used by itself, but its shielding effectiveness is much lower than that of copper or copper/nickel. It is also an excellent base if the article is to be painted. Electroless shielding is applied in general to both sides of a plastic enclosure, which assures the best protection against the transmission of electromagnetic radiation (Fig. 14.11). Still, techniques have been developed to apply the electroless coatings only to the inside of the enclosure, to allow the use of color-molded finishes.

Shielding effectiveness is the ability of the method used to absorb or reflect the unwanted "noise" signal, typically over a frequency range of 14 KHz to 1

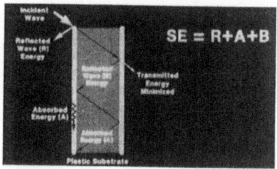

Fig. 14.11—Double-sided shielding.

Table 14.9
EMI Shielding Effectiveness

Attenuation	Shielding effectiveness
0 to 10 dB	Very little
10 to 30 dB	Minimal
30 to 60 dB	Average
60 to 90 dB	Above average
90 to 120 dB and above	Maximum to state-of-the-art

GHz. This is called *attenuation*, which is a function of the electrical conductivity of the shield used, measured in decibels (dB). Attenuation as a function of the thickness of the metal deposits is shown in Fig. 14.10. Shielding values for various attenuation levels are shown in Table 14.9.

The electrical conductivity required to satisfy the minimal functional requirements are expressed in surface resistance (ohms/square) as follows:

- EMI shielding—Less than 1 ohm/square
- RFI (radio frequency interference)—Less than 10 ohms/square
- ESD (electrostatic discharge)—50 to 100 ohms/square

This allows performance to be measured in surface resistance as well as dB.

Some of the other technologies presently being used to shield plastic housings are given in Table 14.10.

Table 14.10
Coating Techniques for EMI Shielding
Of Plastic Housing

Method	Advantages	Disadvantages
Zinc arc spray	Good conductivity Hard, dense coat Effective over a wide frequency range	Need special equipment Adhesive problems Coating cracks May distort the housing
Conductive paints Silver	Good conductivity Conventional equipment Resists flaking Conductive oxide Easy to apply	Expensive
Nickel	Conventional equipment Resists flaking Economical Easy to apply	Multiple coats needed for good conductivity Thickness problems Questionable attenuation

Copper	Conventional equipment
	
	Resists flaking
	Economical
	Easy to apply
		Multiple coats needed for
		  good conductivity
		Thickness problems
		Oxidation reduces
		  conductivity

Vacuum metallizing	Good adhesion
	Good for all plastics
	Not limited to simple
	  designs
	Familiar technology
		Size limited to vacuum
		  chamber
		Base coat needed
		Expensive special
		  equipment
		Low thickness

Cathode sputtering	Good conductivity Good adhesion	Expensive equipment Microscopic cracking May distort housing High power needed
Foil application	Die cut to part shape Good conductivity Good for experimentation	Complex parts are difficult to coat Labor intensive
Conductive plastics	No secondary operation	Material expensive Poor attenuation Surface usually poor
Silver reduction	Good conductivity Low initial cost	Tendency to oxidize Difficult to mask Multiple-step process

Electroless plating (copper
  and/or nickel)
	Uniform thickness
	Good for all size and shape
	  parts
	Good conductivity
	Resists chipping
	Can be electroplated for
	  metallic appearance
	Can be painted
		Limited to certain resins

As can be seen above, besides electroless plating, numerous other methods of shielding are available. The success of plating on plastics technology is dependent on the industry and its requirements.

REFERENCES

1. J.L. Adcock, *Electroplating Plastics*, American Electroplaters and Surface Finishers Society.
2. "Design and Converting Techniques for Plating Cycolac Brand ABS," EP-3510, Marbon Division, Borg-Warner Corporation, Technical Bulletin, (May 1967).
3. "Glossary of Terms," Thomas H. Ferrigno, ed.; Fillers and Additives Committee, Society of Plastic Industry.
4. "Plating on Plastics II," Technical paper—Society of Plastics Engineers, Connecticut Section.
5. A. Rantell and A. Holzman, "Mechanism of Activation of Polymer Surfaces by Mixed Stannous Chloride/Palladium Chloride Catalysts," Division of Metal Sciences, Polytechnic of the South Bank, London SE10AA (February 12, 1973).
6. C.H. deMinjer and P.F.J. v.d. Boom, "The Nucleation with $SnCl_2$-$PdCl_2$ Solutions of Glass Before Electroless Plating," Phillips Research Laboratories, Einhaven, Netherlands.
7. R.G. Wedel, *Plating*, **62**(3), 235 (Mar. 1975).
8. R.G. Wedel, ibid., **62**(1), 40 (Jan. 1975).
9. *Metal Finishing Guidebook—Directory Issue*, Metals and Plastics Publications, Inc., One University Plaza, Hackensack, NJ 07601.
10. ASTM B-117, 15-22; ASTM B-368, 164-169, Part 9, American Society for Testing and Materials, Philadelphia, PA.
11. CSA Standard B-125, LRP-15, Canadian Standards Association, Ontario, Canada.
12. E.B. Saubestre, *Plating*, **52**(10), 982 (Oct. 1965).
13. E.B. Saubestre, *Modern Electroplating*, p. 636 (1974).
14. W.P. Innes, *Plating*, **58**(10), 1002 (Oct. 1971).
15. H. Narcus, ibid., **55**(8), 816 (Aug. 1968).
16. "Plating of Plastics with Metals," *Chemical Technology Review*, Noyes Data Corporation, Vol. 27 (1974).
17. "Plating on Plastics," *Extended Abstracts* (1965-1971), The International Nickel Company, Inc.
18. James L. Adcock, *Products Finishing*, p. 51 (May 1982).
19. Gerald Krulik, *Industrial Finishing*, p. 16 (May 1983).
20. Gerald Krulik, *Products Finishing*, p. 49 (Oct. 1983).

Chapter 15
Electroless Plating
Of Gold and Gold Alloys

Yutaka Okinaka

Gold is currently indispensable for the electronics industry, in spite of its high price, because of its unique physical properties. Because of the unpredictability of the price, substitute materials such as palladium and its alloys are now being used in certain applications. It is generally believed, however, that gold is not likely to be replaced completely by other materials, especially where high reliability is a primary consideration. As compared to conventional electrolytic plating or physical deposition techniques such as vacuum evaporation or sputtering, the principle of electroless plating is highly attractive because of the simplicity of the operation requiring no external source of current and no elaborate equipment. In spite of the fact, however, that much has been written in the open literature and a large number of patents exist, the present state of development of electroless gold technology is still far from being comparable to that of the well-established methods of electroless nickel and copper plating. Nevertheless, much progress has been made since the last comprehensive review by this author, which was published in 1974 (1). The purpose of this chapter is to present a summary of the more recent developments in a form that is useful for those who wish to obtain essential information quickly on the individual processes.

Before describing the processes, it is appropriate to make a brief comment on the terminology. In the literature dealing with the plating of precious metals, particularly gold, it has become customary to use the term "electroless plating" to describe three fundamentally different types of chemical plating performed without the use of an external source of current: (a) galvanic displacement, (b) autocatalytic, and (c) substrate-catalyzed processes. This trend has caused some confusion because "electroless plating" is a term originally meant to describe autocatalytic processes only. This chapter will deal primarily with autocatalytic processes because generally, in principle, they are the most preferred type. In autocatalytic processes, by definition, the deposition should continue indefinitely, and thick non-porous deposits should be obtainable. Recent developments, however, appear to indicate that under certain conditions some non-autocatalytic processes are capable of depositing essentially pore-free films with a considerable thickness (up to a few μm). Furthermore, some processes involve two or more of the three types of reactions mentioned above.

Therefore, this chapter will include some discussion of the non-autocatalytic processes as well.

Galvanic displacement (often called "immersion") processes are easy to distinguish from the other two processes, because generally no reducing agent is present in the bath formulation. Solutions for autocatalytic and substrate-catalyzed processes both contain a reducing agent, and it is not generally possible to distinguish between the two processes based on the solution composition alone. The only way of assuring that a given gold process is autocatalytic is to perform an experiment showing that the plating takes place on a gold substrate.

Table 15.1
Borohydride Baths
(Okinaka)

	Bath A*	Bath B*
KAu(CN)$_2$	0.86 g/L (0.003M)	5.8 g/L (0.02M)
KCN	6.5 g/L (0.1M)	6.5 g/L (0.1M)
KOH	11.2 g/L (0.2M)	11.2 g/L (0.2M)
KBH$_4$	10.8 g/L (0.2M)	10.8 g/L (0.2M)
Temperature	70° C	70° C
Rate	(see text)	

Preparation of Stock Solutions (1 liter)**
For Bath A (5X):
1. Dissolve 56 g KOH and 32.5 g KCN in about 600 mL water.
2. Add 54 g KBH$_4$ and stir until dissolution.
3. Dissolve 4.3 g KAu(CN)$_2$ in about 100 mL water.
4. Mix the above two solutions, and dilute to 1 liter.
5. Filter through Whatman 41 filter paper or equivalent to remove particulates.

Dilute 1 volume of this solution with 4 volumes of water to make Bath A.

For Bath B (2.5X):
1. Dissolve 28 g KOH and 16.3 g KCN in about 500 mL of water.
2. Add 27 g KBH$_4$ and stir until dissolution.
3. Dissolve 14.4 g KAu(CN)$_2$ in about 250 mL water.
4. Mix the above two solutions, and dilute to 1 liter.
5. Filter through Whatman 41 filter paper or equivalent.

Dilute 1 volume of this solution with 1.5 volumes of water to make Bath B.

---

*If these baths are stored at room temperature, the plating rate will decrease at the rate of about 2 percent per day.
**The stock solutions can be stored at room temperature for at least two months without causing a significant change in plating rate.

PURE GOLD

The experiences of many investigators, including this author, have shown that most of the older formulations are not autocatalytic (1,2) except those using either an alkali metal borohydride or an amine borane as the reducing agent (3). More recently, several autocatalytic baths using other reducing agents have been reported. This section will first describe the borohydride and amine borane systems and their recent modifications, followed by the newer autocatalytic baths. Non-autocatalytic processes will also be described in particular reference to the plating on nickel substrates.

Two typical bath compositions that have been used most extensively by this author are given in Table 15.1. Concentrated stock solutions can be prepared for both baths by following the directions given in Table 15.1. These stock solutions can be stored at room temperature for at least three months without noticeable change in plating rate. Since a very small amount of hydrogen gas is continuously liberated from the solution due to the hydrolysis of BH_4^-, the cap of the storage bottle should have a pressure vent or be only loosely tightened for safety. The plating rate depends on agitation, temperature, and gold concentration among other variables, including the concentrations of KOH, KCN, and KBH_4 (3). With mild agitation, bath A gives a rate of approximately 1.5 μm/hr at 70°C, while bath B gives 0.5 μm/hr. Bath A gives deposits with acceptable properties only with vigorous agitation, whereas bath B can be used with little or mild agitation. In plating fine line patterns, e.g., <100 μm, considerably faster plating rates are obtained because of enhanced mass transport effects on such geometries (4). The basic chemistry of this system and some practical knowledge that is essential for successful operation of this process will be summarized below.

Chemistry
The chemical reactions involved in the borohydride process are rather unique in that the reducing agent that actually reacts with $Au(CN)_2^-$ is not the BH_4^- itself but the species with the formula BH_3OH^-, which is an intermediate formed during the stepwise hydrolysis of BH_4^- (5):

$$BH_4^- + H_2O \rightarrow BH_3OH^- + H_2 \tag{1}$$
$$BH_3OH^- + H_2O \rightarrow BO_2^- + 3H_2 \tag{2}$$

According to Efimov et al. (6, 7), another hydrolysis intermediate $BH_2(OH)_2^-$ also participates as a reducing agent. Thus, the overall plating reactions are:

$$6Au(CN)_2^- + BH_3OH^- + 6OH^- \rightarrow 6Au + BO_2^- + 12CN^- + 5H_2O \tag{3}$$
$$4Au(CN)_2^- + BH_2(OH)_2^- + 4OH^- \rightarrow 4Au + BO_2^- + 12CN^- + 5H_2O \tag{4}$$

Efimov et al. observed that all hydrogen gas produced during the plating results from the hydrolysis reactions, and the plating reactions *per se*, as expressed in Equations 3 and 4, do not produce hydrogen. It has been shown (2) that the utilization efficiency of borohydride in this plating system is very low (<2%) because the second step of hydrolysis (Eq. 2) proceeds much more rapidly than the plating reaction, and most of the borohydride is lost by hydrolysis.

Another interesting aspect of the chemistry of the borohydride system is that the plot of the plating rate against the concentration of gold shows a maximum at about 0.003M (3). This unusual dependence appears to be due to the competitive adsorption of BH_3OH^- and $Au(CN)_2^-$ for the same site on the gold surface. The latter species, if present at a large concentration, acts as a catalytic poison and tends to prevent the adsorption of the reducing agent.

Material Compatibility

Electroless gold deposition from the borohydride system takes place on noble metals such as Pd, Rh, Ag, and Au itself, as well as on active metals such as Cu, Ni, Co, Fe, and their alloys. Initial reactions on these two classes of metals are different, however. On noble metals the reaction is catalytic from the very beginning, whereas the gold deposition on the active metals is initiated by galvanic displacement, which results in accumulation of ions of those metals in the bath. No adverse effect occurs with copper, whereas the introduction of ions of Ni, Co, and Fe into the solution is highly detrimental—bath decomposition sets in when these metals are present at a concentration as low as 10^{-3}M. With Ni, a considerable decrease in plating rate occurs at even lower concentrations. The incompatibility of nickel is a significant drawback of the borohydride system, because gold is used on nickel in many applications. This aspect will be discussed further in a subsequent section.

Silicon undergoes a displacement reaction, which deposits loosely adherent gold with simultaneous dissolution of silicon. Aluminum is vigorously attacked by the solution because of the high alkalinity.

The following metals are stable in the bath and do not initiate gold deposition: Cr, Mo, W, Ta, Ti, Zr.

Certain organic materials are not compatible and can cause serious problems (4). For example, polyethylene dissolved in the bath inhibits the plating. Positive photoresists and some surfactants are also incompatible. Polypropylene, teflon, and negative photoresists are stable in the bath.

Impurity Effects

Contamination of the bath with any of the incompatible materials listed above should be avoided. Special precaution should be exercised to prevent contamination with trace amounts of Ni, Fe, and Co. Also to be noted is the sensitivity of the system to trace organic contaminants that may be present in the water used to prepare the bath. For example, deionized water not treated with active charcoal causes suppression of the plating. This problem can be eliminated simply by using charcoal-treated deionized water or distilled (over permanganate) water.

In a reasonably clean environment the borohydride bath is sufficiently stable for routine use and can be used for several days (up to a total operating time of about 20 hours) with proper replenishment of bath components (8). The author is aware of reports that the bath is much less stable than he has claimed. Such conflicting results are most likely to be due to the effect of insidious contaminants.

Ali and Christie (2) recently found that EDTA and ethanolamine are effective for stabilizing the bath. Their bath formulation is given in Table 15.2. The function of these stabilizers apparently is to form strong complexes with metallic contaminants, thereby making them less susceptible to reacting with borohydride.

Methods of Increasing Plating Rate and Bath Stability

An increased plating rate can be obtained by (a) increasing the degree of agitation, (b) increasing the bath operating temperature, (c) decreasing the free cyanide concentration, (d) decreasing the KOH concentration, (e) increasing the borohydride concentration, or (f) bringing the gold concentration to the optimum level of 0.003M (3). Attempts to obtain a rate greater than approximately 3 μm/hr on an unpatterned substrate using the original bath solutions would be futile, even with vigorous agitation, because of the spontaneous bath decomposition that sets in under the conditions giving such high rates.

Three different approaches for obtaining enhanced plating rates beyond this limit have been reported recently, and they will be summarized below.

Addition of Depolarizers

It has been shown that the plating rate can be increased up to 10 μm/hr by adding ions such as Pb^{2+} or Tl^+ (9-12). Kasugai (9) gives a bath composition with $PbCl_2$ (Table 15.3). This bath is said to deposit 4 μm in 30 min. The process yields dense and bright deposits, and has been used to plate gold on ceramics metallized with W or Mo that was preplated with Ni and 0.3-μm-thick displacement gold.

Table 15.2
Borohydride Bath
(Ali and Christie)

$KAu(CN)_2$	1.45 g/L (0.005M)
KCN	11 g/L (0.17M)
KOH	11.2 g/L (0.2M)
KBH_4	10.8 g/L (0.2M)
NaEDTA	5 g/L
Ethanolamine	50 mL/L
Temperature	72°C
Rate*	1.5 μm/hr

*With mild agitation. This rate remains constant until 50 percent depletion of the gold or for 8 hours.

Table 15.3
Borohydride Bath
(Kasugai)

Main Solution

$KAu(CN)_2$	5 g/L
KCN	8 g/L
Na citrate	50 g/L
EDTA	5 g/L
$PbCl_2$	0.5 g/L
Gelatin	2 g/L

Reducing Solution

$NaBH_4$	200 g/L
NaOH	120 g/L

Mix 10 volumes of main solution and 1 volume of reducing solution. Plate at 75-78° C with agitation. Plating rate: 4 μm in 30 minutes.

Matsuoka et al. (10) recently developed a bath containing $Au(CN)_4^-$, trivalent gold cyanide complex, as the source of gold with the addition of a small amount of $PbCl_2$. An example of the bath composition is given in Table 15.4. A deposition rate of 2.5 μm/hr *without stirring* has been reported. In view of the diffusion-controlled nature of the reaction, it should be possible to obtain an enhanced plating rate simply by providing agitation. Effects of various variables on the deposition process were studied using electrochemical polarization measurements (10).

Takakura used thallous sulfate as an additive to obtain a rate as high as 10 μm/hr (11). The bath composition is tabulated in Table 15.5. It is stated that when the concentration of Tl_2SO_4 exceeds 100 ppm, bath instability occurs and the deposit becomes discolored.

In addition, Prost-Tournier and Allenmoz (12) claim the ions of Ga, In, Ge, Sn, Sb, and Bi as additives for increasing the plating rate.

The metallic additives such as Tl^+ and Pb^{2+} are known to induce a significant depolarization effect in the electrolytic reduction of $Au(CN)_2^-$ ions through their specific adsorption on gold surface and subsequent underpotential deposition (13). The same phenomenon is believed to account for the effect of these ions on the rate of electroless gold deposition. While these additives are desirable as far as the plating rate is concerned, it should be noted that they are known to cause a significant decrease in bond strength on aging if gold or aluminum wire is bonded to the surface of electrolytically plated gold (14,15). As far as the author is aware, effects of those additives on the strength of bonding to electroless gold have not been reported. It is advised that this effect be investigated before the metallic additives are used for applications involving such wire bonding.

Table 15.4
Borohydride Bath With Trivalent Gold
(Matsuoka, Imanishi, and Hayashi)

$KAu(CN)_4 \cdot nH_2O$	8 g/L
KOH	10-20 g/L
KBH_4	3-5 g/L
$PbCl_2$	0.5-1.0 mg/L
Temperature	80° C
Rate without agitation	<2.5 μm/hr

Addition of Organic Stabilizers

An alternate approach appears to be the addition of organic compounds that stabilize the bath and hence allow it to be operated at a higher temperature. Sasaki (16) claims in his patent the addition of one or more compounds having N-carboxymethyl group(s) alone or together with compound(s) having S-containing group(s). Some examples are given in Table 15.6. Plating rates ranging from 8 to 23 μm/hr are said to be obtainable by operating the baths at 85-90° C. It should be noted, however, that at such high temperatures the loss of borohydride due to hydrolysis is expected to occur rapidly. This condition will not only result in a problem in bath control, but also lead to rapid accumulation of the hydrolysis product, which eventually causes bath instability if the process is used with replenishment of borohydride (8).

Use of Trivalent Gold Cyanide Complex

El-Shazly and Baker (17) developed a bath using $Au(CN)_4^-$ as the source of gold instead of $Au(CN)_2^-$, the usual monovalent gold complex. The reducing agents

Table 15.5
Borohydride Bath
(Takakura)

$KAu(CN)_2$	(Concentration not specified)
KCN	6.5 g/L (0.1M)
KOH	11.2 g/L (0.2M)
Tl_2SO_4	5-100 ppm
KBH_4	5.4-10.8 g/L (0.1-0.2M)
Temperature	70-80° C
Rate	<10 μm/hr

compatible with this bath are borohydride, amine boranes, and cyano-borohydride. The use of trivalent gold allows it to be replenished in the form of $KAuO_2$ or $KAu(OH)_4$, thus avoiding accumulation of excess cyanide ions, which leads to a decline in plating rate in the monovalent gold system. The trivalent gold system is claimed to give higher plating rates (2-8 μm/hr) with improved bath stability. Examples are given in Table 15.7. The trivalent gold bath containing $PbCl_2$ has already been described in the previous section (10).

Plating on Nickel
As mentioned already, nickel is not compatible with the borohydride or amine borane system. Because of frequent demand for plating electroless gold on nickel substrates, the author's thoughts will be presented below as to how one can accomplish this task by using currently available techniques.

Applying a Preplate of Displacement Gold
Displacement gold can be deposited easily on nickel using a bath prepared according to published information (18) or a commercial bath. Generally, a displacement bath is capable of depositing only a very thin layer of gold (<0.3 μm), and the deposit usually does not completely cover the underlying nickel. Therefore, if the borohydride bath is used subsequently for plating autocatalytic gold, it will still become contaminated with dissolved nickel ions. In view of the extreme sensitivity of the borohydride system to nickel contamination, this approach is considered far from satisfactory.

Table 15.6
Borohydride Baths
(Sasaki)

	Example 1	Example 2
KAu(CN)$_2$	5 g/L	8 g/L
KCN	8 g/L	10 g/L
NaOH	20 g/L	20 g/L
Na$_2$EDTA	15 g/L	-
Glycocol	10 g/L	-
Hydroxy-ethylene diamine triacetic acid (3Na)	-	25 g/L
Nitrilotriacetic acid	-	20 g/L
Mercapto-succinic acid	-	3 g/L
NaBH$_4$	25 g/L	25 g/L
Temperature	90°C	90°C
Rate	12 μm/hr	23 μm/hr

Table 15.7
Dimethylamine Borane or Borohydride Baths
With Trivalent Gold Cyanide
(El-Shazly and Baker)

	Example 1	Example 2
Gold, as KAu(CN)₄	4 g/L	3 g/L
KCN	-	10 g/L
KOH	35 g/L	10 g/L
K-citrate (3K)	30 g/L	-
Dimethylamine borane	5 g/L	-
KBH₄	-	1 g/L
pH	11.5-13	-
Temperature	80°C	(60-85°C)*
Rate	4 μm/hr	2 μm/hr

*General preferred range. No specified temperature is given.

Using a Hypophosphite Bath

A hypophosphite electroless gold bath was described by Swan and Gostin as far back as 1961 (19), and several modified baths have since been described (1). As shown by Brenner (20), the Swan-Gostin bath is not autocatalytic, and most likely the modified baths are also not autocatalytic. The hypophosphite system was reinvestigated more recently by Kurnoskin et al. (21) for the reaction mechanism. Using a split cell in which the anodic and cathodic partial reactions proceed in separate compartments, it has been demonstrated that in the initial stages the plating proceeds primarily by galvanic displacement, but at later stages exposed areas of nickel serve as the catalyst surface where hypophosphite is oxidized anodically, causing the cathodic deposition of gold to proceed in the remaining areas of the surface. The gold deposition continues in this manner for many hours, as long as areas of nickel remain exposed. The plating rate, of course, decreases with time as the total area of exposed nickel becomes progressively smaller. Swan and Gostin (19) reported that a maximum gold thickness of 23 μm was obtained on electroless nickel in 15 hours at 93°C with an initial plating rate of 4.8 μm/hr.

Vratny (22) used a hypophosphite bath to deposit gold on electroless nickel, which was plated over aluminum metallization on silicon integrated circuits. Examples of bath compositions used by Swan and Gostin (19) and Vratny (22) are given in Table 15.8. For other hypophosphite formulations, see Ref. 1.

As may be expected, gold deposits obtained from these types of baths are porous, and the diffusion of substrate metal to the gold surface occurs readily (23).

Using a Hydrazine Bath

An electroless gold bath with hydrazine as the reducing agent was described by Gostin and Swan (24) in 1962 (see Table 15.9). Their data show that gold plating continues for many hours, as in the hypophosphite system, yielding a thickness of 25 μm in 20 hours at 92-94° C on nickel. More recently, Moskvichev et al. (25,26) studied the reaction mechanism of this plating system and showed that three different reactions are involved when nickel is the substrate: galvanic displacement (Reaction 5), substrate-catalyzed deposition (Reactions 6 and 7), and autocatalytic deposition (Reaction 8).

$$2Au(CN)_2^- + Ni \rightarrow 2Au + Ni^{2+} + 4CN^- \qquad [5]$$

$$RCOOH + N_2H_4 \rightarrow RCONHNH_2 + H_2O \qquad [6]$$

$$RCONHNH_2 + 2Au(CN)_2^- \rightarrow 2Au + 4CN^- + N_2 + RCOH + 2H^+ \qquad [7]$$

$$2Au(CN)_2^- + N_2H_5^+ + 2H_2O \rightarrow 2Au + 2NH_3OH^+ + H^+ + 4CN^- \qquad [8]$$

Note that the fatty acid participates in the overall reaction via Reaction 6. Moskvichev et al. report that the plating rate of their bath (Table 15.9) is faster in the initial stages where the displacement and substrate-catalyzed reactions are in progress, but slows down considerably after the surface is completely covered with gold and the reaction becomes entirely autocatalytic. On electroless nickel substrates they obtained 3 μm in the first hour, while in subsequent hours the plating continued at a rate of only 1 μm/hr. Nevertheless, deposits obtained with this system are expected to be less porous than those produced by an entirely displacement process or by the hypophosphite process because of the participation of the autocatalytic reaction. No adverse effect of accumulated nickel ions has been reported in this system.

Table 15.8
Hypophosphite Baths

	Swan & Gostin	Vratny
KAu(CN)$_2$, g/L	2	0.5-10
KCN	—	0.1 → 6
NH$_4$Cl, g/L	75	—
Na-citrate (2H$_2$O), g/L	50	—
Na-acetate, g/L	—	1-30
NaHCO$_3$, g/L	—	0.2-10
NaH$_2$PO$_2$·H$_2$O, g/L	10	1-20
pH	7-7.5	4.5-9
Temperature, °C	93±2	18-98
Rate	2.3-5 μm	0.1-0.5 μm
(on electroless Ni)	in 1 hr	in 15 min

Using "High Build" Displacement Bath

A new "high build" displacement process has been reported recently to be capable of producing an essentially pore-free gold layer with a thickness of up to 3 μm (27). The maximum thickness attainable depends on the material and surface morphology of the substrate. A greater thickness can be obtained on materials with rough surfaces, such as metallized ceramics. The bath is operated in the pH range of 0.5-4 at 50-70° C, and it is reported to be insensitive to metallic contamination. No detailed bath composition has been published, however.

Non-Cyanide Baths

The accumulation of free cyanide ions with bath use causes a significant decline in plating rate of the borohydride bath. The trivalent gold cyanide system already mentioned is advantageous in this respect because it can be replenished with gold compounds containing no cyanide. Electroless gold baths that are entirely free of cyanide have been formulated, and they will be described below.

Au(III) Chloride with Weakly Reducing Amine Boranes

Burke et al. (28) combined Au(III) chloride complex, which is very easily reducible to gold metal, with ether-substituted tertiary amine boranes, which are very weak reducing agents, to prepare an autocatalytic bath. The procedure for synthesizing the reducing agent from dimethylsulfide borane and ether-substituted tertiary amine is also patented (29). Little (30) also uses the chloride complex of trivalent gold in combination with a tertiary or secondary amine borane such as trimethylamine borane, methyl morpholinoborane, or diisopropylamine borane. Stabilizers such as mercaptans, cyanohydroquinolines, or iodine compounds are used for plating temperatures above 35° C. Examples of bath formulations are given in Table 15.10.

Table 15.9
Hydrazine Baths

	Gostin & Swan	Moskvichev et al.
KAu(CN)$_2$, g/L	3	7
NH$_4$-citrate, g/L	90	—
Citric acid, g/L	—	30
NH$_4$Cl, g/L	—	90
FeSO$_4$, g/L*	—	1
Hydrazine sulfate, g/L	—	75
Hydrazine hydrate, g/L	0.0002	—
pH	7-7.5	5.8-5.9
Temperature, °C	92-95	95
Rate	1.3-7.3 μm/hr	3 μm (first hr) decreasing to 1 μm/hr*

*For plating on electroless Ni-P.

Au(I) Sulfite Baths

The Au(I) sulfite complex, which is being used successfully for electroplating gold, has been utilized to prepare electroless gold baths in combination with various reducing agents including hypophosphite, formaldehyde, hydrazine, borohydride, and dimethylamine borane (31-33). Three bath formulations are given in Table 15.11 as examples.

It is of interest to note that both hypophosphite and formaldehyde provide autocatalytic systems with Au(I) sulfite complex, but not with the conventional cyanide complex. On the other hand, it has been shown that gold metal is an excellent catalyst for the anodic oxidation of these reducing agents in solutions containing no cyanide species (34,35). Evidently, the catalytic activity of gold is poisoned by $Au(CN)_2^-$ and/or CN^- ions.

Low pH Baths

The high pH (>13) of the borohydride or amine borane baths is objectionable in applications involving materials that are not stable in alkaline media. Autocatalytic systems operated in the acidic pH range are available. The reducing agents used in such systems are hydroxylamine (36), hydrazine (25), and cyanoborohydride (37). The composition of the hydroxylamine bath is

---

Table 15.10
Amine Borane Baths with Au(III) Chloride

(1) Burke, Hough, and Hefferan (28)
Solution A:

$KAuCl_4$	3.0 g/L
KOH	to pH 14

Solution B:

$CH_3O(CH_2CH_2O)_3CH_2CH_2N(CH_3)_2BH_3$	7.1 g/L
KOH	to pH 14

Mix equal parts of Solution A and Solution B.
Temperature: 55° C
Rate: 8.64 mg/cm^2/hr (4.5 μm/hr) on Pd-activated electroless Ni

(2) Little (30)

$KAuCl_4 \cdot 3H_2O$	2 g/L
$Na_3PO_4 \cdot 12H_2O$	20 g/L
2-Mercaptobenzothiazole	1.2 mg/L
Trimethylamine borane	2 g/L
pH	11.9
Temperature	50° C
Rate (Pd-activated Ni)	0.64 μm/hr

Table 15.11
Au(I) Sulfite Baths

	Reducing Agent		
	Hypophosphite	Borohydride	Formaldehyde
$Na_1Au(SO_1)_2$, g/L	3	1.5	0.6
Na_2SO_1, g/L	15	15	15
1,2-diamino-ethane, g/L	1	1	—
KBr, g/L	1	1	—
Na_4EDTA, g/L	1	1	—
Na-citrate, g/L	—	—	5
NH_4Cl, c/L	—	—	7
Na-hypo-phosphite, g/L	4	—	—
$NaBH_4$, g/L	—	0.6	—
Formaldehyde (37% solution), g/L	—	—	1
pH	9	10	(not given)
Temperature, °C	96-98	96-98	96-98
Rate	0.95 mg/cm³/hr (0.5 μm/hr)	5.2 mg/cm³/hr (2.7 μm/hr)	19 mg/cm³/hr (10 μm/hr)

illustrated in Table 15.12. The fluoride serves as a stabilizer. It is said that no galvanic displacement is involved even in the initial stages of plating when the substrate is either copper or nickel. The hydrazine bath has already been described (Table 15.9). The cyanoborohydride bath of Bellis (37) is operated in the pH range of 1.5 to 5.0 (Table 15.13). At pH <1.5 the bath decomposes, while at pH >5.0 no plating takes place.

Other Baths

Efimov and Gerish (38) describe an electroless bath using hydrazinoborane as the reducing agent (Table 15.14). Other constituents are similar to those used in the conventional borohydride bath. The advantage of this bath is claimed to be the lower operating temperature (58-60° C). The plating rate on Ni-B substrates is 4 μm/hr during the initial 15 minute period, which thereafter decreases to 1.5-2 μm/hr.

Andrascek et al. (39) uses compounds with enol groups such as sodium ascorbate as reducing agents in an autocatalytic bath operated at pH 8. An example of the bath composition is given in Table 15.15. The bath is operated at 63° C and plates gold on copper, nickel, iron, and their alloys.

Table 15.12
Hydroxylamine Baths

	Example 1	Example 2
$NH_4Au(CN)_2$	0.015M	—
$KAu(CN)_2$	—	0.02M
Succinic acid	0.250M	—
KF	0.120M	—
Acrylic acid	0.125M	—
Na_2EDTA	0.010M	—
NH_4Cl	1.200M	—
Citric acid	—	0.10M
KHF_2	—	0.12M
KCl	—	2M
Hydroxylamine sulfate	0.025M	—
Hydroxylamine hydrochloride	—	0.06M
pH	2.3	2.8
Temperature, °C	85	70
Rate	1.2 μm/hr	0.8 μm/hr

GOLD ALLOYS

Au-Ag Alloy

Homogeneous Au-Ag alloys of any composition can be plated by operating the borohydride gold bath with continuous addition of $KAg(CN)_2$ and excess free cyanide (40). The continuous addition method must be used in order to maintain a constant ratio of Au to Ag in the bath and to achieve a uniform composition of the alloy deposit. The reason is that the silver complex is much more readily reducible than the gold complex. An example of bath composition and plating conditions is tabulated in Table 15.16.

Au-Cu Alloy

Molenaar (41) found that gold-copper alloys with a wide range of Au to Cu ratio can be plated from the baths prepared essentially by adding $Au(CN)_2^-$ to the conventional electroless copper plating bath containing EDTA and formaldehyde. Deposits with gold content ranging from 5 to >99.5 wt.% were obtained. Pure gold cannot be deposited from a formaldehyde bath containing $Au(CN)_2^-$, because cyanide poisons the catalytic activity of gold for the anodic oxidation of formaldehyde. When copper is present in the deposit even in a very small quantity, it serves as a catalyst on which formaldehyde is oxidized, resulting in the deposition of both copper and gold. The alloy obtained by this process has been shown to consist of homogeneous mixed crystals of Cu and

Table 15.13
Cyanoborohydride Baths

	Example 1	Example 2
KAu(CN)$_2$, g/L	7	10
Citric acid, g/L	30	30
Thiodiglycolic acid or mercapto-acetic acid, g/L	—	0.01
NaBH$_3$CN, g/L	2	2
pH	3.5	3.5
Temperature, °C	90	90
Rate (on Cu or Ni)	2.5 μm/hr	(not given)

Au with a characteristic lattice constant for each alloy composition. The lattice constant varies linearly with the composition of the alloy. This is in contrast to gold-copper alloys obtained by the conventional electrodeposition method, which yields deposits consisting partly of crystals of the individual metals and partly of Cu-Au mixed crystals. Bath compositions are illustrated in Table 15.17.

Au-Sn Alloy
An electroless gold-tin alloy plating bath has recently been described (42). It employs stannous chloride as the reducing agent. The deposition of tin metal proceeds via a disproportionation reaction of the divalent tin to form Sn(O) and Sn(IV). The plating is carried out at room temperature, and a small amount of toluene is said to accelerate the deposition by a factor of five, yielding a plating rate of 5 μm/hr. The tin content of the deposit can be varied between 5 and 60 percent. Concentration ranges for various bath constituents are tabulated in Table 15.18.

Table 15.14
Hydrazinoborane Bath

KAu(CN)$_2$, g/L	6
KCN, g/L	6.5
KOH, g/L	7-9
N$_2$H$_4$·BH$_3$, g/L	0.6
Temperature, °C	58-60
Rate	<4 μm/hr
(on Cu, Ni, Kovar)	

Table 15.15
Ascorbate Bath

KAu(CN)$_2$, g/L	3
Na$_2$HPO$_4$·2H$_2$O, g/L	32
Citric acid, g/L	1
Na$_4$EDTA·H$_2$O, g/L	3
Na-L(+) ascorbate, g/L	6
pH	8
Temperature, °C	63
Rate	?

APPLICATIONS

Some applications of electroless gold and gold alloys have been reported in the open literature, and they will be described briefly in the following paragraphs.

Borohydride gold has been used to plate fine line conductor patterns and beam leads on silicon integrated circuits (43). The beam leads were 75 μm wide and 10-12 μm thick. The patterns were formed either by vacuum evaporation of titanium followed by palladium through a metal shadow mask or by photolithography and etching. The capability of the borohydride process to plate 6000 discrete beams on each wafer with a thickness non-uniformity of <10% has been demonstrated. For patterned substrates a plating rate as high as 6 μm/hr can be obtained with the original borohydride bath.

Procedures for depositing gold also from the borohydride bath on III-V semiconducting crystals such as GaAs have been developed and used routinely in the fabrication of GaAs microwave field-effect transistors (44). The substrate is activated prior to the electroless plating by employing a solution containing PdCl$_2$ and HF.

Ohmic contacts on n-GaAs were formed using electroless Pd/Sn/Au films (45). In this process the first Pd layer, 250 angstroms thick, is formed simply by extending the activation time to >30 minutes in the PdCl$_2$-HF solution, and the Sn layer, 1000 angstroms thick, is deposited using an electroless tin plating process, which is then followed by the borohydride gold, 1500 angstroms thick. The material is then annealed at 420°C to form an alloy interface.

A modified acidic hydroxylamine bath of Dettke and Stein (36) has recently been used to deposit gold on p-InP activated with Pd using a localized photoactivation method (46). An alkaline bath cannot be used in this application because the active InP layer of the device is attacked by an alkali.

Commercially available aluminized polyvinylidene fluoride (PVDF) films used for making piezoelectric devices are subject to air oxidation, which makes it difficult to form a good electrical contact to the surface. To avoid this problem, Schiavone (47) plated adherent electroless gold using a borohydride bath

Table 15.16
Au-Ag Alloy Bath

$KAu(CN)_2$	0.026M
$KAg(CN)_2$	0.007M
KCN	0.35M
KOH	0.2M
KBH_4	0.2M
pH	13.5-14
Temperature, °C	75
Rate (on Cu)	0.4 μm/hr
Deposit composition	70-80 wt% Au

Deposit Compositions at Various Au/Ag Ratios in Solution

Au:Ag (Mol ratio)	Au in deposit (wt%)
26:10	20-30
26:5	30-40
26:1	40-50
26:0.5	70-90

directly on the PVDF film after etching in an organic solvent, sensitizing with $SnCl_2$, and activating first with $AgNO_3$ and then with $PdCl_2$.

The interior surface of wave guide tubes with complex shapes made of an aluminum alloy was first subjected to a double zincate treatment, followed by electroless nickel, copper, silver, and finally gold plating (48,49). In this process the electroless copper is plated for promoting adhesion of electroless silver, and the conductivity is provided primarily by the top two layers, i.e., silver and gold. The gold is plated to prevent oxidation of the silver.

Tungsten-metallized ceramics for semiconductor packaging have been plated with gold using a borohydride bath (50). It was necessary to preactivate the tungsten with gold and, after the electroless plating, to anneal in hydrogen atmosphere to promote adhesion. It should be noted that a number of displacement processes to plate gold on tungsten are described in the literature; see, for example, References 27 and 51.

Electroless gold plating of transistor headers has been evaluated, but no advantages were found over electrolytic gold in this application (23). Autocatalytic baths were not stable enough for this purpose, and electroless processes based on displacement or substrate-catalyzed reactions were found to give porous deposits, causing fast diffusion of substrate metals to the gold surface.

Table 15.17
Au-Cu Alloy Bath

$CuSO_4 \cdot 5H_2O$	0.04M
Na_4EDTA	0.072M
NaOH	0.12M
$KAu(CN)_2$	x M (see below)
KCN	0.0015M
Formaldehyde	0.10M
Temperature, °C	50

Compositions and Quantities of Deposited Alloys
At Various Gold Concentrations (x)

$KAu(CN)_2$ x M	Alloy deposited in 2 hrs mg/cm^2	Alloy composition, wt%	
		Au	Cu
0.00017	2.5	5.8	94.2
0.00035	2.8	7.4	92.6
0.00087	3.1	17.3	82.7
0.0017	3.6	48.3	51.7
0.0035	4.5	65.0	35.0
0.007	2.7 (1 hr)	74.9	25.1
0.014	2.3	99.0	1.0

Table 15.18
Au-Sn Alloy Bath

$KAu(CN)_2$, g/L	4-10
KCN, g/L	5-15
KOH, g/L	60-100
$SnCl_2$, g/L	40-60
Toluene, g/L	0.05-0.5
Temperature	Room temp.
Plating rate, $\mu m/hr$	
with no toluene	1.8
with 0.5 g/L toluene	5.0

Indications are that numerous other applications have been tried, but they are not documented in the open literature. In view of the current state of development of electroless gold, such applications are most likely to be on a rather small scale. As is apparent from this review, the basics of electroless gold plating are now much better understood than some ten years ago (1), and several improved processes are now available. A process suitable for continuous operation on a commercial scale is yet to be developed, however, and there is little doubt that such a process would vastly expand the range of application of electroless gold. It is hoped that this review will help facilitate the development of such a practical process.

REFERENCES

1. Y. Okinaka, in *Gold Plating Technology*, edited by F.H. Reid and W. Goldie, Chapter 11, Electrochemical Publications Limited, Ayr, Scotland (1974).
2. H.O. Ali and I.R.A. Christie, *Gold Bull.*, **17,** 4, 118 (1984).
3. Y. Okinaka, *Plating,* **57,** 914 (1970).
4. Y. Okinaka, R. Sard, C. Wolowodiuk, W.H. Craft, and T.F. Retajczyk, J. Electrochem. Soc., **121,** 56 (1974).
5. Y. Okinaka, *ibid.*, **120,** 739 (1973).
6. E.A. Efimov, T.V. Gerish, and I.G. Erusalimchik, *Zaschita Metallov.*, **11,** 3, 383 (1975).
7. E.A. Efimov, T.V. Gerish, and I.G. Erusalimchik, *ibid.*, **12,** 6, 724 (1976).
8. Y. Okinaka and C. Wolowodiuk, *Plating,* **58,** 1080 (1971).
9. A. Kasugai, Japanese patent (Kokai Tokkyo Koho), 80-24914 (1980).
10. M. Matsuoka, S. Imanishi, and T. Hayashi, *Proceedings of the Second Asian Metal Finishing Forum*, Tokyo, June 1-3, 1985, p. 121-124.
11. Y. Takakura, Japanese patent (Kokai Tokkyo Koho), 81-152958 (1981).
12. P. Prost-Tournier and C. Allenmoz, U.S. patent 4,307,136 (1981).
13. J.D.E. McIntyre and W.F. Peck, Jr., *J. Electrochem. Soc.*, **123,** 1800 (1976).
14. D.W. Endicott, H.K. James, and F. Nobel, *Plating and Surf. Finish.*, **68,** 11, 58 (1981).
15. N. Wakabayashi, S. Wakabayashi, and A. Murata, *ibid.*, **69,** 8, 63 (1982).
16. S. Sasaki, Japanese patent (Kokai Tokkyo Koho), 77-124,428 (1977).
17. M.F. El-Shazly and K.D. Baker, U.S. patent 4,337,091 (1982).
18. E.A. Parker, in *Gold Plating Technology*, edited by F.H. Reid and W. Goldie, Chapter 10, Electrochemical Publications Limited, Ayr, Scotland (1974).
19. S.D. Swan and E.L. Gostin, *Metal Finish.*, **59,** 4, 52 (1961).
20. A. Brenner, in *Modern Electroplating*, 2nd Ed., edited by F.A. Lowenheim, John Wiley, New York (1963).
21. G.A. Kurnoskin, I.V. Belova, V.A. D'yakonov, V.A. Plokhov, and V.N. Flerov, *Izv. Vyssh. Uchebn. Zaved., Khim. Khim. Tekhnol.*, **20,** 4, 533 (1977).
22. F. Vratny, U.S. patent 4,154,877 (1977).
23. R.K. Asher, *Plating,* **66,** 10, 46 (1979); *Gold Bulletin*, **13,** 1, 7 (1980).
24. E.L. Gostin and S.D. Swan, U.S. patent 3,032,436 (1962).
25. A.N. Moskvichev, G.A. Kurnoskin, and V.N. Flerov, *Zhur. Prikl. Khim.*, **54,** 9, 2150 (1981).

26. A.N. Moskvichev, G.A. Kurnoskin, V.N. Flerov, and Z.P. Gerasimova, *Izv. Vyssh. Uchebn. Zaved., Khim. Khim. Tekhnol.*, **25**, 9, 1104 (1982).
27. M.F. El-Shazly and K.D. Baker, *Proc. First AES Electroless Plating Symposium*, St. Louis, March 23-24, 1982.
28. A.R. Burke, W.V. Hough, and G.T. Hefferan, U.S. patent 4,142,902 (1979).
29. A.R. Burke and W.V. Hough, U.S. patent 4,080,381 (1978).
30. J.L. Little, British patent application GB 2114159 (1983); *Gold Patent Digest*, **1**, 4, 15 (1983).
31. F. Richter, R. Gesemann, L. Gierth, and E. Hoyer, German (East) patent 150762 (1981).
32. R. Gesemann, F. Richter, L. Gierth, U. Bechtloff, and E. Hoyer, *ibid.*, 160283, (1983).
33. R. Gesemann, F. Richter, L. Gierth, E. Hoyer, and J. Hartung, *ibid.*, 160284 (1983).
34. I. Ohno, O. Wakabayashi, and S. Haruyama, *J. Electrochem. Soc.*, **132**, 2323 (1985).
35. J.E.A.M. Van den Meerakker, *J. Appl. Electrochem.*, **11**, 387 (1981).
36. M. Dettke and L. Stein, German patent DE 3029785 (1982); *Gold Patent Digest*, Pilot issue, p. 9 (1982).
37. H.E. Bellis, U.S. patent 3,697,296 (1972).
38. E.A. Efimov and T.V. Gerish, *Zashch. Met.*, **15**, 2, 240 (1979).
39. E. Andrascek, H. Hadersbeck, and F. Wallenhorst, German patent DE 3237394 (1984); *Gold Patent Digest*, **2**, 3, 14 (1984).
40. J. Jostan and W. Mussinger, German patent DE 3132676 (1983); *Gold Patent Digest*, **1**, 2, 8 (1983).
41. A. Molenaar, *J. Electrochem. Soc.*, **129**, 1917 (1982); European patent application EP 70061 (1983); *Gold Patent Digest*, **1**, 2, 12 (1983).
42. V.P. Pilnikov, USSR patent SU 985136 (1983); *Gold Patent Digest*, **2**, 2, 14 (1984).
43. R. Sard, Y. Okinaka, and H.A. Waggener, *J. Electrochem. Soc.*, **121**, 62 (1974).
44. L.A. D'Asaro, S. Nakahara, and Y. Okinaka, *ibid.*, **127**, 1935 (1980).
45. D. Lamouche, P. Clechet, J.R. Martin, G. Haroutiounian, and J.P. Sandino, *Surface Science*, **161**, L554 (1985).
46. G. Stremsdoerfer, J.R. Martin, M. Garrigues, and J.L. Perossier, *J. Electrochem. Soc.*, **133**, 851 (1986).
47. L.M. Schiavone, *ibid.*, **125**, 522 (1978).
48. Y. Takakura, *Jitsumu Hyomen Gijutsu (Metal Finishing Practice)*, **27**, 1, 31 (1980).
49. Y. Takakura, Japanese patent (Kokai Tokkyo Koho), 81-269 (1981).
50. T. Mesaki, J. Nakamura, and N. Onozaki, Proceedings of the 55th Meeting of the Metal Finishing Society of Japan, p. 110 (1977).
51. R.K. Asher, *Plating and Surf. Finish.*, **66**, 10, 46 (1979); *Gold Bull.*, **13**, 1, 7 (1980).
52. A.A. Halecky and M.F. El-Shazly, German patent DE 3343052 (1984); *Gold Patent Digest*, **2**, 3, 13 (1984).

Chapter 16
Electroless Plating
Of Platinum Group Metals

Yutaka Okinaka and Catherine Wolowodiuk

Interest in the deposition of platinum group metals and alloys arises often in connection with applications to the fabrication of electronic devices and components, and also as a means of providing corrosion protection to basis metals in various other applications. For example, electroplated palladium and its alloys have received considerable attention in recent years as substitute materials for the gold plated on connectors and printed circuit board contacts. Physically deposited palladium and platinum are used as a barrier layer to prevent interaction between the first layer (e.g., Ti) and the top layer of gold in multilayer conductors used on semiconductor devices and circuit boards. There is also an interest in palladium as a material for forming metallic contacts on compound semiconductors such as GaAs and InP. Ruthenium is another metal known to be suitable as a contact finish for certain applications. The use of plated rhodium as a wear-resistant decorative finish is well known. Platinum group metals are also known for their catalytic activities for numerous chemical reactions, and the catalysts containing those metals have been deposited from solutions. The principle of electroless plating would be of interest in all of the above applications. However, as far as the authors are aware, there is no established industrial application of electroless plating of platinum group metals. A survey of the literature shows that development activities in this area are inconspicuous at present. On the other hand, some commercial baths have recently become available for plating palladium and rhodium.

This chapter reviews those electroless (autocatalytic) processes described in the literature for plating platinum group metals and alloys. Displacement processes and specialized deposition techniques such as those developed for catalyst manufacture will not be covered in spite of the fact that these processes often are also called "electroless". An extensive coverage will be given to each known autocatalytic process, including listings of essential bath compositions and operating conditions in the hope that the reader can experiment with the various processes based on the information found in this chapter alone. The following metals and alloys will be covered: Pd, Pd-P, Pd-B, Pd-Ni-P, Pd-Co-P, Pd-Zn-P, Pt, Pt-Rh, Pt-Ir, Pt-Pd, Ru, and Rh.

PURE PALLADIUM, PALLADIUM-PHOSPHORUS, AND PALLADIUM-BORON

Reducing agents used for electroless plating of palladium include hydrazine, hypophosphite, amine borane, and formaldehyde. Deposits produced with hypophosphite or amine borane contain 1-3 percent by wt of phosphorus or boron, but they will be discussed in this section. Alloys with other elements will be treated in the next section.

Hydrazine Baths
Rhoda (1)
The first electroless palladium plating bath was developed by Rhoda in 1958 using hydrazine as the reducing agent. Two bath formulations are given in Table 16.1. Palladium is used in the form of tetraammine chloride, $Pd(NH_3)_4Cl_2$, in these examples, but Rhoda states that amine complexes can be used as well. The solution of the tetraammine complex can be prepared either by dissolving palladium diammine chloride, $Pd(NH_3)_2Cl_2$, in dilute ammonia, or by adding a solution of $PdCl_2$ (e.g., 50 g/L Pd in 2M HCl) to dilute ammonia and heating until the precipitate that forms initially dissolves. The overall plating reaction is believed to be represented by

$$2Pd(NH_3)_4^{2+} + N_2H_4 + 4OH^- \rightarrow 2Pd + 8NH_3 + N_2 + 4H_2O \qquad [1]$$

The plating rate increases linearly with temperature between 40 and 80°C from 3.8 to 14.7 μm/hr. The EDTA salt is added as a stabilizer. Without EDTA, the bath decomposes spontaneously when the temperature exceeds 70°C. If the

Table 16.1
Hydrazine Bath for Pd (1)
(Rhoda)

	Bath A*	Bath B
$Pd(NH_1)_4Cl_2$, g/L as Pd	5.4	7.5
Na_2EDTA, g/L	33.6	8.0
NH_4OH, g/L	350	280
Hydrazine, g/L	0.3	—
Hydrazine (1M), mL/hr	—	8
Temperature, °C	80	35±5
Plating rate, μm/hr	25.4	0.89
Plating area, cm²/L	100	1000

*Bath A for rack plating, Bath B for barrel plating.

bath is allowed to stand idle at the operating temperature, the plating rate decreases rapidly with time. For example, the initial rate of 15 μm/hr decreases to 3.8 μm/hr in two hours, and practically to zero in four hours. This phenomenon is attributed to catalytic decomposition of hydrazine by palladium.

The deposit is reported to be at least 99.4 percent pure. It has a density of 11.96 g/cm^3 and an average Knoop hardness of 257, with a range of 150 to 350 at a 25 g load. The hardness depends on the plating rate, with the higher rates producing the softer and more ductile deposits. Rhoda states that good deposits were obtained on Al, Cr, Co, Au, Fe, Mo, Ni, Pd, Pt, Rh, Ru, Ag, steel, Sn, W, graphite, carbon, and properly activated glass and ceramics. Copper and its alloys were plated with palladium after first plating with a catalytic metal such as a displacement-type gold.

Kawagoshi (2)

Bath instability is a serious drawback of the original Rhoda formulation. In order to obtain an improved stability, Kawagoshi added a small amount of thiourea. The bath composition given in Table 16.2 yields 3 g of palladium deposit on 1 m^2 of a steel substrate in one minute at 80° C. This result corresponds to a plating rate of approximately 16 μm/hr.

Table 16.2
Hydrazine Bath for Pd (2)
(Kawagoshi)

PdCl$_2$, g/L	5
Na$_2$EDTA, g/L	20
Na$_2$CO$_3$, g/L	30
NH$_4$OH (28% NH$_3$), mL/L	100
Thiourea, g/L	0.0006
Hydrazine, g/L	0.3
Temperature, °C	80
Plating rate	0.26 μm/min

Laub and Januschkowetz (3)

The baths developed by these workers contain both a stabilizer and an accelerator. As the stabilizer, a mercaptoformazan with the general formula

$$S=C\begin{smallmatrix} \diagup NH\text{-}NHR \\ \diagdown N=NR \end{smallmatrix}$$

is used. Examples are N,N'-di-β-naphthyl-C-mercaptoformazan, N,N'-di-m-carboxyphenyl-C-mercaptoformazan, N,N'-diphenyl-C-mercaptoformazan,

and N,N'-p-phenyl-sulfonic acid-C-mercaptoformazan. As the accelerator, a
water-soluble benzene derivative with two or three functional groups is added.
The following compounds are listed as examples: 3,4-dimethoxybenzoic acid,
2-hydroxy-3-methylbenzoic acid, 2-hydroxy-4-methylbenzoic acid, 2-hydroxy-
4-methoxybenzoic acid, 3-hydroxy-4-methylsulfonic acid, and 2-amino-5-oxo-
sulfonic acid. Palladium is added to the baths as Pd acetate, $PdCl_2$, or
$Pd(NH_3)_2(NO_2)_2$ with a carboxylic acid and ammonia as complexing agents.
Either hydrazine or sodium hypophosphite can be used as the reducing agent.
Two examples of bath compositions and operating conditions are listed in Table
16.3. Both baths apparently behave similarly under identical operating
conditions. The patent states that pure, bright deposits are obtained on Fe, Ni,
Co, Cu, Cu-alloys, Al, Ag, Au, Pt, and activated plastic materials. It is indicated
that the baths must be replenished periodically during operation to maintain the
given compositions.

In spite of the improvement brought about by the addition of stabilizing
agents, the hydrazine baths suffer from a serious drawback: the plating rate
decreases rapidly during bath operation to a much greater degree than expected
from the depletion of palladium in the bath, because of the catalytic
decomposition of hydrazine itself. Hypophosphite baths do not have this
problem and have been investigated more extensively, as described in the next
section.

Table 16.3
Hydrazine or Hypophosphite Bath for Pd (3)
(Laub and Januschkowetz)

	Hydrazine Bath	Hypophosphite Bath
Pd acetate, g/L	8.5	—
$PdCl_2$, g/L	—	7
NH_4 acetate, g/L	70	—
Na succinate, g/L	—	75
NH_4OH (25% NH_3), mL/L	100	100
N,N'-di-o-tolyl-C-mercaptoformazan, g/L	0.02	—
N,N'-di-p-phenylsulfonic acid-C-mercaptoformazan, g/L	—	0.015
3,4-dimethoxybenzoic acid, g/L	10	—
2-amino-5-oxo-sulfonic acid, g/L	—	10
Hydrazine hydrate, 8%, mL/L	10	—
Na hypophosphite, g/L	—	15
pH	8.5-9.0	8.5-9.0
Temperature, °C	65-70	70
Loading, dm^2/L	5	5
Plating rate, μm/hr	15-20	15-20

Hypophosphite Baths
Sergienko (4)

Sergienko was the first to disclose the composition of a hypophosphite bath for electroless palladium plating. An example of specific bath formulation and operating conditions is given in Table 16.4. EDTA and ethylenediamine are said to serve as stabilizers (in addition to forming complexes with palladium) because metallic impurities are also complexed with these agents, minimizing deposition of such impurity metals. The following specific instruction is given in the patent for proper preparation of the bath—"Add $PdCl_2$, ethylenediamine, and disodium EDTA to water and dissolve the solids. Permit the solution to stand until chelation is complete. Chelation can be accomplished in approximately 24 hours if the temperature of the solution is maintained at approximately 160° F (71° C). After chelation is complete, cool the solution to 20° C, add sodium hypophosphite, and adjust the pH to 8.5. Heat the bath through a water jacket to 71° C for plating." The bath can be used, stored, and reused. It can be replenished with $PdCl_2$, which may be added directly to the bath, and with acid. No information is given on plating rate or deposit properties.

Table 16.4
Hypophosphite Bath for Pd (4)
(Sergienko)

$PdCl_2$, g/L	10.0
Na_2EDTA, g/L	19.0
Ethylenediamine, g/L	25.6
NaH_2PO_2, g/L	4.1
pH at 20° C	
(with HCl)	8.5
Temperature, ° C	71

Pearlstein and Weightman (5,6)

Independently of Sergienko (4), these investigators developed a hypophosphite bath using ammonia as the complexing agent and made a detailed study of the process. Palladium is present in this bath as palladous ammine complex, which is prepared by adding slowly, with agitation, a stock solution of $PdCl_2$ (20 g dissolved in 40 mL of 38% HCl and the solution diluted to one liter) into an appropriate volume of NH_4OH. This solution is allowed to stand at room temperature for at least 20 hours and filtered before use. Ammonium chloride is added to stabilize the bath for extended use, although it lowers the plating rate. Hypophosphite is added last, and the bath is brought to final volume. A recommended bath composition is given in Table 16.5.

Behavior of this bath is well documented in the original publications (5,6). The effect of Pd concentration on deposition rate was determined by depleting the

Table 16.5
Hypophosphite Bath for Pd (5)
(Pearlstein and Weightman)

$PdCl_2$, g/L	2
HCl (38%), mL/L	4
NH_4OH (28% NH_3), mL/L	160
NH_4Cl, g/L	27
$NaH_2PO_2 \cdot H_2O$, g/L	10
pH	9.8±2.0
Temperature, °C	50-60
Plating rate	2.5 μm/hr

bath up to 90 percent. The rate decreased almost linearly with the extent of depletion. When the bath was replenished by addition of Pd-ammine complex, the rate was restored almost to the original value, even without addition of hypophosphite. The determination of hypophosphite utilization efficiency showed that only about 31 percent is consumed for the reduction of palladium and the remainder simply decomposes to evolve gaseous hydrogen from the bulk of the solution. The plating rate increases with temperature and hypophosphite concentration up to 60° C and 20 g/L, respectively, above which the bath becomes unstable. Electroless palladium plates spontaneously on copper, brass, gold, steel, or electroless nickel, but there is an initial incubation period ranging from 20 sec to 1.5 min, depending on the substrate. Pretreatment with a mixture of 0.1 g/L $PdCl_2$ and 0.5 mL/L HCl (38%) for 30 sec at room temperature followed by DI water rinse reduces the incubation time. The plating is almost instantaneous on freshly plated nickel or non-metallic surfaces activated by using a $SnCl_2$-$PdCl_2$ process. The palladium deposit produced by using this bath has a hardness of approximately 165 kg/mm^2 on the Vickers scale. The existence of stresses in the deposit was revealed in a microscopic observation of cross sections of the deposit showing a crack pattern. It should be noted that the deposit produced by this process contains about 1.5 percent phosphorus.

The Pearlstein-Weightman bath was used by Hsu and Buxbaum (7) to coat zirconium with an adherent, 5-μm-thick palladium deposit. Zirconium, which is an important metal used in the nuclear and chemical industries, often requires a coating of thin metal films for corrosion protection or for modification of surface properties. A series of special pretreatment steps, which are required to produce deposits without flaking, peeling, or blistering, are described in the original article (7).

Pearlstein and Weightman also obtained various palladium alloys by operating the same basic bath with the addition of suitable alloying elements, as will be discussed in a subsequent section.

Zayats et al. (8)

These investigators found that the addition of a small amount of thiosulfate is a very effective way of stabilizing the bath. Within a limited range of thiosulfate concentration, the bath yields, at otherwise identical conditions, a plating rate much greater than that obtained with the additives used in the two baths described above. A recommended bath composition is given in Table 16.6. The thiosulfate concentration must be controlled carefully, because when it is $<10^{-4}$M, the solution is unstable, whereas at $>3 \times 10^{-3}$M no plating takes place, even if the temperature is raised to 80-90° C. A peculiar dependence of plating rate on Pd concentration was noted; namely, the rate initially increases linearly with Pd concentration, but it reaches a maximum at 0.03M and then falls to zero at 0.06M. This dependence is attributed to rapid consumption of hypophosphite at high Pd concentrations, but this interpretation appears to require a further study. With filtration, the bath is completely stable until exhaustion under the recommended set of conditions. It decomposes spontaneously at 65° C.

Table 16.6
Hypophosphite Bath for Pd (8)
(Zayats et al.)

$PdCl_2$	0.02-0.025 M
	(3.6-4.4 g/L)
NH_4OH (28% NH_3)	0.7-1.4M
	(10-20 mL/L)
$NaH_2PO_2 \cdot H_2O$	0.1-0.2M
	(10.6-21.2 g/L)
$Na_2S_2O_3 \cdot 5H_2O$	1.5-1.8 $\times 10^{-4}$M
	(0.0372-0.0447 g/L)
pH	8-10
Temperature	40-50° C
Plating rate	2-3 μm/hr (40° C)

Vereshchinskii et al. (9,10)

A detailed study for optimizing the ammoniacal hypophosphite bath was conducted by these investigators. The recommended bath composition and operating conditions are shown in Table 16.7. The solution is prepared by first dissolving $PdCl_2$ in 10M HCl (10 g salt per 50 mL HCl) and then adding sodium pyrophosphate and NH_4OH into the palladium solution. The pyrophosphate stablizes the bath and tends to decrease the plating rate. NH_4F, an accelerator, is added to compensate for the decrease of plating rate. The molar ratio equal to one is recommended for palladium to hypophosphite concentrations. The solution that has been used for several hours can be stored for two months without any loss of stability in a proper vessel made of glass, quartz, polyethylene, or Plexiglass.

Table 16.7
Hypophosphite Bath for Pd (9)
(Vereshchinskii et al.)

PdCl$_2$	0.05-0.055M
	(8.9-9.8 g/L)
Na$_4$P$_2$O$_7$·10H$_2$O	0.11-0.12M
	(49.1-53.5 g/L)
NH$_4$OH (25% NH$_1$)	0.1M
	(8 mL/L)*
NH$_4$F	0.3-0.4M
	(11.1-14.8 g/L)
NaH$_2$PO$_2$·H$_2$O	0.05M
	(5.3 g/L)
pH	10.0
Temperature	45-50° C

*The NH$_4$OH concentration of 8M given in Ref. 8 should read 8 mL/L.
See Ref. 9.

The hypophosphite bath is sensitive to impurities such as Zn, Fe, Ni, Cu, and certain organic compounds, and this aspect was studied in detail. For example, the bath stability at 50° C decreases seven times with 0.4 mg/L Zn, three times with the same amount of Fe, and four times with 0.5 g/L saccharin or gelatin.

The phosphorus content of a 5- to 10-μm-thick deposit was reported to be 1 to 2.5 percent by wt. Hydrogen was practically absent in the deposit. The internal stress was low, and no cracking was observed. The mean hardness value was 360 kg/mm^2 as compared to 200-270 kg/mm^2 for electroplated palladium. The porosity determined with a "Walker reagent" for a 2-3 μm-thick deposit on a nickel substrate was 3-4 pores per 100 cm^2. A bright, smooth palladium coating was obtained from the recommended bath on a variety of substrates including Ni, Cu, Ag, Co, Pt, brass, bronze, Kovar, graphite, and glass (roughened, degreased, and activated).

Laub and Januschkowetz (3)

According to the patent granted to these investigators, either hydrazine or hypophosphite can be used in their baths containing a mercaptoformazan (stabilizer) and a benzene derivative with certain functional groups (accelerator). An example is included in Table 16.3 in the previous section.

Mizumoto et al. (11)

Most recently, these workers investigated the behavior of an ammoniacal hypophosphite bath containing thiodiglycolic acid (TDG) as a stabilizer. Their basic bath composition is shown in Table 16.8. The initial incubation period observed with a Pt sheet substrate coated with electroplated palladium was

Table 16.8
Hypophosphite Bath for Pd (11)
(Mizumoto et al.)

PdCl$_2$	0.01M
	(1.78 g/L)
NH$_4$OH (28% NH$_3$)	2.96M
	(200 mL/L)
Na$_4$EDTA·2H$_2$O	0.01M
	(3.7 g/L)
Thiodiglycolic acid	1.33 x 10^{-4}M
	(20 mg/L)
NaH$_2$PO$_2$·H$_2$O	0.06M
	(6.36 g/L)
Temperature	40° C
Plating rate	1.3 mg/cm^2/hr
	(1.1 μm/hr)*

---

*With 0.03M PdCl$_2$, the rate increases to 2.3 mg/cm^2/hr (2.0 μm/hr).

about 5 min. At 30° C, it was as long as 20 min. The length of incubation period is shorter at higher palladium and hypophosphite concentrations. It becomes nearly constant when the molar ratio of NaH$_2$PO$_2$ to PdCl$_2$ exceeds a value of 3 or 4. A unique feature of this bath appears to be the insensitivity of the plating rate to the variation of stabilizer concentration—the rate is reported to be constant between 5 and 200 mg/L TDG. The rate is also insensitive to the variation of NH$_4$OH concentration between 50 and 200 mL/L. These properties are important when the bath is to be used over an extended period of time. This bath yields bright, silvery white palladium with a phosphorus content less than 2 percent at 1 mg/cm^2 (0.9 μm). The phosphorus content is greater in the initial stages of deposition.

The stability of contact resistance of electroless Pd plated on copper was compared with that of semi-bright electroplated nickel and electroplated cobalt-hardened gold. In a 500-hr indoor exposure test and a 50-hr steam treatment, the resistance stability of electroless Pd was significantly better than that of the nickel, and it was comparable to the stability of the go'd. This property is important in electronic contact applications.

Tertiary Amine Borane Baths

Amine boranes have been known as reducing agents for electroless deposition of other metals, especially nickel and gold. Dimethylamine borane (DMAB) is best known. For electroless deposition of palladium, however, the reducing power of this compound is too strong to formulate a stable bath.

Hough et al. (12) discovered that tertiary amine boranes with much weaker reducing power are suitable for preparing stable electroless palladium baths. The following three types of tertiary amine boranes have been found to be useful:

(1) trialkylamine boranes, $R_1R_2R_3NBH_3$, where R_1, R_2, and R_3 are either a methyl or an ethyl group;

(2) straight-chain methoxy-substituted dimethylamine boranes, $CH_3(OCH_2CH_2)_nN(CH_3)_2BH_3$, where n = 1 to 4; and

(3) N-alkyl substituted morpholine boranes, in which the alkyl group contains three or less carbon atoms.

The substituted DMAB (#2 above) gives black, spongy deposits. With the other two types of tertiary amine boranes, gray, smooth deposits can be obtained. Examples of bath compositions are given in Table 16.9. Only those giving smooth deposits are included. Among several classes of compounds that stabilize the baths, thio organic compounds, especially 2-mercaptobenzo-thiazole (MBT) and 3,3'-thiodipropionitrile, have been found to be most effective. The baths are stable for several days at 55° C, and indefinitely at 45° C or at ambient temperature.

The deposits obtained with Bath C in Table 16.9 were tested for hardness and analyzed for composition. The Knoop hardness at 25 g load was 718-764, which is much higher than the hardness of pure palladium metal (70-250), electroplated hard gold (max. 300-350), and electroless Ni-P (500). An X-ray examination showed that the deposit consists of amorphous Pd, amorphous B (1-3 percent by wt), crystalline $PdH_{0.706}$ (1-3 percent), and no more than traces of crystalline Pd and B. The adhesion to the substrate was especially strong on electroless nickel.

Table 16.9
Amine Borane Baths for Pd (12)
(Hough et al.)

	Bath A	Bath B	Bath C
$Pd(NH_3)_4Cl_2 \cdot H_2O$, g/L	3.75	—	—
$PdCl_2$, g/L	—	4.0	4.0
NH_4OH, M	0.3	0.8	0.6
Trimethylamine borane, g/L	3.0	—	2.5
N-Methylmorpholine borane, g/L	—	1.0	—
Mercaptobenzothiazole, mg/L	—	30	3.5
pH	11.4	11.0	?
Temperature, °C	50	45	45
Plating rate, mg/cm²/hr	3.6-3.8	1.0	1.8-2.0
Plating rate, μm/hr	3.2-3.4	0.88	1.6-1.8

The bath compositions and operating conditions described in the original patents (12) evidently have been modified in commercial baths to give a plating rate as high as 3 μm/hr at 60-65° C.

Formaldehyde Bath

Abys (13) formulated an acidic (pH $<$2) electroless palladium bath using formaldehyde as the reducing agent. A typical bath composition is given in Table 16.10. Oxalic acid, tartaric acid, or citric acid can be substituted for formic acid. Instead of nitric acid, hydrochloric or sulfuric acid may be used. The deposit was found to be very bright, indicating that saccharin acts as a brightening agent. The plating takes place on the following materials: Cu, Au, Ag, Ni, Pt, brass, Permalloy, Kovar, as well as semiconductors such as GaAs, InP, and Si. The plating reaction appears to be autocatalytic. The process is unique in that formaldehyde is used in an acidic medium. No extensive characterization has been carried out for this system.

Table 16.10
Formaldehyde Bath for Pd (13)
(Abys)

PdCl$_2$	0.1M
	(1.78 g/L)
Formic acid	0.4M
HNO$_3$	1.0M
Formaldehyde	2.0M
Saccharin	0.002M
pH	1.0-1.5
Temperature	50° C
Plating rate	0.15 μm/min

PALLADIUM ALLOYS

Pearlstein and Weightman (5) investigated the possibility of producing electroless palladium alloys with the addition of various metal salts to the hypophosphite-based bath shown in Table 16.5.

Both electroless nickel and cobalt can be deposited independently of palladium from the same basic medium, and the addition of salts of these elements to the palladium bath yields baths producing Pd-Ni-P and Pd-Co-P alloys. However, palladium is preferentially deposited as indicated by the fact that the addition of the alloying elements at ten times the molar concentration of palladium yields deposits containing only 6 percent Ni and 10 percent Co, respectively (see Table 16.11).

**Table 16.11
Electroless Palladium Alloys (5)**

| | Deposit composition, % (wt) | | |
Metal salt added	Alloying metal	Phosphorus	Pd
None	—	1.52	Balance
$NiSO_4 \cdot 6H_2O$ (29.6 g/L)	5.99	2.68	Balance
$CoSO_4 \cdot 6H_2O$ (29.6 g/L)	9.82	2.77	Balance
$ZnSO_4 \cdot 8H_2O$ (36.0 g/L)	35.99	1.24	61.43

On the other hand, the addition of zinc sulfate produces a deposit containing as much as 36 percent Zn. This induced reduction of zinc is of interest in view of the high electronegativity of this element, from which the reduction of zinc is not expected to occur independently of palladium.

The addition of salts of tungsten and rhenium to the bath did not lead to the formation of their alloys.

PURE PLATINUM

Electroless platinum has been plated using hydrazine or borohydride as the reducing agent. However, the operation of those processes appears to be more difficult than that for electroless palladium, and the recent literature does not reveal much development activities in this area. Those few available processes will be reviewed below.

Hydrazine Baths
Rhoda and Vines (14)
The process patented by these investigators is intended primarily to produce catalytically active platinum layers on materials suitable for making, for example, fuel cell electrodes. Examples of such materials include nickel powder and graphite powder compacts. However, with the addition of a suitable stabilizer, the same process is claimed to be usable for producing bright decorative or protective coatings on less noble metals. An example of bath composition yielding such deposits is given in Table 16.12. Ethylamine (or other nitrogen compounds such as EDTA, quinoline, and sulfamate) serves as the stabilizer. The solution of $Na_2Pt(OH)_6$ is prepared by boiling a chloroplatinic acid solution with excess NaOH. Hydrazine is added just before the plating is started, and its concentration is maintained by either continuous or intermittent additions during the plating run. Hydrazine is not stable in this system. This bath was used to plate platinum on copper, but these materials are also claimed to be plateable: Fe, Mo, Ni, Ag, and Ti.

Table 16.12
Hydrazine Bath for Pt (14)
(Rhoda and Vines)

$Na_2Pt(OH)_n$ (as Pt), g/L	10
NaOH, g/L	5
Ethylamine, g/L	10
Hydrazine hydrate, g/L	1*
Temperature, °C	35
Plating rate	12.7 μm/hr

*Hydrazine is added intermittently to maintain this concentration.

Leaman (15)

Leaman prepared hydrazine-based electroless platinum baths in both acidic and alkaline media and investigated their characteristics. The acidic bath is prepared by mixing two solutions, A and B. Four steps are followed to prepare solution A: (1) pure platinum sponge is dissolved in aqua regia; (2) the resulting solution is evaporated to dryness with low heat; (3) the residue is dissolved in 4 percent by vol HCl, and the solution is evaporated to dryness again; and finally (4) the residue is dissolved in 4 percent HCl (100 mL per gram of Pt). To prepare solution B, hydrazine hydrochloride is dissolved simply in water to make a 1 percent solution. The two solutions are mixed by pouring solution B into an equal volume of solution A. The temperature range for optimum plating is 60-70°C. With agitation, the acidic bath yields deposits to thicknesses up to about 1.5 μm with relatively low stress. With agitation, bright and adherent deposits to about 7.6 μm are attained. In addition to the main bath constituents, various sulfonic acids (e.g., sulfosalicylic acid, benzene sulfonic acid, 2,7-naphthalene disulfonic acid, 0.1-1.0 g/L) and non-ionic surfactants may be employed to improve bath stability, provide reduction in stress, and lower the surface tension of the solution to facilitate the removal of evolved nitrogen gas from the surfaces being plated. The acidic bath plates platinum on Au, Pd, noble metal alloys, and non-conductive materials such as ABS plastics. Because of the relatively rapid loss of hydrazine due to hydrolysis, the plating rate decreases quickly with time, especially at the elevated plating temperatures. However, with proper bath formulation and adequate bath loading, almost all of the platinum initially present in solution can be plated out before the bath becomes inactive. Therefore, a bath can be designed specifically for a given application in such a way that a desired deposit thickness is obtained by allowing the plating to continue until nearly all platinum is exhausted. An example given by Leaman shows that an initial plating rate measured in the first minute, 5.3 μm/hr, decreased to zero after 30 minutes, during which time 98 percent of the platinum initially present in the bath plated out, and a final deposit thickness of 1.2 μm was obtained.

The alkaline bath is prepared by mixing solutions A', B', and NH$_4$OH. Solution A' is a platinum solution prepared similarly to solution A for the acidic bath, except that water is added instead of HCl, for the evaporation procedure to accomplish complete removal of free HCl*. The final residue is dissolved in water to give a solution of 10 g/L Pt. Solution B' is a 50 percent by vol aqueous hydrazine solution. The bath is made up by mixing proper volumes of solutions A', B', water, and concentrated NH$_4$OH. The volume of concentrated NH$_4$OH to be added should correspond to 20 percent by vol of the final bath volume. Platinum and hydrazine concentrations can be varied over wide ranges, depending on bath loading. As an example, Leaman's data show that the concentrations of 1 g/L Pt and 4 mL/L 50 percent hydrazine give a 99 percent efficiency (the percentage of platinum plated out with respect to the initial amount of platinum in solution) in a 30 min plating run with a bath loading of 38.7 cm^2/100 mL. The initial plating temperature was 46° C, which was increased to 70-75° C over a period of 10 minutes, and the plating was allowed to continue for 30 minutes. As in the case of the acidic bath, sulfonic acids (e.g., p-aminobenzene sulfonic acid) and non-ionic surfactants can be added to improve the performance of the alkaline bath. The alkaline formulations are claimed to be suitable for plating platinum on metals more noble than Cu as well as properly activated non-conducting materials, such as polypropylene and polysulfone. Leaman notes that the main disadvantage of this bath is the necessity to use it immediately after preparation, and that working with large bath volumes is difficult.

Strejcek (16)

Strejcek's process uses cis-diammine platinum dinitrite as the source of platinum. The overall plating reaction is

$$2(NH_3)_2Pt(NO_2)_2 + N_2H_4 \cdot H_2O \rightarrow 2Pt + 5N_2 + 9H_2O \qquad [2]$$

This reaction accompanies no accumulation of reaction byproducts in the solution. Sodium nitrite and disodium hydrogen phosphate (optional) are used as a stabilizer and accelerator, respectively. A dilute solution of hydrazine hydrate is added dropwise to the bath at pH 7-9 and 60-95° C. The process deposits platinum on metals, plastics, ceramics, and glass. Catalytically active metals (Ni, Pt, Pd) can be plated directly, whereas for non-catalytic metals (Cu, Ag, Au) the plating reaction must be initiated by contacting them with basic materials such as Al or Mg. Non-conducting materials such as plastics can be activated by immersion in an acidic stannous chloride solution.

---

*It is emphasized that the chloroplatanic acid prepared by HCl evaporations as described for the acidic bath cannot be reduced by hydrazine in ammoniacal media. Evidently, different platinum compounds are formed depending on whether HCl is present during evaporation or not.

Strejcek gives a detailed description of his experiments. Only one example will be given here. Sodium nitrite, 10 g, was dissolved in 100 mL of water, and the solution was heated to boiling. To the boiling solution, 0.124 g of cis-$(NH_3)_2Pt(NO_2)_2$ is added. After complete dissolution (about 5 min), the temperature was lowered to and kept at 75° C. The following substrates were placed in this solution: (a) a Cu wire precleaned with concentrated HCl, (b) a Ni sheet pretreated with dilute HNO_3, and (c) a PVC rod degreased with tetrachloroethylene and activated by immersion in an acidic $SnCl_2$ solution for 5 minutes. A 2 percent hydrazine hydrate solution was added dropwise to the above solution with continuous agitation. The plating on copper was initiated by contacting an aluminum foil. A total of 13 mL of the hydrazine solution was added in a period of 250 minutes. On all substrates, a bright platinum layer was formed. The deposit thicknesses were 1.2 μm on substrate (a), 0.75 μm on (b), and 0.8 μm on (c). The total amount of platinum plated out corresponded to 81 percent of the platinum present initially in the bath.

Torikai et al. (17)

These inventors also used nitro or nitroammine complexes of platinum to prepare electroless platinum baths containing hydrazine. Hydroxylamine was found to be a very effective stabilizer. Two bath compositions are given in Table 16.13. The baths can be operated at relatively low temperatures, and they are said to have good stability and high utilization efficiency for platinum. The

Table 16.13
Hydrazine Bath for Pt (17)
(Torikai et al.)

	Bath A	Bath B
$(NH_3)_2Pt(NO_2)_2$, g	0.5	—
$K_2Pt(NO_2)_4$, g	—	0.6
NH_4OH (28% NH_3), mL	50	20
H_2O, mL	250	100

To the above mixture, add the two components listed below.

	Bath A	Bath B
$NH_2OH \cdot HCl$ (50% soln.), mL	10	—
$NH_2OH \cdot HCl$ (solid), g	—	0.2
Hydrazine hydrate (80% N_2H_4), mL	5	3
H_2O (to a total volume of), mL	400	200
pH	11.7	11.8
Temperature, °C	40-50	40-50
Plating rate	3 μm/2 hr	2.5 μm/2 hr
Pt utilization in 2 hr	98%	98%

following are listed as plateable substrates: Cu, Ni, Fe, their alloys, Ti, Ta, ABS plastics, polyamide, polycarbonate, glass, ceramics, and ion exchange membranes used for water electrolysis. The deposit is ductile and claimed to be especially suitable for use on flexible materials such as the ion exchange membrane electrode.

Borohydride Baths

Valsiuniene et al. (18) developed an electroless platinum bath using ethylenediamine as the complexing agent and sodium borohydride as the reducing agent. Rhodanine (2-mercapto-4-hydroxythiazole) was used as a stabilizer. Table 16.14 lists a typical bath composition and plating conditions. To prepare the bath, all components except $NaBH_4$ were first dissolved in water, and the solution was heated to and maintained at 70°C for 15 minutes for complete complexation. The reducing agent, which had been purified by three recrystallizations from 1M NaOH, was then added. Agitation was provided only by the hydrogen gas evolving during plating. Additional agitation and increased bath loading increase the plating rate and further improve the brightness of the deposit. The bath is made up with Pt in the +4 state, but as soon as borohydride is added, it is reduced to the +2 state by consuming 1/8 mole of BH_4^- for each mole of Pt. The bath is reported to be stable and suitable for plating platinum on copper and properly sensitized and activated glass or plastics. To initiate the plating on copper, it is necessary to contact it with aluminum for 3-5 seconds. No pores were detected at a thickness of 0.5 μm. Good-quality, well-adherent deposits up to thicknesses of 5 to 7 μm were obtained.

The same authors (19) found that the addition of 0.5 x 10^{-4} to 1 x 10^{-4}M disodium cadmium EDTA improves the brightness of the deposit. With rhodanine, the bath can be used until nearly all platinum is plated out. It is reported that sodium diethyldithiocarbamate and 2-mercaptobenzothiazole are also suitable as the stabilizer.

Table 16.14
Borohydride Bath for Pt (18)
(Valsiuniene et al.)

Na_2PtCl_6	0.0051M (2.3 g/L, 1 g/L as Pt)
Ethylenediamine	0.5M (30 g/L)
NaOH	1.0M (40 g/L)
Rhodanine	0.0007M (0.1 g/L)
$NaBH_4$	0.013M (0.5 g/L)
Temperature	70°C
Loading	50-100 cm²/L
Plating rate	1.5 μm/hr

The effect of the addition of thallium chloride (0.0002M) to the same bath was also studied (20). Thallium codeposits with Pt in the form of metal. The codeposition of Tl up to 16 percent is reported for 0.5-μm-thick deposits.

PLATINUM ALLOYS

Only a few electroless platinum alloy baths are known in the literature. Rhoda and Vines (13) give bath compositions to plate Pt-Rh and Pt-Ir alloys. For example, Pt-Rh (10 percent by wt) alloy can be plated from a bath with the same basic composition as that given in Table 16.12, except that the concentration of 10 g/L Pt for $Na_2Pt(OH)_6$ is replaced by 9 g/L, and that 1 g/L Rh is added in the form of $(NH_4)_3RhCl_6$. A plating rate of 6.4 μm/hr at 35° C is given. Pt-Ir alloys with up to 10 percent Ir were also plated from similar baths. Palladium cannot be plated from this type of bath because it precipitates out in the solution.

Torikai et al. (17) states that a deposit of Pt-Pd alloy can be obtained by adding $K_2Pd(NO_2)_4$ to their electroless platinum bath of the type shown in Table 16.13. However, no alloy composition is given. They also state that Pt-Ir and Pt-Rh alloys can be plated from similar baths, but no detailed information is available.

RUTHENIUM

Torikai et al. (21) describes in a recent patent an electroless ruthenium bath using Ru-nitrosylammine complexes in combination with hydrazine. Hydroxylamine is added as a stabilizer as is done to the similar electroless platinum bath described already (Table 16.13). The baths contain both $[Ru(NO)(OH)(NH_3)_4]^{2+}$ and $[Ru(NO)(NH_3)_5]^{3+}$. Four methods of preparation of solutions containing these species are summarized in Table 16.15. The active ruthenium species are either added as their chloride salts or formed *in situ* from other ruthenium salts such as $RuCl_3$ or $K_2[Ru(NO)Cl_5]$ with $NaNO_2$ and NH_4OH.

Two examples of actual plating procedures described by the inventors will be given below. In Example 1, a bath was prepared by adding 10 mL of water and 0.5 mL of a 40 percent hydrazine hydrate solution to 40 mL of solution A, Table 16.15. A degreased and cleaned copper sheet (2 cm x 4 cm) was activated by treating in a 5 percent $PdCl_2$ solution in 2M HCl for 30 seconds, and immersed in the bath at pH 11.8 and 40° C for three hours. A 5 μm thick ruthenium deposit was obtained with 90 percent Ru utilization. In Example 2, a bath was prepared by adding 150 mL of water and 5 mL of 40 percent hydrazine hydrate to 70 mL of solution D. A sheet of ABS plastic (5 cm x 5 cm x 2 mm) plated with electroless nickel was used as the substrate. It was activated with palladium as in Example 1 above, and immersed in the bath at pH 12.0 and 40-50° C for 3 hours. The thickness of the Ru deposit obtained was 3 μm, and the Ru utilization was 95 percent.

Table 16.15
Preparation of Ruthenium
Nirosylammine Solutions (21)
(To make 500 mL)

	Sol. A	Sol. B	Sol. C	Sol. D
$RuCl_3 \cdot 3H_2O$, g	2.6	2.6	—	—
$K_2[Ru(NO)Cl_5]$, g	—	—	3.8	—
$[Ru(NO)(NH_1)_5]Cl_3$, g	—	—	—	3.2
$NaNO_2$, g	2.5	2.5	—	—
NH_4OH				
(28% NH_1), mL	20.0	20.0	20.0	—
Hydroxylamine HCl, g	—	1.0	1.0	1.5
pH 12 Buffer*	—	—	—	30 mL

*0.1M NaOH plus 0.1M Na_2CO_3

Solution A
Dissolve $RuCl_3 \cdot 3H_2O$ in 50 mL H_2O. Add a few drops of HCl and $NaNO_2$ in small increments and boil. Add NH_4OH and heat to boil. Dilute to 500 mL after cooling.

Solution B
As above, except that NH_4OH and hydroxylamine are added at the same time instead of NH_4OH alone.

Solution C
Add hydroxylamine after boiling mixture of Ru-salt and NH_4OH. Then dilute to 500 mL.

Solution D
Add Ru-salt in the buffer and heat for dissolution. Add hydroxylamine and dilute to 500 mL.

The bath is stable, especially in the presence of hydroxylamine. It becomes unstable at pH >13 and at temperatures above 60° C. A pH range of 10-12 and a temperature range of 35-50° C are recommended. The plating reaction produces N_2 gas and small amounts of H_2 and NH_3. The process is claimed to be suitable for plating Ru on Cu, Ni, Fe, their alloys, Ti, Ta, and non-conducting materials such as ABS, polyamide, polycarbonate, glass, ceramics, and especially ion exchange membranes used for industrial electrolysis with solid polymer electrolytes. The substrates must be activated; for example, metallic substrates can be activated by immersion in a solution containing a salt of Pd, Pt, Ru, Au, or Ag, followed by immersion in a borohydride solution, if necessary. The non-conducting materials are activated by an ordinary sensitization and activation procedure. Ion exchange membranes are first roughened and coated with an initial layer of thin (0.1-0.2 μm) ruthenium by immersion in a solution of ruthenium salt followed by reduction with a borohydride solution. The inventors note that the low operating temperature makes this process especially useful for ruthenium deposition on materials that are unstable at high temperatures.

RHODIUM

Strejcek (16) describes an example of an electroless rhodium bath using hydrazine as the reducing agent. To prepare a rhodium solution, 0.1 g of $RhCl_3 \cdot 3H_2O$ was dissolved in 100 mL of water, and a large excess of $NaNO_2$ (10 g) was added. It was then heated to 95 to 98° C. After about 30 minutes, the color of the solution changed from red to pale yellow. After cooling, 5 mL of concentrated NH_4OH was added to form $(NH_3)_n Rh(NO_2)_x$. A copper wire (with aluminum foil contacting) and a nickel sheet were hung in this solution. With simultaneous heating and continuous agitation, a 2 percent solution of $N_2H_4 \cdot H_2O$ was added dropwise. At 60° C a bright deposit of rhodium was reported to form.

CONCLUDING REMARKS

An attempt was made to review the state of the art of electroless plating of platinum group metals with reasonably detailed descriptions of practical procedures for solution preparation and plating conditions. Wherever possible, critical remarks were included. However, it should be noted that much of the information on individual processes were obtained from the patent literature, and basic understanding of those processes is often lacking. Furthermore, with a few exceptions, detailed information on process characteristics and deposit properties is not available. Therefore, it was difficult to make critical judgement on the practical usefulness of various processes without first-hand experiences. It is hoped, however, that this review will serve as a consolidated source of scattered information on electroless plating of platinum group metals and alloys, and that it will stimulate further work to improve existing processes as well as to develop new processes useful for today's sophisticated applications.

REFERENCES

1. R.N. Rhoda, *Trans. Inst. Metal Finish.*, **36,** 82 (1959); U.S. patent 2,915,406 (1959).
2. S. Kawagoshi, Japanese patent (Kokai Tokkyo Koho) 77-733 (1977).
3. H. Laub and H. Januschkowetz, Ger. Offen. DE 2841584 (1980).
4. A. Sergienko, U.S. patent 3,418,143 (1968).
5. F. Pearlstein and R.F. Weightman, *Plating*, **56,** 10, 1158 (1969).
6. F. Pearlstein, in *Modern Electroplating*, edited by F.A. Lowenheim, Chap. 31, pp. 739-41, John Wiley, New York (1974).

7. C. Hsu and R.E. Buxbaum, *J. Electrochem. Soc.*, **132,** 2419 (1985).
8. A.I. Zayats, I.A. Stepanowa, and A.V. Gorodyskii, *Zashch. Metal.*, **9,** 1, 116 (1973).
9. S. Yu. Vereshchinskii, S.B. Kalmykova, and N.V. Korovin, *ibid.*, **9,** 1, 117 (1973).
10. N.V. Korovin, S.B. Kalmykova, and S. Yu. Vereshchinskii, U.S.S.R. patent SU291991 (1971).
11. S. Mizumoto, H. Nawabune, M. Haga, and K. Tsuji, *Extended Abstracts of Papers Presented at the 73rd Tech. Conf., Metal Finish. Soc. Japan,* **27B-8,** 116 (1986).
12. W.V. Hough, J.L. Little, and K.V. Warheit, U.S. patent 4,255,194 (1981); 4,279,951 (1981).
13. J.A. Abys, U.S. patent 4,424,241 (1984).
14. R.N. Rhoda and R.F. Vines, U.S. patent 3,486,928 (1969).
15. F.H. Leaman, *Plating,* **59,** 5, 440 (1972).
16. J. Strejcek, Ger. Offen. DE 2607988 (1977).
17. E. Torikai, K. Takenaka, and Y. Kawami, Japanese patent (Kokai Tokkyo Koho) 84-80764 (1984).
18. J. Valsiuniene, J. Vinkevicius, and A. Yu. Prokopchik, *Liet. TSR Mokslu Akad. Darb.,* **Ser. B,** 5, 25 (1976).
19. I. Vinkevicius, J. Valsiuniene, and A. Yu. Prokopchik, Issled. Obl. Elektro-osazhdeniya Met., Mater. Resp. Konf. Elektrokhim. Lit. SSR, 14th, 210 (1976).
20. A. Yu. Prokopchik, J. Vinkevicius, and J. Butkevicius, *Liet. TSR Mokslu Akad. Darb.,* **Ser. B,** 1, 3 (1977).
21. E. Torikai, Y. Kawami, and K. Takenaka, Japanese patent (Kokai Tokkyo Koho) 84-80766 (1984).

Chapter 17
Electroless Plating Of Silver

N. Koura

It is frequently required to coat insulators such as glass and ceramics with metals. Many methods are available for coating, e.g., baking, chemical vapor deposition, ion-sputtering, and chemical plating. The chemical plating method is very effective because the apparatus is simple, it can be done on a complex substrate, and it is suitable for mass production. This method can be classified into two categories: (1) the galvanic exchange deposition method, where the potential difference between metals is utilized; and (2) the electroless plating method, where reducing agents are used. In this chapter, only electroless plating will be reviewed.

The silvering reaction is a well-known example of the electroless plating of silver. This reaction was devised by Drayton (1) in 1830 and developed by Liebig (2). This method is normally used to make mirrors. On the other hand, electroless plating methods, which are utilized in the printed circuit industry, have come into the limelight with the development of the electronics industry. The electroless silver plating method (including the silvering reaction) is important because any substance, i.e., both metals and insulators, can be coated with silver by using this technique. Here the fundamental research and applications of this method will be reviewed.

PLATING PROCESS
AND THE REACTION MECHANISM

The substrate pretreatment for electroless silver plating affects early stages of the plating itself, and it plays an important role in the success of the plating process.

The reactions that occur in the pretreatment and the plating processes are separately summarized.

Pretreatment Process
The treatment for glass will be described first because electroless silver plating is usually carried out on glass. The plating is done soon after the pretreatment (3). In the pretreatment process, the glass is degreased with acetone and alcohol,

immersed in a $SnCl_2$ solution for 1 to 2 minutes, then rinsed in warm water. A $SnCl_2$ solution has been used as the pretreatment bath for a long time (4) and is thought to have two effects (1). Fine silver particles, which have a positive charge in the solution, adhere easily to the glass because the SnO_2 generated in the solution from the hydrolysis of Sn^{2+} adsorbs on the surface of the glass, and the negative charge on the surface increases. Ag^+ is reduced by Sn^{2+} and is then adsorbed on the surface. Recently, the mechanism was discovered (5). The behavior of Sn on glass was analyzed with X-ray photoelectron spectroscopy by Pederson (6) as follows:

$$-\overset{|}{\underset{|}{Si}}-OH + SnCl_3^- + H_2O \rightarrow \overset{|}{\underset{|}{Si}}-O-Sn-OH + 2H^+ + 3Cl^- \qquad [1]$$

$$\underset{Si}{\diagdown}\overset{-OH}{\underset{-OH}{}} + SnCl_3^- \rightarrow \underset{Si}{\diagdown}\overset{O}{\underset{O}{}}Sn + 2H^+ + 3Cl^- \qquad [2]$$

$$O \underset{=Si-OH}{\overset{=Si-OH}{\diagdown}} + SnCl_3^- \rightarrow O \underset{Si-O}{\overset{Si-O}{\diagdown}}Sn + 2H^+ + 3Cl^- \qquad [3]$$

First, the exchange reaction of the Sn^2 ion occurs on the surface of the glass according to Eqs. 1-3. Then silver deposits onto the glass in the early stages of the plating reaction, as shown in Eq. 4:

$$\underset{Si}{\diagdown}\overset{O}{\underset{O}{}}Sn^2 + 2Ag^+ \rightarrow \underset{Si}{\diagdown}\overset{O}{\underset{O}{}}Sn^4\overset{Ag}{\underset{Ag}{\diagdown}} \qquad [4]$$

In this manner, the deposition of silver onto the pretreated glass surface is induced early in the reaction. Then the plating reaction progresses according to the theory stated in the next section. Geelen (7) also proposed a similar mechanism by using the model as shown in Fig. 17.1.

Plating Reaction

It is well known that the silvering bath is highly unstable. The bath decomposes as soon as the silvering reaction starts, and becomes muddy. Therefore it was believed that the silvering occurs through the adhesion of silver particles with positive charges generated in the solution (8). With this in mind, attempts were made to improve the silvering bath by stabilizing the fine silver particles in the solution with a protective colloid (9). Gelatin, arabic gum, organic acids, inorganic salts of zinc and lead, and copper sulfate were used as protective colloid reagents.

There are many reports (10,11) on the reaction mechanism of electroless plating. One study states that the plating reaction progresses with a combination of a cathodic reduction of a cation, i.e, $M^{n+} + ne \rightarrow M$, and an anodic oxidation of

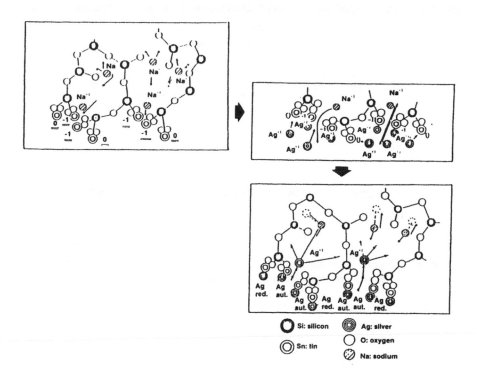

Fig. 17.1—Model of the early stages of silver plating reaction on a glass substrate (7).

the reducing agent, i.e., R → O + ne. In this case, it is thought that a metal ion is cathodically reduced to metal on the surface, which is activated by a catalyst.

The above theory of electroless plating is applicable to silver plating because silver is a metal on which an autocatalytic reaction occurs (12) as shown in Table 17.1. The concept of potential-pH diagram is very important in this theory. The potential-pH diagrams for $Ag-NH_3-H_2O$ and $Ag-CN-H_2O$ systems and various reducing agents are shown in Figs. 17.2, 17.3, 17.4 and Table 17.2 (11,13). As the potential of silver is very noble, as can be seen in Fig. 17.2, many reducing agents can be used in electroless silver plating. There are 14 kinds of commercially available reducing agents: formalin; dextrose; Rochelle salts; Rochelle salts + silver nitrate; glyoxal; hydrazine sulfate; a boiled mixture solution of Rochelle salts and crystallized sugar; sugar inverted by nitric acid (14); KBH_4 or DMAB; aldonic acid and aldonic lactone (2,15); cobalt ion (16); sodium sulfide (17); triethanol amine (18); and $CH_2OH(CHOH)_nCH_2OH$ (n = 1-6) (19).

Several examples of the reaction with representative reducing agents are described below. Ag_2O precipitation in the bath must be avoided because the electroless silver plating is generally carried out in a basic solution. A

Table 17.1
Metals Plated Autocatalytically (12)

			Group			
Period	**VB**	**VI B**	**VIII**			**IB**
3	V	Cr	Fe	Co	Ni	Cu
4			Ru	Rh	Pd	Ag
5			(Os)	(Ir)	Pt	Au

( ) = no report.

complexing agent such as ammonia is added to the solution. The complex formation reactions are thought to be as follows (14):

$$2AgNO_3 + 2NH_4OH = Ag_2O + 2NH_4NO_3 + H_2O \qquad [5]$$

$$Ag_2O + 4NH_4OH = 2[Ag(NH_3)_2]OH + 3H_2O \qquad [6]$$

$$[Ag(NH_3)_2]OH + NH_4NO_3 \rightleftharpoons [Ag(NH_3)_2]NO_3 + NH_4OH \qquad [7]$$

The silver plating is carried out by the reaction of the ammonia complex with aldehyde, as in Eq. 8 (14). The electromotive force is 0.971V at pH 10.0 (20).

$$2[Ag(NH_3)_2]OH + RCHO \rightarrow 2Ag + 4NH_3 + RCOOH + H_2O \qquad [8]$$

There are many studies on the reaction with Rochelle salt as the reducing agent. They are as follows:

$$2Ag(NH_3)_2^+ + 2OH^- \xrightarrow{Ag} Ag_2O + 4NH_3 + H_2O \qquad [9]$$

$$3Ag_2O + C_4O_6H_4^{2-} + 2OH^- \rightarrow 6Ag + 2C_2O_4^{2-} + 3H_2O \qquad [10]\ (21)$$

$$2Ag(NH_3)_2NO_3 + KNaC_4O_6H_4 + H_2O$$
$$= Ag_2O + KNO_3 + NaNO_3 + (NH_4)_2C_4O_6H_4 \qquad [11]$$

$$4Ag_2O + (NH_4)_2C_4O_6H_4$$
$$= 8Ag + (NH_4)_2C_2O_4 + CO_2 + 2H_2O \qquad [12]\ (1)$$

$$4AgNO_3 + 4NH_3 + C_4H_4NaKO_6 + H_2O$$
$$= 4Ag + 4NH_4NO_3 + C_3H_2KNaO_5 + CO_2 \qquad [13]\ (1)$$

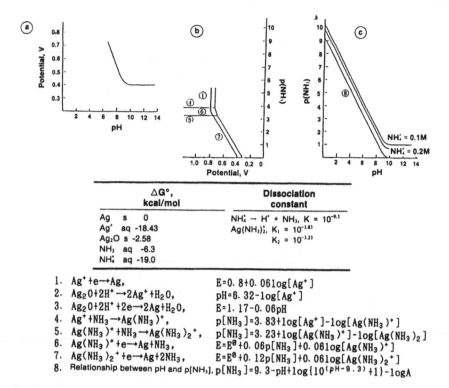

Fig. 17.2—Potential-pH diagram of Ag-NH₃-H₂O system (13).

The equations and table within the figure:

	ΔG°, kcal/mol
Ag s	0
Ag⁺ aq	-18.43
Ag₂O s	-2.58
NH₃ aq	-6.3
NH₄⁺ aq	-19.0

Dissociation constant

$$NH_4^+ \rightarrow H^+ + NH_3, \quad K = 10^{-9.3}$$
$$Ag(NH_3)_2^+, \quad K_1 = 10^{-3.83}$$
$$K_2 = 10^{-3.23}$$

1. $Ag^+ + e \rightarrow Ag$, $E = 0.8 + 0.06\log[Ag^+]$
2. $Ag_2O + 2H^+ \rightarrow 2Ag^+ + H_2O$, $pH = 6.32 - \log[Ag^+]$
3. $Ag_2O + 2H^+ + 2e \rightarrow 2Ag + H_2O$, $E = 1.17 - 0.06pH$
4. $Ag^+ + NH_3 \rightarrow Ag(NH_3)^+$, $p[NH_3] = 3.83 + \log[Ag^+] - \log[Ag(NH_3)^+]$
5. $Ag(NH_3)^+ + NH_3 \rightarrow Ag(NH_3)_2^+$, $p[NH_3] = 3.23 + \log[Ag(NH_3)^+] - \log[Ag(NH_3)_2]$
6. $Ag(NH_3)^+ + e \rightarrow Ag + NH_3$, $E = E^\theta + 0.06p[NH_3] + 0.06\log[Ag(NH_3)^+]$
7. $Ag(NH_3)_2^+ + e \rightarrow Ag + 2NH_3$, $E = E^\theta + 0.12p[NH_3] + 0.06\log[Ag(NH_3)_2^+]$
8. Relationship between pH and p[NH₃], $p[NH_3] = 9.3 - pH + \log\{10^{(pH-9.3)} + 1\} - \log A$

In Eq. 14, glyoxal (CHO-CHO) generated in the bath acts as a reducing agent. Silver carbonate is filtered off.

The following plating reactions are reported for hydrazine (22) and hydrazineborane (23):

$$4[Ag(NH_3)_2]NO_3 + N_2H_4 \rightarrow 4NH_4NO_3 + 4NH_3 + N_2 + 4Ag \qquad [15]$$

$$N_2H_4BH_3 + 3Ag^+ + 4OH^- \rightarrow N_2H_4 + B(OH)_4^- + 3/2H_2 + 3Ag \qquad [16]$$

The plating rates for various reducing agents are shown in Figs. 17.5-17.7 (20,24,25). The rate for borohydride is somewhat faster than that for other

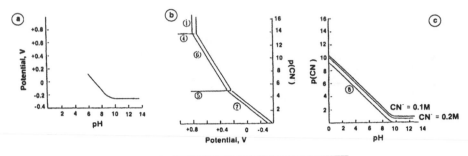

	ΔG°, kcal/mol	Dissociation constant
Ag s	0	HCN, K = $10^{-9.4}$
Ag⁺ aq	18.43	Ag(CN)₂⁻, K = 3.8 × 10^{-19}
Ag₂O s	2.58	
CN⁻ aq	39.6	
AgCN s	39.2	
Ag(CN)₂⁻ aq	72.05	

1. $Ag^+ + e \rightarrow Ag$, $E = 0.8 + 0.06\log[Ag^+]$
2. $Ag_2O + 2H^+ \rightarrow 2Ag^+ + H_2O$, $pH = 6.32 - \log[Ag^+]$
3. $Ag_2O + 2H^+ + 2e \rightarrow 2Ag + H_2O$, $E = 1.17 - 0.06pH$
4. $Ag^+ + CN^- \rightarrow AgCN$, $p[CN^-] = 13.8 + \log[Ag^+]$
5. $AgCN + CN^- \rightarrow Ag(CN)_2^-$, $p[CN^-] = 5.0 - \log[Ag(CN)_2^-]$
6. $AgCN + e \rightarrow Ag + CN^-$, $E = -0.017 + 0.06p[CN^-]$
7. $Ag(CN)_2^- + e \rightarrow Ag + 2CN^-$, $E = -0.31 + 0.12p[CN^-] + 0.06\log[Ag(CN)_2^-]$
8. $H^+ + CN^- \rightarrow HCN$, $p[CN^-] = 9.4 - pH + \log\{10^{(pH-9.4)} + 1\} - \log A$

Fig. 17.3—Potential-pH diagram of Ag-CN-H₂O system (13).

reagents. Moreover, hydrazine is often used in the spray method for mirror production, as the deposition rate is fast (24).

The activation energies of the silver plating reaction for the glucose method (26) and the Rochelle salt method are 27.1 kcal/mol and 22.0 kcal/mol, respectively. Activation energies of electroless nickel plating for the alkane succinic acid bath and the KBH₄ bath are 16.9 kcal/mol and 9.9 kcal/mol, respectively. These values indicate that the deposition rate for the electroless silver plating is strongly influenced by temperature.

The polarization curves for the electroless silver plating bath that uses these reducing agents are shown in Figs. 17.8-17.11 (21,29,30). A polarization curve for the solution containing both silver and formalin cannot be determined, because the solution is unstable (31).

The effect of pretreatment and the mechanism of electroless silver plating can be studied from the partial anodic and cathodic polarization curves (32). Both the partial cathodic polarization curve for a solution without Rochelle salt as the reducing agent and the partial anodic curve for a solution without silver nitrate are shown in Fig. 17.12. The 3,5-diiodotyrosine added bath (DIT bath), which is

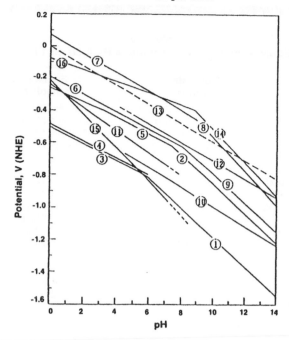

Fig. 17.4—Potential-pH diagram of various reducing agents (11). (For Explanation of the numbers (1) to (16), see Table 17.2)

the long-life bath described later, is used. The result for the electrode pretreated with a SnCl₂ solution is shown with a solid line and that for the non-treated one is shown with a dotted line. The difference between the partial cathodic polarization curves in Fig. 17.12 is minimal. On the other hand, the difference between the partial anodic polarization curves of the treated and the untreated electrodes is clearly seen. In the case of the treated electrode, an anodic current appears also in the range of 100-600 mV vs. SCE. At the same time, the rest potential of the untreated electrode becomes 25 mV vs. SCE, and that of the treated one becomes -81 mV vs. SCE. The difference in the rest potential is attributed to the adsorption of Sn^{2+} ions on the electrode. Namely, it is thought that the anodic current due to oxidation of the Sn^{2+} ion flows at potentials of 100-600 mV vs. SCE. When the anodic and cathodic polarization curves for the treated electrode are combined, a point of intersection appears at 0.02 mA/cm² of current density. When this value of 0.02 mA/cm² is converted to a plating rate, the value is equivalent to 0.04 mg/cm²/30 min. This converted value is close to the plating rate that is determined from the deposition quantity in Fig. 17.13. In the case of the untreated electrode, the current density at the intersection is

Table 17.2
Equilibrium Potential of Various Reductants (11)

No.*	Electrode reaction	Equilibrium potential (E)
1	$H_2PO_2^- + 3OH^- = HPO_3^{2-} + 2H_2O + 2e$	-0.31 @ 0.09 pH
2	$HPO_3^{2-} + 3OH^- = PO_4^{3-} + 2H_2O + 2e$	0.14 @ 0.09 pH
3	$H_2PO_3^- + H_2O = H_2PO_2^- + 2H^+ + 2e$	-0.504 @ 0.06 pH
4	$H_3PO_2 + H_2O = H_3PO_3 + 2H^+ + 2e$	0.499 @ 0.06 pH
5	$H_3PO_3 + H_2O = H_3PO_4 + 2H^+ + 2e$	-0.276 @ 0.06 pH
6	$HCOOH = CO_2 + 2H^+ + 2e$	-0.199 @ 0.06 pH
7	$HCHO + H_2O = HCOOH + 2H^+ + 2e$	0.056 @ 0.06 pH
8	$2HCHO + 4OH^- = 2HCOO^- + H_2 + 2H_2O + 2e$	0.32 @ 0.12 pH
9	$HCOO^- + 3OH^- = CO_3^{2-} + 2H_2O + 2e$	0.25 @ 0.09 pH
10	$BH_4^- + 8OH^- = BO_2^- + 6H_2O + 8e$	-0.45 @ 0.06 pH
11	$N_2H_5^+ = N_2 + 5H^+ + 4e$	-0.23 @ 0.075 pH
12	$CN^- + 2OH^- = CNO^- + H_2O + 6e$	-0.13 @ 0.06 pH
13	$H_2 = 2H^+ + 2e$	0.000 @ 0.06 pH
14	$S_2O_4^{2-} + 4OH^- = 2SO_3^{2-} + 2H_2O + 2e$	0.56 @ 0.12 pH
15	$S_2O_6^{2-} + 2H_2O = 2SO_4^{2-} + 4H^+ + 2e$	-0.22 @ 0.12 pH
16	$HS_2O_4^- + 2H_2O = 2H_2SO_3 + H^+ + 2e$	-0.056 @ 0.03 pH

*Numbers correspond to those in Fig. 17.4.

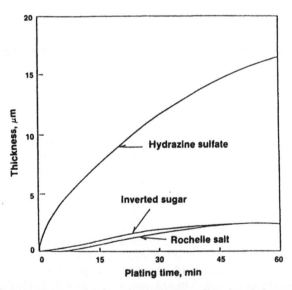

Fig. 17.5—Plating rates for solutions using hydrazine sulfate, inverted sugar or Rochelle salt as reducing agents.

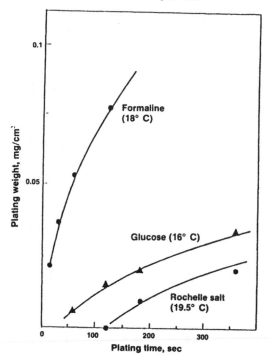

Fig. 17.6—Plating rates for solutions using formaline, glucose or Rochelle salt as reducing agents.

extremely small. Therefore, in the case of the DIT bath, it is thought that the electroless silver plating reaction proceeds only on the substrate surface treated with the $SnCl_2$ solution and does not occur on the untreated surface.

Reaction in Non-Aqueous Solution

The silvering reaction in a non-aqueous solution is also reported (33). Figure 17.14 represents the reaction mechanism. A silver soap, which forms the micelle, is stable in the solution. Since the soap adheres to the substrate surface, the silvering reaction progresses.

Stability of the Plating Bath

It is very important to stabilize the bath even for the electroless plating of Cu and Ni. As stated above, the electroless silver plating bath is very unstable and short-lived. Therefore, if the bath were made more stable, it would become more useful. Many attempts have been made to improve the bath stability. It has been determined that a small amount of 3-iodotyrosine or 3,5-diiodotyrosine added to the bath works well as a stabilizer (34).

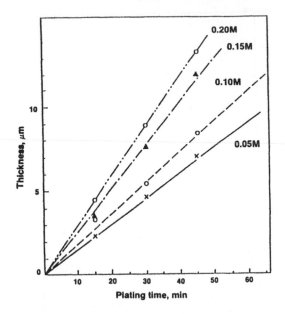

Fig. 17.7—Plating rates for bath containing various concentrations of KBH₄ (25). Bath composition = 0.05M NaAg(CN)₂, 0.10M NaCN, 0.40M NaOH, temperature 75 ±2° C.

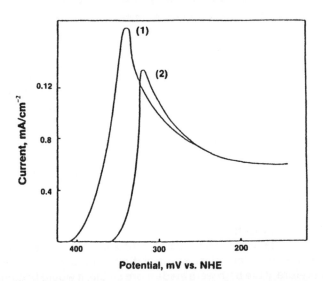

Fig. 17.8—Cathodic polarization curves for Ag-NH₃-tartaric acid solution (21). Curve 1 = 0.02M AgNO₃, 0.15M NH₃, 0.1M NaOH; Curve 2 = 0.02M AgNO₃, 0.15M NH₃, 0.1M NaOH + tartaric acid. Sweep rate = 0.2 V/min.

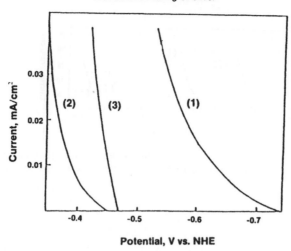

Fig. 17.9—Partial anodic polarization curves for N₂H₄ solution with various CN⁻ concentrations (29). Bath composition = 6M N₂H₄ + 1.5M NaOH at 40° C. Curve 1 = OmM NaCN; Curve 2 = 5 mM NaCN; Curve 3 = 40 mM NaCN.

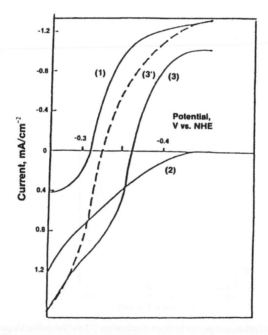

Fig. 17.10—Polarization curves for Ag-CN⁻-N₂H₄ solutions (29). (1): partial cathodic polarization curve for solution containing 0.03M AgNO₃, 0.1M NaCN, and 1.0M NaOH; (2): partial anodic polarization curve for solution containing 6M N₂H₄, 0.04M NaCN, and 1.0M NaOH; (3): total curve (experimental) for solution containing 0.03M AgNO₃, 0.01M NaCN, 6M N₂H₄, and 1.0M NaOH; (3'): total curve (calculated).

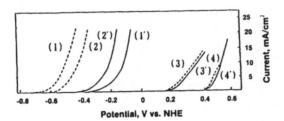

Fig. 17.11—Partial anodic polarization curves for various reductants (30): formaline (1, 1'); BH₄ ion (2, 2'); ascorbic acid (3, 3'); Fe²⁺ ion (4, 4'). Dotted line = activated electrode; solid line = inactive electrode.

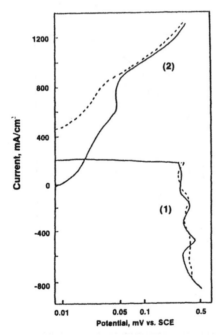

Fig. 17.12—Partial polarization curves for various electrodes: (1) partial cathodic polarization curve of electroless silver plating solution without Rochelle salt—AgNO₃ (8.8 x 10⁻³ M), ethylenediamine (5.4 x 10⁻² M), DIT (4 x 10⁻⁵ M), 35° C, pH 10.0; (2) partial anodic polarization curve for electroless silver plating solution without silver nitrate—ethylenediamine (5.4 x 10⁻² M), Rochelle salt (3.5 x 10⁻² M), DIT (4 x 10⁻⁵ M), 35° C, pH 10.0. Solid line = Pt electrode sensitized with SnCl₂ solution; dotted line = Pt electrode, not sensitized.

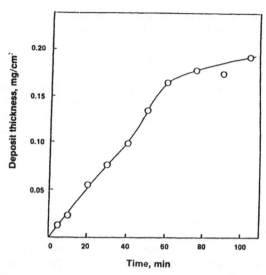

Fig. 17.13—Relationship between amount of silver deposit and plating time. Bath composition: AgNO$_3$ (8.8 x 10^{-1} M), ethylenediamine (5.4 x 10^{-2} M), Rochelle salt (3.4 x 10^{-2} M), DIT (4 x 10^{-5} M), pH 10.0, 35° C.

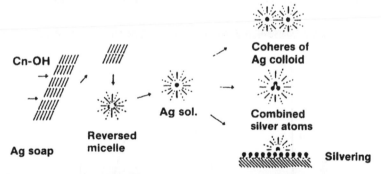

Fig. 17.14—Silvering reaction mechanism from nonaqueous solution (32).

The bath conditions are as follows:

Silver nitrate 3 x 10^{-1} M   Ethylenediamine 1.8 x 10^{-2} M
Rochelle salt 3.5 x 10^{-2} M   3,5-diiodotyrosine 4 x 10^{-5} M
pH 10.0-10.5   Temperature 35-40° C

It was found that the plating rate and the bath stability were greatly affected by the pH of the bath, therefore the factors affecting the bath pH were studied (34). The bath pH was found to change with the addition of silver nitrate. The change was remarkable at pH values <10.5. The complex formation between silver ion and ethylenediamine (en) is believed to be as follows:

$$Ag^+ + 2H_2O + 2en \rightleftharpoons [Ag(OH)_2en_2]^- + 2H^+ \qquad [17]$$

The equilibrium consideration of this reaction shows that the bath pH must change with concentrations of Ag^+ and ethylenediamine. On the other hand, even with the stable bath (DIT bath), the plating reaction almost stopped after 24 hours and the pH of the bath dropped. It was thought that the pH change strongly affected the life of the bath, because a slight drop in the bath pH reduced the plating rate significantly, as seen in Fig. 17.15. When a bath of small temporal change was used (Table 17.3), the same plating quantity was obtained even after 24 hours. The life of this bath increased to one week by adjusting the bath pH occasionally.

The life of the DIT bath was also studied from partial polarization curves (32). As seen in Fig. 17.16, the DIT controls the cathodic reaction. This effect may contribute to the bath stability. Moreover, the temporal changes of rest potential and bath pH were measured for the DIT bath (Fig. 17.17a), for the bath without Rochelle salt (Fig. 17.17b), and for the bath without silver nitrate (Fig. 17.17c). Those pH values remained unchanged. In the case of Fig. 17.17b, the rest potential moved to a noble potential and the color of the solution gradually changed from colorless to yellow. The color change suggested that the fine

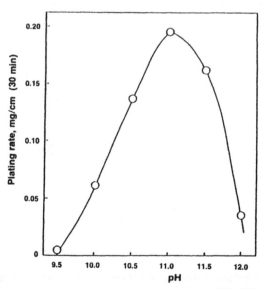

Fig. 17.15—Electroless plating rate of silver as a function of bath pH: $AgNO_3$ (2.9 x 10^{-3} M), ethylenediamine (1.8 x 10^{-2} M), Rochelle salt (3.5 x 10^{-2} M), 3,5-diiodotyrosine (4 x 10^{-5} M), 35° C.

Table 17.3
Deposit Thickness as a Function
Of Bath pH and Plating Time

Bath*	pH	Deposit thickness, mg/cm^2	Silver content, %
Initial	9.97	0.115	
After 24 hr	9.95	0.092	96.0

Bath composition: AgNO$_1$ (8.8 x 10^{-1} M); Ethylenediamine (5.4 x 10^{-2} M); Rochelle salt (3.5 x 10^{-2} M); 3,5-diiodotyrosine (4 x 10^{-3} M); temperature 35° C.

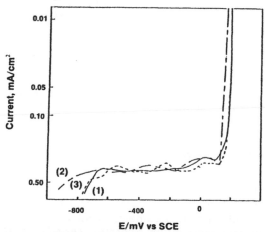

Fig. 17.16—Effects of various additives on the partial cathodic polarization curve for the Pt electrode in electroless silver plating solution without Rochelle salt. Bath composition: AgNO$_3$ (8.8 x 10^{-3} M), ethylenediamine (5.4 x 10^{-2} M), pH 10.0, 35° C. (1) No additive; (2) DIT (4 x 10^{-3} M); (3) KI (4 x 10^{-1} M).

silver particles were growing in the bath (35). Based upon the theory of Vaskelis (36) concerning the electrode potential and the particle size in an electrolyte, it was also believed that the potential becomes noble when the particle size increases. The temporal changes for the non-additive bath (Fig. 17.17b), the KI-added bath, and the DIT bath (Fig. 17.17a) were 2.1, 1.9, and 1.1 mV per 24 hours, respectively. From these facts, it can be stated that DIT restricts the reaction at the electrode surface by controlling the partial cathodic reaction, and stabilizes the bath by controlling the growth of fine silver particles in the solution.

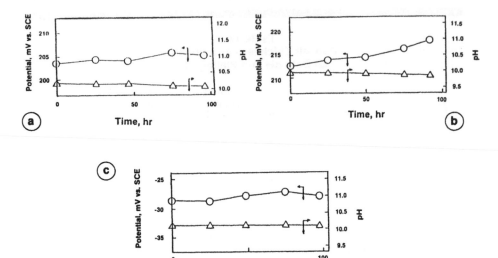

Fig. 17.17—Change of rest potential and bath pH with plating time. Bath composition: (a) AgNO₃ (8.8 x 10⁻¹ M), ethylenediamine (5.4 x 10⁻² M). DIT (4 x 10⁻⁵ M); (b) AgNO₃ (8.8 x 10⁻¹ M), ethylenediamine (5.4 x 10⁻² M), Rochelle salt (3.5 x 10⁻² M), DIT (4 x 10⁻⁵ M); (c) ethylenediamine (5.4 x 10⁻² M), Rochelle salt (3.5 x 10⁻² M), DIT (4 x 10⁻⁵ M).

Other stabilizers reported include: metal ions (Cu^{2+}, Ni^{2+}, Co^{2+}, Zn^{2+},) (37); Na-2,3-mercaptopropane sulfonate ($NaBH_4$, N_2H_4,CN⁻ system bath) (38); cystine, cysteine, dimethyldithio carbamate, NaSCN (N_2H_4, NH_3 bath) (39); RCONHR′ (e.g., dodecylammonium acetate), [$AgNO_3$, $(NH_4)_2Fe(SO_4)_2$, $Fe(NO_3)_3$, citric acid system bath] (40), and CaO (41).

PLATED FILM

There are many reports about the characteristics and the protective nature of the plated film that is obtained by the silvering method.

Dense plating is obtained by using glucose and formalin as the reducing agent. The X-ray diffraction pattern for the chemical plating method (the silvering method) is similar to those of a metal silver plate and a silver film obtained by the vapor deposition method as shown in Fig. 17.18 (42) and Table 17.4 (20). The diffraction pattern changes slightly with different reductants (Fig. 17.19). Etching and sensitizing during pretreatment are important in order to obtain good adherent plating (43).

The thickness of the coated silver film was measured by quantitative analysis with KSCN (44), the β-ray method (45), and the iodine-ring method (46). The plated silver film was usually protected by a polymer film such as lacquer. Multiple platings of Cu or Ni improve durability (14).

PRACTICE OF ELECTROLESS SILVER PLATING

Bath Composition

The immersion and the spray methods are known as the commercialized electroless silver plating technique. As it is impossible to review all reports with respect to the bath composition for the immersion method, some typical examples are presented in this chapter. (A and B denote a silver ion solution and a reducing agent solution, respectively.)

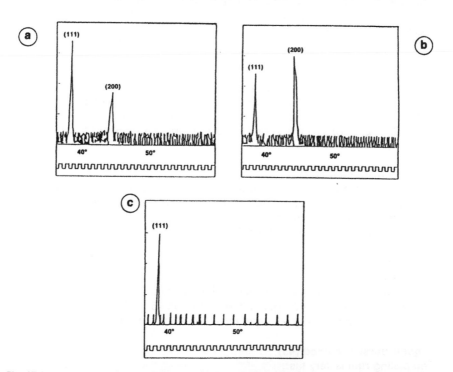

Fig. 17.18—X-ray diffraction patterns of various silver films (42). (a) Ag film obtained by chemical plating (thickness—1000 angstroms); (b) Metal Ag film (thickness—1 mm); (c) Ag film obtained by vapor deposition method (thickness—1000 angstroms).

Table 17.4
X-ray Diffraction Patterns
Of Various Silver Films

Plane orientation	Intensity		
	Metal Ag film	Chemical plating	Vapor deposition
(111)	100	100	100
(200)	40	41	13.5
(220)	25	23	—

Glucose method (47)

Although the operation is complex, a good plating is obtained.

A: Silver nitrate 3.5 g, ammonia solution proper quantity, water 60 mL, sodium hydroxide solution 2.5 g/100 mL.

B: Glucose 45 g, tartaric acid 4 g, water 1 L, alcohol 100 mL.

Rochelle salt method (48)

Because handling Rochelle salt is easy, this method is widely used.

A: Silver nitrate 454 g, ammonia solution 355 mL, water 5.45 L.

B: Rochelle salt: 1590 g, Epsom salt 114 g, water 3.64 L.

Distilled water of 3.61 L is added to the mixture of A (256 mL) and B (256 mL) solutions.

Formaldehyde method (1)

The deposition rate is fast, but a cloudy film is obtained or peeling occurs.

A: Silver nitrate 20 g, ammonia solution proper quantity, water 1 L.

B: Formaldehyde 40 mL, water 200 mL.

A and B are mixed at the ratio of 5:1.

Hydrazine method (48)

This is used for the spray method.

A: Silver nitrate 114 g, ammonia solution 227 mL.

B: Hydrazine sulfate 42.5 g, ammonia solution 45.5 mL.

A and B are diluted to 4.55 L and mixed in a ratio of 1:1.

Organic borane method (49)

The plating rate is very fast.

A: $Na[Ag(CN)_2]$ 1.83 g/L, NaCN 1.0 g/L, NaOH 0.75 g/L.

B: DMAB (dimethylamine borane) 2.0 g/L.

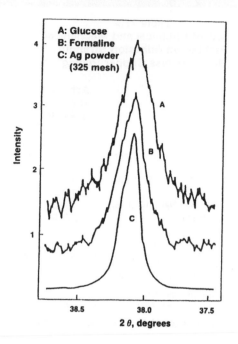

Fig. 17.19—X-ray diffraction patterns of the silvering films obtained by using (a) glucose, (b) formaline, and (c) Ag powder (325 mesh) as reductants (20).

Aldonic acid and lactone methods (15)

The deposition rate is fast. The yield of silver is very high.

A: $Ag(NH_3)_2^+$ 5×10^{-3} M, NaOH 4.0×10^{-3} M.
B: δ-gluconolactone 0.5–2.0×10^{-3} M.

In the silvering method, the silver ion solution and the reducing agent solution are prepared separately, then combined just before plating. In order to prepare the silver ion solution with ammonia as a complexing agent, a black-brown precipitate is produced upon addition of ammonia solution to a silver nitrate solution; then the ammonia solution is added until this precipitate almost dissolves. This solution is dark brown and clear. Reducing agents are prepared by the specific methods.

In the spray method, various guns have been devised (50). For example, the gun may have two tubes. The silver ion solution and the reducing agent are sprayed separately, and mixed on the substrate.

Other distinctive baths are as follows: long life bath (34); good adherence (51) or high speed (52); for semiconductors (53); mixture of silver and copper ions (54); for substrate pretreated by hydrazine or dextrine (55); for plating on Si (56);

Table 17.5
Effects of Chemical and Mechanical[a]
Etching on Adhesion Strength
Of Electroless Copper Plating[b]

Chemical etchant	Adhesion strength[c], g/4 mm^2
None	499.4
	(613.2)[d,e]
HCl (20 wt. %), 100° C, 15 min	501.8
	(467.9)
NaOH (10 wt. %), 100° C, 15 min	498.3
	(750.0)
NaOH (10 wt. %) + NaCl (10 wt. %) 100° C, 15 min	780.2
	(933.0)

[a]Mechanical etching (Al$_2$O$_3$ powder treatment): alumina powder (0.5 wt. %); ultrasonic agitation for 15 min.
[b]Undercoated with Ag (0.05 μm) by electroless plating.
[c]L-type tensile strength.
[d]( ) = with mechanical etching.
[e]When Ag was not undercoated, the value was 206.0.

with cyanides of the platinum group (57); electrolyzed Rochelle salt solution (58).

The displacement plating method is usually used in order to electroless-plate Ag on metals (59). For example, the bath composition for Ag plating on a Cu alloy consists of silver nitrate (7.5 g/L), sodium thiosulfate (105 g/L) and ammonia (75.0 g/L). There are many other reports (60).

Applications

There are applications for electroless silver plating in many fields. They are broadly divided into two categories:

1. Optics and decoration.
2. Electrical conductivity or for undercoating prior to other plating.

A good example of the first category is a mirror. There are many reports about this process (61). The second application makes the best use of the characteristics of electroless silver plating, including silvering, which is used for all insulators because the plated layer has electrical conductivity. For example, the plating is used as an undercoating (62) for electroless gold plating, for Ag plating on ceramics, for Sn-Pb alloy plating (63), and electrolytic plating (64).

When electroless plated silver was used as the undercoating on 96 percent Al$_2$O$_3$ ceramics, an improvement in the adhesion of copper plating on the ceramics was obtained (32). SEM photographs confirmed that electroless silver plating was useful in obtaining smooth and uniform electroless copper plating. Furthermore, the adhesion was greatly increased by etching the ceramics by both mechanical and chemical techniques, as shown in Table 17.5.

In other cases, silver plating on ceramics is used for SEM measurement of the surface (65). The plating is also used for quantitative analysis of trace arsenic (66), for determination of the pore diameter of a membrane (67), and for recovery of mercury.

REFERENCES

1. K. Nakanishi, *Kagaku no Ryoiki*, **4**, 604 (1950).
2. B. Liebig, *Ann. Chem. Phar. XIV*, 140 (1835).
3. *Encyclopedia of Chem. Technology*, **12**, 443 (1961), Maruzen.
4. H. von Wartenberg, *Z. anog. allgen. Chem.*, **199**, 185 (1930).
5. C.H. Minjer, *J. Electrochem. Soc.*, **120**, 1644 (1983); M. Tsukahara, T. Kishi, H. Yamamoto, T. Nagai, *J. Met. Fin. Soc. Japan*, **23**, 2, 83 (1972).
6. L.R. Pederson, *Sol. Energy Mater.*, **6**, 2, 221 (1982).
7. A. van Geelen, *Silicates Ind.*, **35**, 4, 93 (1970).
8. J. Loiseleur, *Compt. rend.*, **209**, 993 (1939).
9. S. Miyagi, T. Inagaki, *Kogyo Kagaku Zasshi*, **41**, 122 (1938); V. Kohlschutter, H. Schacht, *Z. Elektrochem.*, 19 (1913).
10. P. Bindra, J. Tweedie, *J. Electrochem. Soc.*, **130**, 1112 (1983).
11. I. Ohno, S. Haruyama, *Bull. Jpn. Inst. Met.*, **20**, 12, 979 (1981).
12. D.J. Levy, *Tech. Proc. Am. Electroplater's Soc.*, **50**, 29 (1963).
13. T. Ishibashi, *Bull. Himeji Inst. Tech.*, **15**, 16 (1962).
14. *Handbook of Metal Finishing*, Ed. Met. Fin. Soc. Jpn., 525 (1972).
15. S. Christain, S. Joseph F., Ger. Offen. 2,162,338 (1972).
16. A. Vaskelis, O. Diemontaite, USSR 709,713 (1980).
17. V.G. Vorobev, USSR 163,549 (1963).
18. R.D. Barnard, *Ind. Eng. Chem.*, **34**, 637 (1942).
19. B. Harry, U.S. patent 3,983,266 (1976).
20. S. Morimoto, Y. Moori, *Int. Congr. Glass,* (Paper), 10th, No. 8, 87 (1974).
21. A. Vaskelis, O. Diemontaite, *Liet. TSR Mokslu Akad. Darb.*, Ser. B, 5, 13 (1984).
22. I.I. Berisova, *Steklo i Keram.*, **18**, 12, 8 (1961).
23. A. Vaskelis, O. Diemontaite, *Liet. TSR Mokslu Akad. Darb.*, Ser. B, 6, 9 (1976).
24. R.J. Heritage, J.R. Balmer, *Met. Fin.*, **51**, 75 (1953).
25. Y. Takakura, *Jitsumu Hyoumen Gijutsu*, **27**, 31 (1980).
26. P. Mayaux, S. Descureux, *Silicates Ind.*, **33**, 381 (1968).
27. A.P. Modi, S. Ghosh, *J. Ind. Chem. Soc.*, **45**, 5, 446 (1968).
28. M. Matsuoka, T. Hayashi, *Denki Kagaku*, **42**, 8, 424 (1974).
29. A. Vaskelis, O. Diemontaite, *Liet. TSR Mokslu Akad. Darb.*, Ser. B, 5, 3 (1976).

30. G.K. Levchuck, *Dokl. Akad. Nauk. BSSR*, **25**, 10, 922 (1981).
31. A. Vaskelis, O. Diemontaite, *Liet. TSR Mokslu Akad. Darb.*, Ser. B, 4, 9 (1968).
32. A. Kubota, N. Koura, *J. Met. Fin. Soc. Jpn.*, **37**, 694 (1986).
33. R. Gotou, *Surface*, **18**, 404 (1980).
34. A. Kubota, N. Koura, *J. Met. Fin. Soc. Jpn.*, **37**, 131 (1986).
35. T. Asaoka, *Colloid Chem.*, 42 (1975), Sankyo.
36. A. Vaskelis, *Elektrokhimiya*, **14**, 1770 (1978).
37. T. Hata, T. Hanada, U.S. patent 3,337,350 (1967).
38. A. Vaskelis, O. Diemontaite, *Liet. TSR Mokslu Akad. Darb.*, Ser. B, 4, 23 (1975).
39. A. Vaskelis, O. Diemontaite, *ibid.*, 4, 3 (1982).
40. G.M. Fofanav, *Zashch. Met.*, **15**, 1, 125 (1979).
41. V.M. Vimokurov, B.N. Moskvin, *Optiko-Mekhan. Prom.*, **9**, 6-7, 17 (1939).
42. H. Fukyo, *Bull. Tokyo Inst. Tech.*, **89**, 69 (1968).
43. G. Rozovskii, V. Radziuniene, A. Venckiene, *Elektrokhim. Method.*, 1, 40 (1970).
44. P. Sen, P.S. Rao, *Glass Ind.*, **38**, 89 (1957).
45. V.A. Dubrovskii, *Steklo i Keram.*, **13**, 4, 14 (1956).
46. H. Fry, *Glass Ind.*, **39**, 328 (1958).
47. T. Ishibashi, *Less. Metal Finish*, **9**, 42 (1968).
48. S.K. Gupata, *Cent. Glass Ceram. Res. Int. Bull.*, **14**, 18 (1967).
49. F. Pearlstein, *Plating*, **58**, 1024 (1971).
50. I. Cross, *Met. Finish.*, **48**, 77 (1950).
51. Y. Mori, Jap. patent appl. 7,606,135 (1976).
52. S. Noguchi, Jap. patent appl. 7,593,234 (1975).
53. S. Horst, Ger. Offen. 2,063,334 (1972).
54. T. Mukaiyama, Jap. patent appl. 7,216,819 (1972).
55. S. Werner, Ger. Offen. 2,639,287 (1978).
56. J. Valsiuniene, *Galvanicheeskie Khim. Pokrytiya Dragotsennymi Redk. Met. Mater. Semin.*, 49 (1978).
57. Max Ernes, Ger. Offen. 950,230 (1956).
58. D.J. Hood, U.S. patent 2,120,203.
59. T. Kanbe, H. Ise, *Ind. Met. Sur.*, **7**, 64 (1979).
60. V.A. Bochkarev, E.Z. Napukh, *Elektrokhimiya*, **18**, 6, 843 (1982).
61. G. Maurice, Ger. Offen. 2,153,837 (1971).
62. F. Helmut, V. James, U.S. patent 4,091,128 (1978).
63. Compegnie generale d'electroceramique, Ger. Offen. 1,108,183 (1957).
64. H. Narcus, to be published in *Trans. Electrochem. Soc.*
65. V.J. Pakolenko, *Zanod. Lab.*, **48**, 10, 89 (1982).
66. L.V. Markova, *Zh. Anal. Khim.*, **25**, 8, 1620 (1970).
67. K. Wekua, B. Hoffmann, *Furbe u. Lack*, **58**, 250 (1952).
68. K. Hirabayashi, Jap. patent appl. 77,105,578 (1977).

Chapter 18
Electroless Cobalt and Cobalt Alloys

Section I
W.H. Safranek, CEF

Interest in electroless cobalt was first aroused with the reports of Brenner and Ridell (1) who used sodium hypophosphite as the reducing agent. They concluded that deposition from acid solutions was impractical and recommended a solution containing cobalt chloride, sodium hypophosphite, sodium citrate and ammonium chloride, operated at a pH of 9 to 10, controlled by additions of ammonium hydroxide. Ions such as Cd^{2+}, Zn^{2+}, Mg^{2+}, and Fe^{2+}, decreased the rate of deposition; CN^- and SCN^- ions retarded deposition completely.

Table 18.1 (1-19) summarizes the reducing agents disclosed for depositing cobalt. Results vary appreciably. For example, bright and dull coatings were obtained with 1.2 and 1.8 M concentrations of hydrazine sulfate, while deposition rates were 2.8 and 4.5 μm/hr, respectively (14).

Commercial electroless cobalt deposits are made in alkaline baths although an acid bath using dimethylamine borane as the reducing agent reportedly produces bright deposits containing 1.7 percent boron, about 1.0 percent carbon and 0.05 percent nickel at a rate of about 13 μm/hr at 70° C (4).

Electroless cobalt can provide effective protection against corrosion of steel. Applications for electroless cobalt involve its magnetic properties, wear resistance, electrical conductivity or thermal conductivity.

MAGNETIC PROPERTIES

The range of magnetic properties that can be developed and the potential usefulness of cobalt-phosphorus alloys have resulted from extensive research and manufacturing development. Methods for measuring magnetic properties have been compiled in a single volume (20).

Editor's Note:

Chapter 18, "Electroless Cobalt Deposition," has been divided into two parts. Section I covers the literature dealing with electroless cobalt plating during the time period 1947 to 1969, whereas Section II is concerned with the subject for the years 1969 to the present. Examples of electroless cobalt plating baths, the resulting deposits, and the properties of these deposits are discussed in both sections.

The preponderance of information contained in the chapter as a whole deals with the magnetic properties of electroless cobalt alloys and their applications. This is merely a reflection of the nature of the industry and content of the available literature.

Table 18.1
Reducing Agents Used
For Electroless Cobalt
And Cobalt Alloys

Agent	Reference
Amine boranes	2-4
Borohydrides	5-7
Cobalt boranes	8-11
Dimethylborane	12,13
Hydrazine	14-18
2-oxazolidinone	19
Sodium hypophosphite	1

The magnetic properties of the alloys are critically dependent upon preparation conditions and are a complicated function of the phosphorus content, crystallite size, and the orientation of the deposits, which, in turn, are controlled by the deposition variables, primarily solution composition, pH and temperature. For example, coercivity can range from a low value of only 1 oersted for electroless cobalt with a low phosphorus content to more than 1000 oersteds for alloys with a fine-grained structure and containing a relatively high phosphorus content (21).

On the basis of single domain theory, coercivity depends not only on crystallite or grain size, but also the separation between grains and size distribution from crystal-, shape-, and strain-anisotropy energy of the individual grains, modified by coupling effects between grains.

Reviews of chemically deposited magnetic alloys, including cobalt alloys, have appeared in Russian (22), English (23) and Japanese (24). The properties required for various computer components have been discussed (25). Magnetic properties and recording behavior of chemically deposited films have been correlated (26).

The magnetic properties of a material are based on its hysteresis loop (Fig. 18.1) wherein the field (H) applied to magnetic elements is plotted along the horizontal axis. The output or resulting flux density or induction change (B) is plotted along the vertical axis. Remanence (B_r) is defined as the flux remaining in the material when the magnetic field is reduced to zero. The magnetic element in its rest state is always in a position of $+B_r$ or $-B_r$. This property, coupled with the ability to sense whether the magnetic element is in the + or - state, enables the element to perform as a binary memory. Coercive force (H_c) is defined as the applied field which is necessary to maintain the magnetic element at zero flux density. The coercive force is essentially a measure of the amount of energy which must be applied to the magnetic material in order to change its state from $+B_r$ to $-B_r$, or from $-B_r$ to $+B_r$. The maximum value along the B axis which can be obtained by applying large fields is B_m, which is a measure of the total amount of flux density or induction available in the magnetic element for the maximum field applied. The squareness of the hysteresis loop is a property of some importance

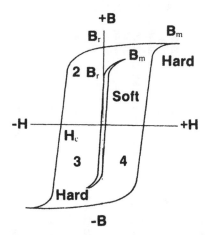

Fig. 18.1—Magnetization curves for hard and soft magnetic materials.

in magnetic memory applications and is defined as B_r/B_m. Occasionally, workers in this field have determined the saturation magnetization (M_s), which has been measured between 0 and 160 emu/g. From the second quadrant of the hysteresis loop (Fig. 18.1) the maximum energy product. $(BH)_{max}$, of the material and other characteristics are determined.

Requirements for high-density data storage are: (a) a nearly rectangular B-H loop, (b) a coercive force generally higher than 200 oersteds, and (c) a relatively low ratio of remanent magnetization (B_r) to coercive force (H_c). Substantial increases in the packing density are obtained in a magnetic recording system by the use of high-coercivity metal films. Of several techniques considered, electroless deposition was reported in 1967 to be the most successful for producing films with these magnetic properties (28).

Besides crystallite size and orientation, which can be influenced by the nature of the substrate, and the phosphorus content of the alloy, the deposit thickness and residual stress affect magnetic properties. Table 18.2 (3, 15, 16, 27-53) summarizes the role of these and other factors.

Data for Cobalt-Phosphorus Alloys

In general, by increasing the concentration of hypophosphite and hydrogen ions, and decreasing temperature and thickness in the plating solution, the cobalt deposit exhibits an increasing coercive force (38). A low coercivity, <12 oersteds, was reported for films obtained at a high temperature (92° C), which tends to limit the phosphorus content of the alloy to only 1.0 or 1.5 percent (49). Table 18.3 (42, 49, 54, 55) shows coercivity increasing to as much as 1200 oersteds and remanence decreasing to 0.8 kilogauss for 600- to 800-angstrom films deposited in solutions operated at lower temperatures.

An increasing coercive force of Co-P films was associated with an increase in

Table 18.2
Factors Affecting Magnetic Properties

Factor	Coercivity, H_c	Remanent flux density, B_r	Maximum flux density, B_m	Squareness of hysteresis loop, B_r/B_m	Magnetic moment	Refs.
Substrate	Slight	—	—	—	—	27-32
Pretreatment procedures	Can be significant	—	—	—	—	33,34
Cobalt ion concentration in solution, increasing	Decreases	Increases	Increases	No change	—	35,36
Hypophosphite ion concentration, increasing	Increases to a max. which varies with thickness	Decreases	Decreases	Decreases	—	37,38
Addition agents	Variable	—	—	—	Variable	16, 38-46
pH, increasing	Increases to a max. at pH 8.1	No change	No change	Increases	Max. at pH 8.1-8.8	15,28, 31,38 47,48
Temperature, increasing	Decreases	No change	—	No change	—	28,38
Agitation	Variable	—	—	—	—	28
Applied magnetic field	Variable	No change	No change	No change	—	49
Film thickness, increasing above 500 Å	Decreases	Decreases	Decreases	Decreases	No change	28,47 50,51
Phosphorus content, increasing	Increases to a max. at 3.8-4.2%				Decreases	28
Grain size, increasing	Increases to a max. at 300-450 Å				—	15,30 38,52,53
Stress	Variable	—	—		—	28
Heat treatment	Variable	—	—	No change	—	49

Table 18.3
Magnetic Property Data for Electroless Cobalt
As a Function of Solution Temperature[a]

Temp., °C	Coercivity, Oe	Remanent flux density, K gauss	Squareness of hystersis loop, B_r/B_m	Phosphorus content, percent	Film thickness, Å
50	1000-1200	0.8	—	5.0-5.8	600-800
75	640-800	7-10	0.750.90	2.0-2.4	9000-30,000
80[b]	115-344	11.6-14.6	0.65-0.79	About 2	4600-5400
92	1-11[c]	11.0-14.7	0.93-1.0	1.0-1.5	1200-6500

[a]Data for solutions containing 0.075 to 0.11M $CoSO_4$ or $CoCl_2$, 0.085 to 0.33M $Na_4H_2PO_2 \cdot H_2O$, 0.22 to 0.85M NH_4Cl or $(NH_4)_2SO_4$, and 0.06 to 0.3M $Na_4C_6H_5O$ with a pH of 8.0 to 8.7 (42,49,54,55).
[b]A saturation magnetic moment of 100 emu/g was reported for a film with a coercivity of 250 Oe, which was deposited in a solution heated to 88° C.
[c]Higher coercivities, up to 350 Oe, were reported for films deposited in solutions having a high pH of 9.5 (49).

the pH from 7.4 to 8.1 of electroless solutions containing boric acid (15,56). A maximum of about 750 oersteds was reported for solutions with a pH of 8.0 to 8.2. A higher pH above 8.3 resulted in a much lower coercivity of about 100 oersteds. With another bath that contained no boric acid, a film with a coercivity of 120 oersteds was produced when the pH was adjusted to 7.7, whereas a maximum coercivity of 875 oersteds was recorded when the pH was raised to 9.4 (55).

The coercive force for cobalt films produced in solutions prepared with cobalt sulfate tends to be slightly higher than the coercivity of the cobalt-phosphorus alloy deposited in cobalt chloride baths. This effect is illustrated in Fig. 18.2 which also shows a maximum coercivity at a film thickness of 500 to 700 angstroms. According to Figure 18.3 coercivity and flux density decreased as film thickness increased to 14,000 angstroms. This decrease in coercivity may be associated, at least in part, with a decreasing phosphorus content with increasing film thickness (37,38).

Electroless cobalt films containing 3.5 to 4.2 percent phosphorus, which were deposited in 0.1 or 0.3 M cobalt sulfamate solutions at 80° C, also decreased in coercivity as film thickness increased from 1500 to 8000 angstroms (51). Deposits on Kapton exhibited a decrease to only 100 oersteds from a high of 800 oersteds. The B_r-B_m ratio also decreased from about 0.65 to the range of 0.5 to 0.57.

Increasing the phosphorus content of the alloy by increasing the hypophosphite concentration of the bath increases coercivity to a maximum value that depends on film thickness (37). For films deposited in a cobalt sulfate bath with a pH of 8.3 (85° C), the maximum (1200 oersteds) occurred at a thickness of 560 angstroms when the phosphorus content was 3.5 percent, but shifted at 2250 angstroms to 4.2 percent phosphorus.

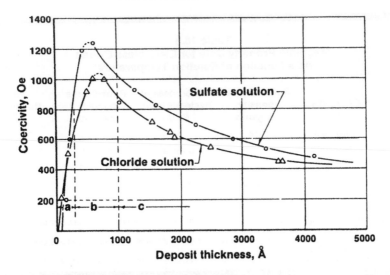

Fig. 18.2—Coercivity as a function of film thickness (54).

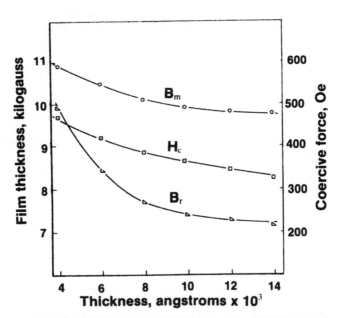

Fig. 18.3—Coercive force (H$_c$), remanence (B$_r$), and maximum flux density (B$_m$) of cobalt deposits as a function of thickness (57).

Addition agents containing a C=S group such as dimethylthiourea and thioacetamide decreased the coercive force of the films, whereas those containing a C=O group such as urea and acetamide increased or had little effect on the coercive force of similar films (42). Agents with the carbon-sulfur double bond produced a significant effect on grain size, crystal orientation, and coercive force. The magnitude of the change in coercivity was dependent on the concentration of the addition agent. The solution employed for these studies consisted of 0.075 M cobalt chloride, 0.33 M sodium hypophosphite, 0.22 M ammonium chloride, and 0.06 M citrate ions and was maintained at 80° C.

Relationships between coercivity and grain size are complex, but one report showed a maximum of about 700 oersteds for films with a grain size of 300 to 400 angstroms (58). Lower coercive force values were reported for coarser-grained deposits. Another report showed coercivity decreasing from >650 to 500 oersteds as grain size increased from 1000 to 2000 angstroms (57).

The application of a magnetic field during deposition induced a uniaxial anisotropy in films exhibiting a very low coercive force (49).

The magnetic properties of typical deposits were unchanged when held at 100° C for 1100 hours (59) but annealing at temperatures in the range of 350 to 500° C increased the coercivity of low-coercivity films and decreased the coercivity of high-coercivity deposits (60). After annealing, coercive force was approximately 300 oersteds for both classes. A transition from a supersaturated solution to a two-phase structure including Co_2P probably was responsible for the shift in the coercive force of the low-coercivity films. This transition occurs at about 320° C (61).

The saturation magnetization of Co-P films decreased with increasing phosphorus content from 156 emu/g at 2.7 percent phosphorus to 114 emu/g at 3.45 percent phosphorus and then increased to 127 emu/g at 4.2 percent phosphorus (37).

Ternary Alloy Deposits

A low coercivity of 1.5 to 5 oersteds was reported for cobalt-nickel-phosphorus films deposited at a high temperature (90° C). In one case, film thickness was 2000 angstroms and the nickel and phosphorus contents were about 40 and 1 to 2 percent, respectively (62). In another, thickness was 5000 angstroms (59). In a nickel-cobalt deposit on glass, which had a thickness of 2000 to 3000 angstroms and a phosphorus content of 9 to 11 percent, coercivity was 3 to 5 oersteds (63). The temperature of this solution was 54° C. All of these films were obtained in a sulfate solution containing sodium and/or ammonium ions, sodium citrate and phosphorous or hypophosphorus acid.

Thicker, nickel-rich deposits exhibited even lower coercivity values. For example, a 10,000-angstrom deposit containing 33 to 45 percent cobalt had a coercive force below 1.0 oersted (64), and a 20-μm thick deposit containing 23 percent cobalt had a coercivity of 0.14 oersted (65).

Films with a nickel content below about 40 percent exhibited higher coercivities, ranging from 270 to 2000 oersteds, as shown in Table 18.4 (62-73). Figure 18.4 shows increasing coercivity to about 2000 oersteds as the cobalt

Table 18.4
Magnetic Property Data for Ternary Alloy Deposits

Tertiary metal, percent	Phosphorus content, percent	Coercive force (H_c), Oe	Saturation magnetization, emu/g	Film thickness, Å	Refs.
Nickel, 15-40	2-12	270-2000	90-110	500-2500	63,66, 67,68
Nickel[a], 50-75	1-11	0.1-15	40-80	2000-80,000	62-67
Iron, 1-5	About 4	600-1480	110	500-10,000	69
Iron, 6-40	0.3-3.7	50-150	125-189	500-10,000	69
Rhenium, 30	2	0-290[b]	—	—	70
Tungsten, 9	2	250-400	—	—	70
Zinc[c], 3.5-4	—	680-900	—	1000-5000	71
Zinc[d], 4-8	4	1000-1080	—	20,000	72
Tin[e], 1-3	2	280	—	—	73

[a] Remanent flux density shifted from 0.3 to 7.0 kilogausses with a shift in the cobalt and phosphorus concentrations from 3.75 and 5.5 percent, respectively, to 23 and 6.9 percent, respectively. Saturation flux density changed from 3.0 to 6.0 kilogausses with this change in composition.

[b] Coercivity increased from 0 to 290 oersteds as film thickness was increased.

[c] For films with a thickness >300 angstroms, B_r/B_m ratios (squareness) were in the range of 0.70 to 0.75.

[d] Squareness of the film was 0.7.

[e] Squareness of the film was 0.6.

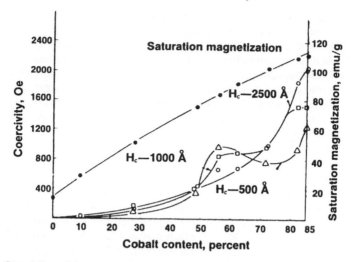

Fig. 18.4—Coercivity and saturation magnetization as a function of the cobalt content of cobalt-nickel-films deposited in solutions containing citrate ions (67).

content of 500-angstrom-thick deposits increased to 85 percent. These alloy films, obtained in solutions containing citrate ions, also exhibited increasing saturation magnetization with increasing cobalt content.

Small concentrations (20 to 30 mg/L) of lead, copper or aluminum in sulfate-malonate-malate-succinate baths greatly increased the coercivity of cobalt-nickel- phosphorus alloy deposits which normally ranged from 580 to 620 oersteds; 1 g/L of zinc had a similar effect, but 50 to 100 mg/L of molybdenum, iron or silver reduced coercivity considerably (74).

With good control of impurities, films with a thickness of 0.08 μm exhibited a B_r/B_m ratio of 0.80 to 0.84 (74). Such films are applied on magnetic disks.

Low-coercivity, cobalt-nickel-phosphorus deposits from sulfate-citrate-phosphorous acid baths had good stability at temperatures of 25 and 65° C during 1500-hour tests (59). A 50 percent increase in coercivity was noted for films with a thickness of 400 to 3600 angstroms during heating for 1500 hours at 100° C, but values for thicker deposits changed very slightly at 100° C.

Annealing at 300° C increased coercive force and magnetization of 16 to 18-μm-thick deposits from a sulfate-acetate-sodium hypophosphite solution at 87° C (75). Heating at 400° C further increased coercivity from 6.5 to 12.7 oersteds and magnetization from 115 to 150 gauss.

Iron-alloy films with a high cobalt content (>95 percent) also exhibited a high coercivity (600 to 1400 oersteds), whereas films with an iron content of 6 to 40 percent exhibited an appreciably lower coercivity (50 to 150 oersteds) (69). The higher values in each range corresponded to 500-angstrom films and the lower values to 2000-angstrom films. The phosphorus content of the alloy decreased rapidly from about 4 to only 0.3 percent as the iron content increased from 6 to 25 percent. These films were deposited in 80° C solutions with a pH of 8 and containing 0.09 M cobalt sulfate, 0.38 M sodium hypophosphite, 0.1 M sodium citrate, 0.3 M ammonium sulfate, and up to 0.07 M ferrous sulfate.

Thin, cobalt-rhenium-phosphorus films exhibited low coercivities (Table 18.4) and ternary alloy deposits containing tin or tungsten were intermediate from 250 to 400 oersteds. The rhenium and tungsten alloys were deposited in dilute (0.01 M) chloride solutions containing citric acid and sodium hypophosphite with a pH of 8.8 or 8.9 (70).

Despite the relatively high temperatures (80 to 90° C) of the cobalt-zinc-phosphorus baths, deposits from these solutions had a high coercivity ranging up to 1080 oersteds (72). High coercive force, above 1000 oersteds, was reported for cobalt-phosphorus films containing aluminum (76).

Among the electroless cobalt-phosphorus alloys adopted or developed for high density recordings, one containing 24 percent nickel and 1.3 percent tungsten has been reported to have a coercivity of 700 oersteds and a B_r/B_m ratio of 0.5. Deposits were obtained in a sulfate bath at 85° C with a pH of 9.0 (77).

The codeposition of 1 to 1.5 percent manganese with 3 percent phosphorus and 20 percent nickel, balance cobalt, increased coercivity to a maximum of about 2200 oersteds (78). Coercive force values declined rapidly to 1500 oersteds, or less, when the manganese content was raised to 3 percent or reduced to 0.5 percent. The 5-μm-thick alloy deposits were produced in a sulfate bath at 85° C with a pH of 9.6. The optimum manganous sulfate concentration was 0.04 M.

MECHANICAL PROPERTIES

Stress

Stress in electroless deposits has an appreciable influence on magnetic properties (27,79). High stress, ranging from about 5 to 38.5 kg/mm^2 (7,000 to 55,000 psi), has been reported for cobalt-phosphorus films. (27) The higher values corresponded to thin, 500-angstrom films, whereas 1000- to 2000-angstrom films exhibited a stress of 5 to 10 kg/mm^2 (7,000 to 14,000 psi). Concentrated solutions produced films with high stress, in comparison with dilute solutions. The influence of film thickness and solution concentration is shown in Figure 18.5.

Exposure to a warm, humid atmosphere (80 percent RH) increased stress in 600- to 800-angstrom films deposited in solutions containing 0.11 M cobalt chloride or 0.13 M cobalt sulfate, 0.18 or 0.19 M sodium hypophosphite, 0.5 M ammonium sulfate and 0.12 M sodium citrate (54). Solution pH and temperature were 8.7 and 50° C, respectively. The phosphorus content ranged from 5.0 to 5.8 percent. Stress was increased from about 20 to 32 kg/mm^2 after 60 hours of exposure to the humid atmosphere and the magnetic moment of the cobalt-phosphorus alloy also was increased, which was attributed to the increase in

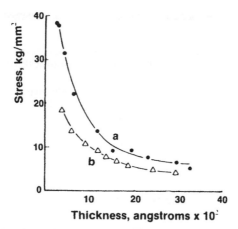

Fig. 18.5—Average stress vs. thickness for electroless cobalt. (a) = deposits from concentrated solutions; (b) from dilute solution (27).

surface area caused by corrosion. An overlay of rhodium is sometimes adopted to prevent such corrosion.

Microstrain and hardness of electroless cobalt films increased as their phosphorus contents increased above 2 percent, as shown in Table 18.5. Microstrain was determined by X-ray analysis. The pH of the 88° C solution ranged from 7.4 to 8.2, to vary the phosphorus content. The bath contained 0.085 M cobalt sulfate, 0.19 M sodium hypophosphite, 0.3 M ammonium sulfate, and 0.27 M sodium citrate.

Hardness

The hardness of electroless cobalt ranges from 120 to about 750 kg/mm^2, depending on the phosphorus content and other factors. An increase in the hypophosphite concentration in the plating bath and an increase in the phosphorus content of the alloy increased hardness to the higher end of this range (35,80). Heat treating for 1 or 2 hours at 300 to 500° C increased hardness to 1050 kg/mm^2 for alloy containing about 5 percent phosphorus, as indicated in Fig. 18.6. Data in Fig. 18.6 relate to 30- to 35-μm coatings deposited on steel in 90 to 92° C solutions containing 0.16 M cobalt chloride, 0.19 M sodium hypophosphite, 0.095 M ammonium chloride, and 0.33 M sodium citrate with a pH of 9 to 10.

Cobalt-nickel alloy deposits with a phosphorus content of 3.6 percent and a hardness of 450 to 480 kg/mm^2 exhibited better resistance to wear than electroless nickel or cobalt coatings, when thickness was adjusted to 13 to 14 μm (82). However, both the nickel and cobalt coatings showed less wear than the alloy when thickness was in the range of 41 to 44 μm, as shown in Table 18.6.

Cobalt-nickel-phosphorus alloy with an initial hardness of about 500 kg/mm^2 was increased in hardness to 1000 kg/mm^2 by heating in nitrogen at 500° C (83).

Table 18.5
Microstrain and Hardness of Cobalt Films
As a Function of Phosphorus Content (80)

Approx. P content, percent	Fiber axis	Particle size D (Å)	Microstrain, $(\epsilon\omega)^{1/2}$, percent	Hardness, kg/mm^2 (1-g load)
1	(0001)	200	0.15	120
2	(1010)	100	0.11	160
3	(1010)	100	0.15	320
5	(1010)	100	0.22	380

Cobalt alloy deposits containing 32 percent rhenium, 2.6 percent boron and 10 percent oxygen (in the form of oxides) had a hardness of 350 kg/mm^2 (84). These coatings were deposited in a chloride bath at 60° C containing ethylenediamine, sodium hydroxide, sodium borohydride and perrhenate ions. Heating 0.5 hour at 400° C raised the hardness to 750 kg/mm^2.

The hardness of cobalt-nickel alloy containing 3 to 7 percent boron ranged from 400 to 700 kg/mm^2, depending on the boron content (85). Electrical resistance was 60 to 80 μohm-cm. Heat treatment (in vacuum) at 400 to 600° C increased hardness to 1200 kg/mm^2 and decreased specific resistance to about 20 μohm-cm.

STRUCTURE

Cobalt-Phosphorus Alloys
Examination by electron diffraction revealed a hexagonal-close-packed crystalline form for cobalt in many cobalt coatings. With increasing phosphorus content in the film, the face-centered-cubic structure becomes dominant in the mixture of hexagonal-close-packed and face-centered-cubic cobalt (61).

X-ray examination has shown a fine-grained structure for high-coercive-force films. The high coercivity of such films is related to their semiparticulate nature and depends primarily on particle size and secondarily on particle interaction and their crystallographic interactions (60).

Cobalt-phosphorus films nucleate as discrete particles with a circular area. Crystallites range in size from 200 to 1000 angstroms, as a rule (37), but larger crystals were observed in deposits thicker than 700 angstroms (57). Another report (58) correlated grain size with the pH of the electroless solution. Deposits from a bath at pH 7.5 had an average grain size of 600 angstroms whereas films produced when the pH was raised to 8.4 exhibited an average grain size of 300 angstroms. Crystallites in the 200- to 400-angstrom range were reported to have their C axes parallel to the polycarbonate or polyamide substrate (86).

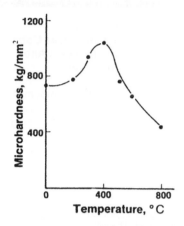

Fig. 18.6—Microhardness for 30- or 35-μm-thick electroless cobalt coatings containing 5 percent phosphorus as a function of heat treatment temperature (81). Microhardness was measured with a 50-g load.

The phosphorus content of the deposit correlates with crystallographic orientation. Below about 2 percent phosphorus, a random form was noted by one investigator (87). At phosphorus contents in the range of 2 to 5 percent, an acicular structure with a (1010) fiber axis was reported (80). The particle size appeared unchanged, however.

X-ray and electron-beam diffraction structural analysis reveals a random distribution of hexagonal crystallites with their longitudinal axis parallel with the plane of the substrate (28,37,53,69,88). These high-coercivity films had a stacking fault frequency of about 0.13 with crystallite sizes of 200 to 700 angstroms (89). In lower coercivity films (about 300 oersteds), orientation of the longitudinal axis was predominantly perpendicular to the film plane. Films with intermediate values of coercivity (300 to 600 oersteds) showed a random distribution (52,89).

Magnetization decreased with increasing temperature to about 20° C, increased abruptly at 320° C, then remained constant (61). Differential thermal analysis and X-ray diffraction analysis indicated that the transition at 320° C was accompanied by a breakdown of the solid solution and formation of Co_2P. Cobalt-phosphorus films contained hcp and fcc cobalt and Co_2P after heating to 500° C (90). Hexagonal-close-packed crystals were not present above 550° C. At 700° C, the Co_2P decomposed. Coercivity increased sharply at 350 to 450° C.

Metallographic examination of cobalt-phosphorus deposits containing 2 to 6 percent phosphorus reveals a microstructure consisting of laminations oriented parallel to the supporting metal. This layer-like structure has been attributed to the nonuniform distribution of phosphorus (91). Heat treating at 800° C recrystallized the metal and obliterated the layered structure (81).

Table 18.6
Loss in Weight of Electroless Cobalt,
Nickel and Cobalt-Nickel Alloy
During Wear

	Weight loss, mg*	
Alloy	13-14 μm coatings	41-44 μm coatings
Cobalt-phosphorus	8.4	15.8
Nickel-phosphorus	11.2	12.8
Cobalt-nickel- phosphorus	7.1	97.7

*Loss during 1000 cycles with a 1000-g load on a Taber Abraser CS17F wheel.

Ternary Alloy Deposits

Cobalt films containing about 4 percent iron had only the hcp phase with random crystallite orientation (69). Films with a higher iron content showed an increasing alignment of the hexagonal longitudinal axis of the crystallites perpendicular to the plane of the film. With 20 to 50 percent iron contents, mixtures of hcp and fcc phases were found. This indicates a supersaturated solution of iron in cobalt, since under equilibrium conditions the composition should be below these values. Crystallite sizes were estimated at between 300 and 1000 angstroms for most of these films, except for the alloys with a 40-45 percent iron content, which contained very small crystallites (<100 angstroms) and a large proportion of crystallites too small for resolution by X-ray analysis.

As nickel was introduced into cobalt-phosphorus alloy, the fcc phase appeared and there was an increasing orientation of the hexagonal (002) and cubic (111) axis perpendicular to the substrate (66,67). For high nickel contents, $CoNi_3$ appeared, again with the hexagonal axis (002) perpendicular to the film plane (66). Crystallite size varied directly with the rate of deposition. It ranged from <200 to 700 angstroms for both cobalt and the high nickel-cobalt alloy films, but exceeded 1000 angstroms for a 73 Co-25 Ni-2 P composition. Cobalt-nickel films with 9 percent phosphorus and a thickness of about 500 angstroms appeared amorphous and exhibited a maximum coercivity of 200 oersteds (92).

Both Co_2P and Ni_2P were identified in cobalt-nickel films with a thickness of 24,000 angstroms (93). Annealing in the range of 100 to 225° C increased the concentration of Co_2P. Recrystallization occurred at a temperature above 200° C. Annealing the films at 300° C caused a decrease in coercivity, caused by the destruction of the preferred orientation of the defects (22). Increasing the temperature to 400° C caused an increase in coercivity, which may relate to coagulation of phosphides.

Cobalt-zinc-phosphorus films were discontinuous, with a particle diameter of 64 angstroms after 10 seconds of plating, and 105 angstroms after 20 seconds (71).

REFERENCES

1. A. Brenner and G.E. Riddell, *RP1725, Journal Research National Bureau of Standards,* **37**(1), 31 (1946); *RP1835,* ibid., **39**(5), 385-395 (1947). A. Brenner, D.E. Couch and E.K. Williams, *RP2061,* ibid., **44**(1), 109 (1950).
2. L.M. Weisenberger, U.S. patent 3,431,120 (1969). Assigned to Allied Research Products, Inc.
3. T. Berzins, U.S. patent 3,338,726 (1967). Assigned to E.I. du Pont de Nemours & Company.
4. F. Pearlstein and R.F. Weightman, *J. Electrochem. Soc.,* **121,** 1023 (1974).
5. V. Cupr, *Povrchove Upravy,* **7**(5-6), 167 (1967).
6. K. Lang and H.-G. Klein, U.S. patent 3,373,054 (1968). Assigned to Farbenfabriken Bayer AG.
7. K. Lang and H.-G. Klein, Belgian patent 650,755 (1965). Assigned to Farbenfabriken Bayer AG.
8. British patent 836,480 (1960). Assigned to E.I. du Pont de Nemours & Company.
9. E. Zirngiebl and H.-G. Klein, U.S. patent 3,140,188 (1964). Assigned to Farbenfabriken Bayer AG.
10. R.M. Hoke, U.S. patent 3,150,994 (1964). Assigned to Callery Chemical Company.
11. J.S. Mathias and J.J. McGee, U.S. patent 3,416,932 (1968). Assigned to Sperry Rand Corporation.
12. R.M. Hoke, U.S. patent 2,990,296 (1961). Assigned to Corning Glass Works.
13. J.M. Mochel, U.S. patent 2,968,578 (1961). Assigned to Corning Glass Works.
14. N.I. Kozlova and N.V. Korovin, *Zh. Prikl. Khimii,* **40**(2), 445 (1967); *J. Appl. Chem.,* **40**(2), 426 (1967).
15. M. Aspland, G.A. Jones and B.K. Middleton, *IEEE Transactions on Magnetics,* **MAG-5,** (3), 314 (1969).
16. O. Takano, T. Shigeta and S. Ishibashi, *J. Met. Fin. Soc. Jpn.,* **18**(8), 7 (1967).
17. O. Takano, T. Shigeta and S. Ishibashi, ibid., **18**(8), 299 (1967).
18. M.P. Makowski, U.S. patent 3,416,955 (1968). Assigned to Clevite Corporation.
19. E.D. Prueter and W.E. Walles, U.S. patent 3,472,665 (1969). Assigned to the Dow Chemical Company.
20. K.L. Chpra, *Thin Film Phenomena,* McGraw-Hill, New York, NY, 1969.
21. V. Zentner, *Journees Intern. Appl. Cobalt,* Brussels, Belgium, p. 152 (1965).

22. V.V. Bondar, Itogi Nauki. *Elektrokhim.*, **56** (1966—pub. 1968).
23. F.R. Morral, *Plating*, **59**, 131 (1972).
24. O. Takano, *J. Met. Fin. Soc. Jpn.*, **28**, 541 (1977).
25. J.S. Sallo, *Plating*, **54**(3), 257 (1967).
26. D.E. Speliotis, J.R. Morrison and J.S. Judge, *IEEE Transactions on Magnetics*, **MAG-1**, (4), 348 (1965).
27. H.E. Austin and R.D. Fisher, *J. Electrochem. Soc.*, **116**(2), 185 (1969).
28. I. Blackie, N. Truman and P.A. Walker, *Magnetic Materials and Their Applications*, The Institution of Electrical Engineers, Conference Publication No. 33, p. 207 (1967).
29. R. Bate, *J. Appl. Physics*, **37**(3), 1164 (1966).
30. R.D. Fisher and S.D. Taylor, ibid., **37**(6), 2512 (1966).
31. I.F. Kostenich, *Mater. Nauch. Konf. Sovnarkhoz, Nizhne-volzhsk. Ekon. Raiona*, Volograd Politekh. Inst., Volograd, USSR, **2**, 207 (1965).
32. V. Morton and R.D. Fisher, *J. Electrochem. Soc.*, **116**(2), 188 (1969).
33. C. Ruscior, M. Suciu and Gh. Calugaru, *An. Stiint. Univ., "Al. I. Cuza," Is ai, Sec. 1b*, 12, 51 (1966).
34. J.P. Levy, British patent 1,003,575 (1965). Assigned to Sperry Gyroscope Company Ltd.
35. L. Cadorna, P. Cavallotti and G. Salvago, *Electrochimica Metallorum*, **1**(2), 177 (1966).
36. O. Takano, K. Deguchi and S. Ishibashi, *J. Met. Fin. Soc. Jpn.*, **18**(2), 46 (1967).
37. J.S. Judge, J.R. Morrison and D.E. Speliotis, *J. Electrochem. Soc.*, **113**(6), 547 (1966).
38. M.G. Miksic et al., ibid., **113**(4), 360 (1960).
39. G.S. Alberts, R.H. Wright and C.C. Parker, ibid., **113**(7), 687 (1966).
40. O. Takano et al., *J. Met. Fin. Soc. Jpn.*, **18**(1), 2 (1967).
41. O. Takano, S. Yoshida and S. Ishibashi, ibid., **18**(3), 99 (1967).
42. R.D. Fisher and W.H. Chilton, *Plating*, **54**(5), 537 (1967). Discussion by Weil, ibid., **55**, 263 (1968).
43. E.E. Saranov, N.K. Bulatov and S.G. Mokrushin, *Izvestiya Vysshikh Uchebnynkh Zavedeniy SSSR, Khimiya i Khimicheskaya Tekhnologiya*, **10**(4), 403 (1967).
44. O. Takano, S. Masai and S. Ishibashi, *J. Met. Fin. Soc. Jpn.*, **18**(5), ι80 (1967).
45. W.W. Pendleton, U.S. patent 3,446,660 (1969). Assigned to Anaconda Wire and Cable Company.
46. F.W. Schneble Jr. et al., U.S. patent 3,403,035 (1968). Assigned to Process Research Company.
47. I.D. Bursuc and Gh. Calugaru, *Thin Solid Films*, **3**(1), R5 (1969).
48. O. Takano and S. Ishibashi, *J. Met. Fin. Soc. Jpn.*, **19**(4), 139 (1968).
49. L.D. Ransom and V. Zentner, *J. Electrochem. Soc.*, **111**(12), 1423 (1964).
50. O. Takano and S. Ishibashi, *J. Met. Fin. Soc. Jpn.*, **17**(8), 299 (1966).
51. G. Asti et al., *Surface Tech.*, **10**, 171 (1980).
52. G.A. Jones and B.K. Middleton, *J. Materials Science*, **3**(5), 519 (1968).

53. K. Aoki et al., *J. Met. Fin. Soc. Jpn.*, **19**(8), 301 (1968).
54. J.S. Judge et al., *J. Electrochem. Soc.*, **112**(7), 681 (1965).
55. Y. Moradzadeh, ibid., **112**(9), 891 (1965).
56. Gh. Calugaru, *Thin Solid Films*, **3**(5), R27 (1969).
57. R.D. Fisher and W.H. Chilton, *J. Electrochem. Soc.*, **109**(6), 485 (1962).
58. M. Aspland, G.A. Jones and B.K. Middleton, *Trans. IEEE, Magnetics*, **MAG-7,** 215 (1971).
59. G.W. Lawless, *J. Electrochem. Soc.*, **115**(6), 620 (1968).
60. D.E. Speliotis, J.S. Judge and J.R. Morrison, *J. Appl. Physics,* **37**(3), 1158 (1966).
61. T. Kanbe and K. Kanematsu, *J. Physical Soc. Jpn.*, **24**(6), 1396 (1968).
62. G.W. Lawless and R.D. Fisher, *Plating*, **54**(6), 709 (1967).
63. J.C. Hendy, H.D. Richards and A.W. Simpson, *J. Materials Science*, **1,** 127 (1966).
64. J.O. Holmen and J.S. Sallo, Proc. Intermag. Conf., Washington, D.C. (1964).
65. K.M. Gorbunova and A.A. Nikiforova, *Physicochemical Principles of Nickel Plating*, USSR Academy of Sciences, Moscow, 1960. Translated by the Israel Program for Scientific Translations, Jerusalem (1963).
66. J.S. Judge, J.R. Morrison and D.E. Speliotis, *J. Appl. Physics*, **36**(3), Part 2, 948 (1965).
67. D.E. Speliotis, J.S. Judge and J.R. Morrison, *Proc. Second AES Plating in the Electronics Industry Symp.*, p. 301 (1969).
68. O. Takano and H. Matsuda, *J. Met. Fin. Soc. Jpn.*, **31,** 146 (1980).
69. J.R. DePew and D.E. Speliotis, *Plating*, **54**(6), 705 (1967).
70. F. Pearlstein and R.F. Weightman, ibid., **54**(6), 714 (1967).
71. R.D. Fisher, *IEEE Transactions on Magnetics*, **MAG-2**, (4), 681 (1966).
72. O. Takano, H. Matsuda and M. Yoneda, *J. Met. Fin. Soc. Jpn.*, **32,** 610 (1981).
73. H. Matsuda and O. Takano, *Proc. Interfinish '80*, p. 142 (1980).
74. F. Goto, Y. Suganama and T. Osaka, *J. Met. Fin. Soc. Jpn.*, **33,** 414 (1982).
75. N.A. Korenev, *Ivz. Akad. Nauk. SSSR, SER Fiz.*, **29**(4), 650 (1965); *Bull. Acad. Sci., USSR Physical Series*, **29**(4), 651 (1965).
76. J.S. Mathias and J.J. McGee, *Proc. AES Plating in the Electronics Industry Symp.* (1966).
77. T. Osaka and N. Kasei, *J. Met. Fin. Soc. Jpn.*, **32,** 309 (1981).
78. T. Osaka et al., *J. Electrochem. Soc.*, **130,** 790 (1983).
79. V. Zentner, *Plating*, **52,** 868 (1965).
80. A.S. Frieze, R. Sard and R. Weil, *J. Electrochem. Soc.*, **115**(6), 586 (1968).
81. S.A. Shatsova, S.A. Andreeva and N.V. Koroleva, *Zaschita Metallov*, **3**(6), 744 (1967).
82. Unpublished data from research sponsored by the Cobalt Information Center, Brussels (1960).
83. O. Takano and S. Ishibashi, *J. Met. Fin. Soc. Jpn.*, **19,** 489 (1968).
84. E.P. Rinkyavichene, A. Yu Youiavichyus and Ya. I. Val'syunene, *Issld. V. ob. Elektroosazhdeniya Met., Vilnius,* 163 (1977).

85. M. Matsuoka and T. Hayashi, *J. Met. Fin. Soc. Jpn.*, **31,** 567 (1980).
86. J.R. DePew, *J. Electrochem. Soc.*, **120,** 1187 (1973).
87. A. Amendola, G.W. Brock and R. Travieso, *Lubrication Engineering*, **26,** 1 (1970).
88. J. Bagrowski and M. Lauriente, *J. Electrochem. Soc.*, **109**(10), 986 (1962).
89. G. Bate, *IEEE Transactions on Magnetics*, **MAG-1,** (3), 193 (1965).
90. V.P. Moiseev, G.A. Sadakov and K.M. Gorbunova, *Zh. Fiz. Khimii*, **42**(11), 2751 (1968).
91. A.M. Kharitonyuk, *Metal Science and Heat Treatment*, (3), 234 (1966); *Metallovdeniye i Obrabotka Metallov*, (3) 57 (1966).
92. H.D. Richards, *British Journal of Applied Physics*, **17,** 879 (1966).
93. E.H. Schmidt and J.O. Holmen, *J. Appl. Physics*, **37**(3), 1378 (1966).

Electroless Cobalt and Cobalt Alloys

Section II
Peter Berkenkotter and Donald Stephens

The object of this section is to review progress in the practice of electroless cobalt plating for the years 1969 to the present. Several reviews have appeared during this period, but these have been too narrow in scope to constitute a general reference source (1-3).

In selecting the articles for review, material has been included on the basis of perceived workability, and therefore some "solutions in search of a problem" are presented. The electroless process itself had such a status for about 35 years (4). Articles on electroless Co plating beyond the scope of this section have been listed as additional references.

REDUCING AGENTS

Hypophosphite has remained the reducing agent of choice for the electroless deposition of cobalt and its alloys, mainly because of its low cost and docile behavior as compared to other reducing agents. A major problem is the tendency of some reducing agents to decompose in the absence of a plating process. Even in those cases where films from other reducing agents have shown promising properties, such as low-coercivity magnetic films, electroplating has proved to be more economical. Research continues on non-hypophosphite baths, however, and these efforts are reported here.

An extensive study of the anodic oxidation of the reductants used in electroless plating has been carried out (5) on several pure metals, including cobalt (see Table 18.7). By measuring the current-potential curves with variation in temperature, reagent concentration, and pH, it was possible to calculate the polarization curves for the oxidation of reductants, the Arrhenius activation energies at constant potential, and the reaction orders for reductants and OH (Table 18.8).

The current-potential curves for the decomposition of reductants on Co were complicated, but the reaction orders with respect to reductant and hydroxyl concentrations were found to vary between 0.4 to 0.8 and 0.2 to 0.4, respectively. Hydrogen evolution on Co at open circuit potentials was greater than at more cathodic potentials. This was true also for Ni, Pd, and Pt. Hydrogen evolution increased, however, on Cu, Ag, and Au with increasing potential.

Since the metals used in this study contained no phosphorus, and since codeposited phosphorus is thought to inhibit the electroless process (6), it is difficult to assess the significance of the hypophosphite-related data to the Co-P process.

Table 18.7
Electroless Cobalt Reductants
And Operating Conditions

Component	Bath				
	A	B	C	D	E
Na₃ citrate, m/L	0.2	—	—	0.2	—
Na₂ EDTA, m/L	—	0.175	0.175	--	0.175
H₃BO₃, m/L	0.5	—	—	0.5	—
Reductant, m/L	NaH₂PO₂ 0-0.4	HCHO 0-1.0	NaBH₄ 0.01-0.1	DMAB 0-8 g/L	NH₂NH₂ 0-2.0
pH	7.0-9.5	11.0-13.2	11.5-13.8	6.0-6.8	11.0-13.0
Temp., °C	17-77	5-57	4-47	14-56	25-66
Atmosphere	Air	N₂	N₂	N₂	N₂

Table 18.8
Reactions of Various Electroless Cobalt Reductants

Reductant	Conc.[a]	pH	E*[b]	E_a[c]	log i*[d]	E_{ref}[c]
NaH₂PO₂	0.2	9.0	-0.854	21.0	11.0	-0.80
HCHO	0.1	12.5	0.450	7.4	4.8	0.45
NaBH₄	0.03	12.5	-1.180	9.6	8.2	-0.95
DMAB	2.0 g/L	7.0	-0.832	8.8	6.3	-0.55
NH₂NH₂	1.0	12.0	-0.940	7.4	5.7	-0.75

[a]Concentration measured in mol/L.
[b]E*—Potential vs. SCE for a current density i = 1 x 10⁻⁴ A/cm .
[c]E_a—Apparent activation energy, kcal/mole.
[d]log i*—Log (pre-exponential factor)
[c]E_{ref}—Potential of Arrhenius experiment vs. SCE

Hydrazine

Works relating to the use of hydrazine as a reducing agent for cobalt have been somewhat inaccessible, and no known commercial uses have appeared. The films generated are unique in that products from the reducing agent are not codeposited with the Co; i.e., pure Co is deposited.

Cobalt and cobalt/nickel alloys were plated using the baths described in Table 18.9 (7,8). The cobalt deposit yields α-Co with a grain size of 200 to 400 angstroms (about twice the size of a typical cobalt-phosphorus crystallite), with coercivity (H_c) = 100 Oersteds, remanent flux density (B_r) between 11,000 to 13,000 Gauss, and a squareness of 0.4. The plate becomes dull above 2 μm.

Table 18.9
Cobalt and Cobalt/Nickel
Alloy Baths

Component	Bath	
	Co	Co + Ni
Na tartrate	0.4M	0.4M
$N_2H_2 \cdot HCl$	1.0M	1.0M
Thiourea	2-4 ppm	3 ppm
Co	0.05-0.08M	—
Co + Ni	—	0.05M
pH (NaOH)	12.0	12.0
Temp., °C	90	90

Cobalt and nickel can be alloyed at any ratio by varying the ratio of the metals in the bath.

Sodium Borohydride

The catalytic influence of Co-B on the oxidation of $NaBH_4$ and the reaction orders for several bath components were determined in a study using $NaBH_4$ (9). Co-B was plated onto activated Pt gauze, which was then immersed in a $NaBH_4$ solution (Table 18.10). The reaction order was found to be zero order with respect to OH^- and BH_4^-; the presence of $D_2O\text{-}OD^-$ led to about 50 percent deuterium appearing in the H_2 generated by the decomposition, and some deuterium was found in the originally isotopically-pure BH_4^-. A strong deuterium isotope effect was found, indicating that hydrogen is probably involved in a rate-limiting step of the hydrolysis of BH_4.

Table 18.11 shows the composition of a bath for depositing a Co-B film at room temperature (10). While the plating rate changes with temperature and reagent concentrations, the composition of the resulting amorphous films does not; the composition was not explicitly stated, however.

Thermocalorimetry up to 700° C gives three transitions: the first, at 200° C, is attributed to a transition to a hcp structure; this is followed by the formation of CoB at 320° C; and finally by the formation of an fcc structure and/or CoB_2 at 460° C.

The magnetic properties of the as-plated film are almost entirely independent of the plating conditions and result in a saturated magnetization of 30 to 50 emu/g (Co alone is 160 emu/g) and coercivities of less than 50 Oe. After annealing, however, the magnetic properties show a dependence on plating conditions, as well as on the annealing temperature. It is possible to obtain 74 emu saturation magnetization at a coercivity of 250 Oe after 1 hour at 400° C, with these values changing to 60 Oe at 100 emu after 5 hours at the same temperature.

Table 18.10
Catalytic Influence of Co-B
On NaBH₄ Oxidation

	Concentration	
Component	Bath	Test solution
CoCl₂, M	0.10	—
Na citrate, M	0.10	—
NaBH₄, M	0.05	0.0027-0.0962
NaOH, M	—	1.85-0.095
pH (NH₄OH)	11.2	—
Temp., °C	22	22

Table 18.11
Borohydride Electroless
Cobalt Bath

Component	Concentration
CoCl₂·6H₂O	0.0126-0.0315M
Na citrate·2H₂O	0.0126M
NaBH₄	0.0031-0.0142M
pH (NH₄OH)	10.2-11.3
Temp., °C	13-28
	(21 linear rate)

Cobalt electrodeposition was compared to electrodeposition augmented with NaBH₄; the solutions are shown in Table 18.12 (11). With NaBH₄ augmentation, the deposition rate was increased considerably and the surface of the Co was reported to be "changed."

The use of $Na_2S_2O_3$, $K_2S_2O_5$, or H_2SeO_3 has been reported to stabilize a mixed borohydride-hydrazine bath, as shown in Table 18.13 (12). Bright, adherent plate was obtained on Cu, Fe, and Ti, which contained about 3 percent B. No details are given as to the concentrations of stabilizer necessary or to the stabilizers achieved.

Alkyl Amine Boranes

Co-Ni alloys can be plated at room temperature with N,N dimethylamine borane (DMAB) as the reductant, using the solutions given in Table 18.14 (13). Extremely short bath lifetimes of about 30 minutes can be expected; the amount of boron in the deposit was not reported.

Table 18.12
Composition of Sodium
Borohydride-Augmented
Electroless Cobalt Bath

Component	Concentration A	Concentration B
$CoCl_2 \cdot 6H_2O$, g/L	20	10
Na_1 citrate·3 H_2O, g/L	100	—
$NH_2CH_2CH_2NH_2$, g/L	60	—
Na_2 EDTA, g/L	—	35
NaOH, g/L	40	40
$NaBH_4$, g/L	1	1
Temp., °C	60-80	60-80

Table 18.13
Composition of Mixed
Borohydride-Hydrazine Bath

Component	Concentration
$CoCl_2 \cdot 6H_2O$	20 g/L
Na_1 citrate	100 g/L
NH_2NH_2	60 g/L
NH_4Cl	10 g/L
NaOH	40 g/L
$NaBH_4$	1 g/L
Temp., °C	60

Table 18.14
Bath for Co-Ni Alloy
Deposition at
Room Temperature

Component	Concentration
Co + Ni	25 g/L $MCl_2 \cdot 6H_2O$
Co	10-50 percent
$Na_4P_2O_7 \cdot 10H_2O$	50 g/L
DMAB	10 g/L
pH (NH_4OH)	10.5

Table 18.15 shows a mixture of pyridine, citric acid and $CrCl_3$ that was found to stabilize a DMAB bath (14). These inhibitors are not codeposited into the plate, unlike heavy metal and sulfur compounds.

The introduction of hypophosphite into DMAB baths will inhibit plating under certain circumstances such as the solutions shown in Table 18.16 (15). The addition of hypophosphite to an acid bath (Bath A) slows the plating process, and the mixed potential of the deposited metal electrode becomes more noble. Increasing the hypophosphite concentration eventually results in the cessation of the deposition process.

Formaldehyde

Table 18.17 shows the novel use of formaldehyde as a cobalt reductant (16). At sufficiently high OH⁻ concentrations, Co forms the soluble $[Co(OH)_4]^{-2}$ complex,

Table 18.15
Mixture of Stabilizers
for DMAB Bath

Component	Concentration
$CoSO_4$	0.2M
Glycolic acid	0.0003M
Lactic acid	0.00016M
Succinic acid	0.055M
$(CH_3CH_2)_2NH \cdot BH_4$	4M
Pyridine	0.0063M
$CrCl_3$	0.009M
Citric acid	0.014M
pH	6.0-6.5
Temp., °C	70-80

Table 18.16
Composition of DMAB Baths
Used to Study Influence
Of Hypophosphite Ions

Component	Bath A	B	C
$CoSO_4 \cdot 7H_2$, g/L	25	30	30
DMAB, g/L	4	4	—
$NaH_2PO_2 \cdot H_2O$, g/L	—	—	20
Na_2 succinate	25	—	—
Na_3 citrate·$2H_2O$	—	80	80
NH_4Cl	—	60	60
NH_4OH	—	60	60
pH	5.0	9.0	9.0
Temp., °C	70	80	80

and no organic complexing agents are necessary. The bath deposits a film with a metallic luster on $PdCl_2$-$SnCl_2$ activated glass, with an approximate rate of 4 x 10^{-3} μm/min or less. Plating is accompanied by local evolution of hydrogen.

Rate control is best achieved by variation in Co concentration. In addition to reducing cobalt, formaldehyde undergoes the Cannizzaro reaction with second order kinetics, and reagent loss becomes rapid at higher concentrations.

Oxazolidinones
Heating a mixture of a substituted 2-oxazolidinone (for instance, 5-methyl-2-oxazolidinone [MOI]) and an organic cobalt salt deposits a metallic cobalt layer

Table 18.17
Composition of Bath
Containing Formaldehyde

Component	Concentration
$CoCl_2$	0.002-0.010M
HCHO	0.1-0.2M
NaOH	7-9M
Temp., °C	30

on many substrates (17). Typically, a mixture of 5 parts cobalt acetylacetone in 100 parts MOI is baked for 2 hours at 230 to 250° C, to produce a continuous conductive Co coating. Substrates such as glass, transite, amny metals, and some plastics have been found to be suitable substrates. The thickness limit depends only on the amount of mixture originally on the substrate.

Hypophosphite

In a study of bath additives and their influence on Co and Ni plating, Tafel behavior has been reported for the phosphorus codeposited with cobalt (8). A plot of log (deposition of phosphorus) vs. the steady-state plating potential yielded a plausibly straight line, the slope of which allows calculation of the product of the number of electrons involved in phosphorus deposition and the transfer coefficient.

While baths for plating cobalt alone seldom require stabilization, this is not always true for alloy baths. A well-known inhibitor, thiourea, has been found to be an accelerator up to a concentration of 5.8×10^{-6}M; above this concentration, it is inhibitory (9). The amount of phosphorus in the plate drops by 1 percent over this range of concentrations. The hydrogen evolved remained proportional to the Co plated. Imidazole has been proposed as a bath stabilizer (which is not incorporated into the film) (14).

It has been suggested that Co^{+3}, which is usually unintentionally present in cobalt baths due to air oxidation of Co^{+2}, acts as a stabilizer (19). The deposition rate in hypophosphite baths, particularly in the acidic range, is inhibited by this species.

BATH ANALYSIS

While it is beyond the scope of this review to cover all analytical advances in the determination of cobalt and other bath-related components, a few methods have been selected on the basis of utility, or which were developed for bath analysis *per se*.

Classical methods for the determination of Co, Ni and P have been adapted to quantities small enough for the analysis of plated films in the μin. range (20). By dissolving a sufficient area of plated film to yield 2 mg of metal, Co, Ni, and P could be determined to better than ± 10 percent. Cobalt and nickel were determined by pulse polarography. Cobalt can be measured to as little as 0.05 mg in the presence of a 20-fold excess Ni.

By conversion of dissolved phosphorus species to phosphate by means of permanganate oxidation, as little as 1.4 μg P in the plate could be determined by phospho-molybdate colorimetry.

The analysis of Co^{+3} generated by air oxidation of cobalt plating solutions has received scant attention in spite of its presence. One method that should prove adaptable has appeared in the gold plating literature (21). Sufficient cyanide is added to convert any Co^{+3} to the stable $Co(CN)^{-3}$, which is retained on an anion exchange resin. Co^{+2} is not retained, and the difference in cobalt concentration before and after this separation, as measured by, say AAS, is due to Co^{+3}.

The rate of deposition of cobalt from two baths (Table 18.18) onto $SnCl_2$-$PdCl_2$ activated glass was followed by optical density measurements (22). By monitoring what is apparently the transmission through the glass, the OD at 530 μm followed Beer's law to greater than 40 mu in film thickness.

Table 18.18
Composition of Baths
Used to Determine Deposition Rate
By Optical Density Measurement

Component	Bath	
	A	B
$Co(Cl)_2$	0.001-0.010M	0.001-0.008M
NaH_2PO_2	0.1-1.5M	0.1-0.2M
Na_1 citrate	0.3M	—
NH_4Cl	—	2.0M
NaOH	0.01-0.15M	—
pH	12.0-13.2	8.5-9.0
Temp., °C	20-70	15

PROCESS DEVELOPMENTS

Several interesting process developments for cobalt plating on nonconductors have been reported, with the main trend being the elimination of the need for palladium activation. A substitute for the Pd(II)-Sn(IV) activation process in plating nonconductors has been reported in a patent (23). An etched nonconductor is immersed in a cobalt hydroxide colloid for 24 hours at 65° C, followed by an additional 6-days soak at room temperature. The substrate is then rinsed and developed in 1 g/L KBH$_4$ solution at pH 8.5. Plating commences upon immersion in a Co bath. Hopefully, further developments will reduce these process times.

The basic idea of reducing an adsorbed "hydrous oxide" to either a metal or metal boride to form a catalytic surface was further elaborated in two subsequent patents (24,25); after formation of the "hydrous oxide colloid", treatment of the colloid with optimal concentrations of several possible cations enhanced the catalytic effect of cobalt in subsequent nickel plating. In some cases, Mn^{+2}, Cr^{+3}, and ZrO^{+2} were found to be effective in increasing the coverage to 100 percent, and reducing the induction time.

Substrates treated with colloidal metals (including Cu), generated via borohydride reduction, have been found to be catalytically active in a dimethyl-amine borane-cobalt bath, such as that given in Table 18.19 (25,26). Additionally, secondary organic colloids and anti-oxidants have been found to extend the life of the metal colloids.

**Table 18.19
Composition of Bath
Used To Study Colloids
In a DMAB Bath**

Component	Concentration
CoSO$_4$·7H$_2$O	25 g/L
Na$_2$ succinate·6H$_2$O	25 g/L
DMAB	4 g/L
pH	7.6
Temp., °C	50

A method has been developed (26,27) whereby either a negative or positive photoactivated image can be plated on properly prepared substrates. After Sn(IV) sensitization, Pd(II) activation and drying at room temperature, followed by irradiation with UV light and a rinse, the photo-activated regions will either plate or remain passive, depending on the pH of the cobalt bath; the non-activated area will show the opposite behavior (see Table 18.20). In connection with this work, it was found that the life of the Sn(IV) solutions can be extended up to three months (28).

An improvement in adhesion of electroless Co-Ni metal films on nonmetallic substrates is reported (29) to result from additional steps in the activation process; a dip in SnCl$_2$ and a subsequent dip in an Ag salt solution followed by the traditional SnCl$_2$-PdCl$_2$ treatment resulted in spall-resistant thin film resistors.

A process (once used as a magnetic disk overcoat) for generating an oxidation and abrasion resistant surface with good frictional properties utilizes the oxidation of an electroless cobalt alloy layer (30).

**Table 18.20
Composition of Bath
Used to Study
pH Effects**

Component	Concentration
$CoSO_4 \cdot 7H_2O$	35 g/L
$NaH_2PO_2 \cdot H_2O$	40 g/L
Na_3 citrate	35 g/L
NH_4SO_4	70 g/L
Temp., °C	60-65
pH (NH_4OH)	9.1-9.5
	(pos. image)
	6.5
	(neg. image)

To protect an oxidizable substrate, an optional barrier layer of either Ni-P or Cu (0.01 to 0.25 μm) is deposited, followed by a 0.05 to 0.2 μm alloy layer containing at least 85 percent Co. This layer is oxidized at the appropriate temperature and time (3 hr at 200° C or 2 hr at 275° C). To generate optimal surface preparation, buffing may be necessary.

BATH CONTROL TECHNOLOGY

A proposed method for the continuous control of cobalt baths involves the determination of the instantaneous plating rates by means of polarization measurements (32). The method was demonstrated with three different cobalt baths (Table 18.21).

Polarization curves for cobalt and the reductants alone in the bath matrix are illustrated in Fig. 18.7. The intersection of the polarization curve of the cobalt ion and that of a reductant represents the *mixed potential* of the bath when both components are present. (See Ref. 33 for a well worked out exposition of this theory.)

Polarization resistance was determined by means of current measurements, utilizing a potentiostat/triangle wave generator that impressed an 8 mV signal between an auxiliary electrode and the working electrode. In operation, the working electrode is allowed to plate under open circuit conditions. Prior to a measurement, the voltage, relative to a reference electrode, is held constant and the triangle wave superimposed on this dc voltage. As expected, the measured current increased as the concentration of reactants dropped.

This measurement contains components that are the result of surface reactions, surface resistance, and perhaps diffusion resistance; and, since not all of these components are necessarily sensitive to a change in the concentration

Table 18.21
Baths Used to Determine
Instantaneous Plating Rates

Component	Bath		
	A	B	C
CoSO$_4$, M/L	0.10	0.10	0.10
Citrate, M/L	0.20	0.20	0.20
H$_3$BO$_3$, M/L	0.50	0.50	0.50
NaH$_2$PO$_2$	0.10	—	—
DMAB	—	0.05	—
NaBH$_4$	—	—	0.05
pH	6-9	7	7
Temp., °C	80	80	80

of a reactant, an ac method was developed to better isolate the reaction current (34).

Two ac frequencies (0.01 and 10 kHz) were impressed simultaneously between two identical electrodes whose dc voltage is essentially open circuit. From the high frequency component of the signal, the solution resistance can be calculated. This, in turn, allows calculation of the "transfer resistance" from the low frequency current, which, along with an empirical calibration constant, K, yields the current resulting from plating alone.

Both devices gave resistance values that correlated well with the weight gains of plated samples, and therefore proved effective monitors of the extent of the reaction in the bath. The bath polarization curve could not be calculated directly from the polarization curves of the individual components. The plating rate did not follow the trajectory predicted by the movement of the point of intersection of the superimposed polarization curves, which correspond to the concentration changes that would occur as a reaction progresses. A notch-like depression occurred in the observed currents, which was not predicted by the individual polarization curves, as shown in Fig. 18.8. It is proposed that, over the potential region of the notch, metal oxides form that partially block the catalytic surfaces responsible for the oxidation of the reductant, thus lowering the plating rate.

Since the polarization monitor analyzes the total effect of bath changes, and not the change in a specific ion, it shows promise as a means of measuring reductants indirectly. Cobalt concentration and pH are easily monitored by colorimetry and a pH electrode. The remaining variable, the reductant concentration, could be determined from polarization measurements.

For these devices to evolve beyond being a simple monitor for one-shot baths, and to be useful as bath replenishment controller, at least three requirements must be fulfilled:

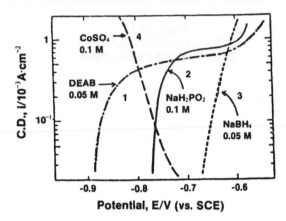

Fig. 18.7—Polarization curves for individual bath components.

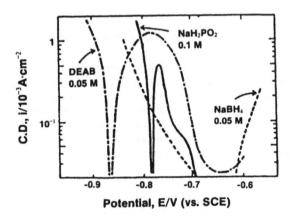

Fig. 18.8—Polarization curves for baths A, B & C.

• The polarization resistance is related to the plating rate by a factor K that must remain reasonably constant as the bath is replenished, even though replenishment changes the bath matrix.

• The bath-to-bath behavior will have to be reproducible enough so that K is reproducible also.

• The inherent precision attainable in the polarization resistance (R_p) for a simple instrument must be high enough for the control of concentrations in the 1 percent range, as is required by some magnetic cobalt baths.

A patent has been granted for a novel hypophosphite analyzer (35). At sufficiently low pH, orthophosphite ion will not reduce Fe^{+3}, whereas hypophosphite will. Fe^{+2}, generated from the oxidation of hypophosphite by Fe^{+3}, reacts with *ortho*-phenanthroline (o-Ph) to form a red complex that may be measured spectrophotometrically:

$$2[Fe(o-Ph)_3]^{+3} + H_2PO_2^- + H_2O \rightarrow 2[Fe(o-Ph)_3]^{+2} + HPO_3^{2-} + 3H^+$$

    Blue   Red

or

$$[Fe(o-Ph)_3]^{+3} + E^- \rightarrow [Fe(o-Ph)_3]^{+2}$$

By control of the transit time of successive samples through tubing, the mixing of reagents, heating, and colorimetry can be done repeatedly, forming the basis for a hypophosphite auto-analyzer/controller.

INITIATION AND FILM STRUCTURE

The initiation of Co-P plating and the structure of the plate resulting thereby has continued to be a topic of investigation, particularly with respect to the magnetic properties of Co-P. Perhaps because of experimental difficulties, no complete studies have been reported concerning the initiation of nucleation and the growth of these nuclei up to the point where the film becomes continuous.

In one comparative study, cobalt films were found to initiate with low nucleation density on a Ni-P substrate, relative to Co-Re sputtered films (36), and the nuclei were found to coalesce when the deposit thickness reached 1.2 μin. Since neither the bath composition nor the means of preplating activation are given, the relevance of these observations to published information on magnetic baths cannot be determined.

Sparse nucleation for Co-P, if unavoidable, constitutes a possible limit to the recording densities achievable by electrolessly-plated media. A model advanced by the authors proposes that, as a consequence of sparse nucleation, the surface of a thinly plated disk will be rough. The 3-D "domes", which result from coalesced nuclei, create roughness amounting to 30 to 50 percent of the film depth. Each dome is regarded as the source of a structural magnetic dipole which, in the aggregate, will lower the S/N ratio. When this film is compared to a sputtered Co-Re film (which grows in 2-D islands up to coalescence), it is seen to be smoother. Curiously, no corroborative S/N data are given.

An independent study of plated film noise in Co-Ni electrolessly-plated disks found a noise contribution due to film structure, and a contribution resulting from the influence of the underlayer (37). A 7-dB larger S/N ratio was found when comparing smooth vs. rough plate on a disk. How the difference in roughness was generated is not stated.

The cited roughness is at least an order of magnitude larger than that reported in Ref. 36 (it is observable with an optical microscope), and the thickness of the plate is twice as great (3.2 to 17 μm). It is not clear that these findings support the model proposed in Ref. 37. The importance of the control of film growth in generating high S/N ratios is, however, demonstrated.

By means of an Au underlayer followed by an electroless Ni strike, an additional 3 dB improvement in S/N ratio was obtained with the smooth disk. By control of plating smoothness and by paying attention to the surface plated upon, S/N could be raised from 30 to 40 dB.

One report on the effects of activation on Co-P plating demonstrated the effect of Pd-augmented initiation (38). Pd pretreatment of the Ni-P undercoat resulted in a 200 Oe increase in coercivity. While no direct evidence is given, this is probably a result of increased nucleation density.

A study reports on the comparative nucleation, chemical composition, and magnetic properties of Co-P and Co-Ni-P films for unspecified baths (39); the influence of pH is particularly emphasized. At a pH of 7.8, Co-Ni-P forms needle-like crystallites that differ from the dome shapes reported for Co-P. As the pH is increased to 9.3, the nuclei become more circular. In the pH interval 7.8 to 8.4, film continuity is said to improve. Over the pH range investigated, the cobalt content of the film remains above 65 percent, and goes through a maximum; the nickel content remains below 25 percent, and phosphorus below 8 percent. Over the same range, the remanent magnetization (B_n) steadily drops from around 1100 gauss to below 500 gauss. Clearly structure, as well as composition, is important in determining remanence.

The recording output for a Co-P disk at 600 Oe and 14,000 gauss-μm was compared to a Co-Ni-P disk at 700 Oe and 10,000 G-μm; the Co-P output fell below that of the Co-Ni-P disk at 26 KFCI and remained somewhat below, to 45 KFCI, possibly because of a coercivity difference.

The magnetic properties of Co-P plated on a substrate composed of aluminum-electroless Ni-copper have been reported (40). Five variations (Table 18.22) on the basic Brenner and Riddell bath (41) were investigated and compared for rate, induction time, magnetic properties, and phosphorus

Table 18.22
Influence of $H_2PO_2^-$
And pH on Magnetic Properties of Co-P

Component	Bath				
	1	2	3	4	5
$H_2PO_2^-$	0.01	0.10	0.01	0.10	0.01
pH (NH_4OH)	8.5	8.5	9.5	9.5	10.5
Rate, μm/min	1.0	1.3	1.8	2.4	3.0
P, at.%	7.8	7.6	8.1	9.6	10.2
emu/cm^3	103	100	106	115	121
Temp., °C	76.3	76.3	76.3	76.3	76.3

content. It was found that grain definition improves with film thickness and grain size increases with thickness. Grain structure is influenced, perhaps strongly, by the induction time as well as the plating rate. It is hypothesized that coercivity is governed by grain size and definition, whereas squareness (S^*) is governed by grain separation and uniformity. Because of the observed growth characteristics, optimizing both coercivity and squareness will require a compromise.

The magnetic properties of Co-P as plated on mylar and Ni-P-plated mylar show distinct differences (42). The coercive force, over a range of 10 to 1000 Oe, was higher for the Co-P with a Ni underlayer. The sum of the integrated intensities for all X-ray reflection lines was also higher, and varied in a linear fashion with thickness. This sum-of-intensities is proposed as a measure of total crystalline order without regard for the particular crystalline structures involved, which may be difficult to assign in some cases.

The lower crystalline order found for the mylar implies that there is more amorphous Co-P in this film, where "amorphous" can include crystalline particles smaller than the resolution of the X-ray wavelength used. It is proposed that the lowered coercivity for the mylar-based films is due to the presence of amorphous/fine particles.

Electroless Co-P, deposited on carbon, glass, and single-crystal Cu, initiated in isolated spots, eventually coalesced into a continuous film (43). All of the above substrates required $SnCl_2$ sensitization and $PdCl_2$ activation, and the initiation may differ from initiation on a catalytic surface, such as Ni-P.

At pH 8.3 the grains, after coalescing, showed visible granularity in growth. X-ray diffraction showed relatively distinct points, indicating a coherent structure. At pH 8.9, a loss of structure was noticable both micrographically and crystallographically.

The low pH behavior indicates that the Co crystals do not coalesce even when the film covers the substrate surface, and that some structural feature of growth isolates these crystals, perhaps Co-P. This "insulation" of Co crystal growth is thought to be responsible for the high coercivity shown by these films. The structures of electroless Ni-P, Ni-Co-P, and Co-P on a formvar substrate have been determined by electron diffractometry (44). The bath compositions and conditions are described in Table 18.23. Ni-P was found to deposit as a strained fcc lattice. The mixed Ni-Co-P bath was assigned a fcc structure and a mixed fcc/hcp structure was deemed plausible for the deposit from the cobalt bath. The grains of the Co-Ni and Co films, with deposited mass/area held constant, increased in size as phosphorus content decreased.

Metallic Co in electrolessly-deposited Co-P, furthermore, was found to form hcp crystallites about 20 angstroms in size (45). No orthorhombic Co_2P was evident and, depending upon P content, the crystallites tended to orient themselves out of the formvar plane. Additionally, it was observed that the crystallites grew in clusters, presumably in a matrix of non-crystalline Co-P.

The annealing behavior of Ni-P, Co-Ni-P, and Co-P films deposited in Table 18.24 baths was examined. The alloys plated out had proportions different from those in Baths A to F in Table 18.25. Several properties of these alloys were determined as a function of annealing temperature, as shown in Table 18.25. We

Table 18.23
Bath Compositions for Examination
Of Structure by Electron Diffraction

	Bath		
Component	Ni	Ni-Co	Co
NiSO$_4$·6H$_2$O, g/L	30	30	—
CoSO$_4$·7H$_2$O, g/L	—	30	35
NaH$_2$PO$_2$·H$_2$O, g/L	10.6	20	5-40
Na$_3$ citrate·2H$_2$O, g/L	100	—	35
NaK tartrate·4H$_2$O, g/L	—	200	—
NH$_4$Cl, g/L	53.6	—	—
(NH$_4$)SO$_4$, g/L	—	50	70
Temp., °C	85	85	85
pH (NaOH)	7-11	7.5-10.5	7-12

Table 18.24
Baths Adopted to Study
Annealing Effects

	Bath					
Component	A	B	C	D	E	F
NiCl$_2$, g/L	30	21	15	9	—	—
CoCl$_2$, g/L	—	9	15	21	30	30
NaH$_2$PO$_2$, g/L	30	30	30	30	30	20
Na$_3$ citrate, g/L	100	100	100	100	100	100
NH$_4$Cl, g/L	50	50	50	50	50	50
pH (25% NH$_4$OH)	8-9.5	8-9.5	8-9.5	8-9.5	8-9.5	9-10
Temp., °C	90	90	90	90	90	90-92

have given the maxima here, and in the last column, the temperature at which the maximum occurred. A qualitative discussion of film morphology and crystalline transitions is given in Ref. 47.

The growth of Co-P on the crystalline faces of single-crystal Cu and Ni has received extensive attention (48-50). It has been found that growth is initially fcc but becomes hcp within 10 angstroms, with orientation usually in the plane formed by the face of the host crystal. At intermediate thicknesses (up to 2000 angstroms), film growth is influenced by the crystal face of the host upon which it is growing, and at relatively large thicknesses the bath becomes influential in the direction of crystalline growth. A detailed analysis is presented in the references.

Table 18.25
Rates, Composition and Effects Of Heat Treatment on Co Plate*

Bath	pH	Rate, μm/hr	Wt. % Ni	Wt. % Co	Wt. % P	H_v, kg/mm^2	T_{Hv}, °C	H_c, Oe	T_{Hc}, °C	B_r, gauss	T, °C
A	8.0	15.7	89.4	—	10.6	700	250	125	350	3500	275
	9.5	12.4	91.6	—	8.4						
B	8.0	12.5	77.1	14.4	8.5	750	250	130	350	7000	200
	9.5	12.1	75.4	17.6	7.0						
C	8.0	11.0	55.6	37.7	6.7	900	250	320	350	9500	270
	9.5	10.0	50.2	43.9	5.9						
D	8.0	11.0	49.4	44.5	6.1	725	250	320	350	10,750	450
	9.5	11.0	27.4	66.9	5.7						
E	8.0	5.9	—	94.7	5.3	775	375	250	470	13,000	470
	9.5	8.9	—	94.2	5.8						
F(47)	—	8.9	—	95.0	4.5-5.0	1070	400	—	—	—	—

*Baths A-F from Table 18.24.

The activation energy, E_a, of 10 to 12 kcal/mole seems typical of commercial cobalt-phosphorus baths, and in the absence of other specific data, can be used to estimate the influence of temperature on the plating; a detailed analysis of crystalline habit vs. film thickness is given in Refs. 48-50.

The relative influence of bath, substrate, and substrate activation on the magnetic properties of Co-P films in the thickness range of 0.15 to 0.8 μm has been treated by the same authors (48-51).

The plating was performed on Pd-activated mylar and kapton, and Ni-P-plated mylar, as well as Cu single crystals and foils. A consistent inverse relationship between loop squareness and film coercivity was noted. Highly-grained Co-P plate gave higher H_c values than smooth bright plate unless the grain was too small, and grain size in turn was highly dependent upon the substrate. It is suggested that surface roughness increases coercivity by forcing an alignment of grains such that magnetic coupling between them is unfavorable.

Interestingly, it is reported that all films generated in this study had a saturation magnetization, M_s, of about 130 emu; the value for pure cobalt is 160 emu. To obtain these values, fields of 20,000 Oe were necessary to obtain saturation. With lesser fields, the apparent squareness was greater (0.78 at 3000 Oe vs. 0.68 at 20,000 Oe). This indicates that some of the cobalt is probably superparamagnetic, i.e., it consists of crystallites too small to maintain a magnetic polarity by themselves. A superior magnetic recording film will have a lower as well as an upper bound to its crystallite size distribution.

The preferred orientation (PO) of crystal growth can be controlled by changing the concentration of bath constituents, and by changing the pH of the bath. The (100) orientation has its hcp axis in the plane of the substrate, whereas the (110) is vertical to it.

MISCELLANEOUS PROPERTIES

The optical and magneto-optical constants have been determined for Co-P alloys over the wavelength range of 0.4 to 2.0 μm (52). The constants for Co-P films were largely independent of substrate and thickness, and were smaller in magnitude compared to the values for evaporated Co.

Co-P alloys were found to deposit on activated mylar with their c-axes (hcp) vertical at low P levels, and to deposit in an increasingly horizontal fashion as the P content in the film increased (1.7 to 3.1 percent). Wear lifetime, as measured with a tungsten carbide ball, improved as the c-axis of the deposit became horizontal (53).

The stress in plated Co-P films deposited on the (111) plane of a silicon single crystal has been determined and found to be tensile (54). The magnitude of the stress is dependent upon bath concentration. For a dilute bath, the stress rises linearly with thickness, but with a concentrated bath, the stress increases rapidly over the first 750 angstroms and then levels off. The stress is greater in deposits from concentrated baths.

Magnetic Recording Baths

Progress in plating magnetic media has probably been slowed somewhat by the emergence of sputtering technology, which is currently in vogue. As production costs become more important in the total cost of magnetic media, interest in plating processes should be rekindled.

The progress achieved has been obscured by the fact that most developments are kept "in-house", perhaps because chemical patents are notoriously difficult to police. The few developments that have found their way to print are reviewed here.

There are two approaches to digital magnetic recording currently under discussion: horizontal and vertical recording. The plating problems presented are somewhat different for these two cases. Ideally, a horizontal film would consist of uniform hcp cobalt crystals aligned in one direction and lying in the plane if the medium. Each crystal would be the size of one magnetic domain and would be separated from its neighbors by about 20 angstroms of nonmagnetic material, e.g., Co-P. Ideal films have proven somewhat elusive in practice.

The ideal vertical film presents an additional problem. As before, uniformity and separation are important, as well as the direction, which now must be vertical with respect to the plane of the medium. In addition, the saturation magnetization (M_s) along the hexagonal (vertical) axis must be reduced. The "vertical" crystal must be more anisotropic than the horizontal film. To achieve verticality, alloying to form ternary and quaternary deposits is therefore critical.

Horizontal Recording Baths

An ammonia-citrate bath, described in Table 18.26, has received much attention (55). The coercivity of deposits produced in these baths is easily controlled by changing the Co^{+2}/citrate ratio. The hypophosphite efficiency (g $H_2PO_2^-$ consumed/g Co^{+2} deposited) goes through a minimum at pH 9.2, demonstrating that these two components are consumed in distinct pathways. The magnetic properties are relatively unaffected by a change in the $H_2PO_2^-/Co^{+2}$ ratio. Ammonium hydroxide acts as a brightener until it approaches a NH_4^+/citrate ratio of 4.4, where the ammonium ions begin to compete with citrate ions in complexing Co^{+2}. As a result of this competition, the plating rate drops. The enigmatic phenomenon of bath loading is found in this bath, and is demonstrated in Table 18.27.

A patent has been granted (56) covering magnetic Co-P and Co-Ni-P plating baths that include phosphate as a component, and suggests the use of ethanolamines as brighteners and wetting agents (see Table 18.28). A subsequent overcoat of nonmagnetic Ni-P was applied and the resulting disks baked 2 hr at 200° C. Any effect of the thermal soak on coercivity is unreported.

Also proposed is the formation of thick, high coercivity films by the deposition of thin, high H_c films, alternating with layers of nonmagnetic Ni-P layers.

Alpha-amino acids, such as glycine and α-alanine, in conjunction with tetraborate, has recently been reported to increase the coercivity of Co-P films (57). A typical formulation is shown in Table 18.29.

Table 18.26
Ammonia-Citrate
Horizontal Recording Bath

Component	Range	Typical
$CoSO_4 \cdot 7H_2O$, g/L	10-53	20
Na_1 citrate $\cdot 2H_2O$, g/L	25-123	50
NH_4Cl, g/L	30-60	40
$NaH_2PO_2 \cdot H_2O$, g/L	20-40	20
pH (NH_4OH)	8.0-10.2	9.2
Temp., °C	70-92	90

Table 18.27
Effects of Bath Loading
On Plating Rate
And $H_2PO_2^-$ Consumption

Load, dm^2/L	Rate, μm/hr	$H_2PO_2^-$, g/g Co-P
0.8	6.7	7.7
1.7	5.1	6.0
3.4	4.9	5.0

Table 18.28
Electroless Cobalt Baths
Prepared with Amines

Component	Bath A	B
$CoSO_4 \cdot 7H_2O$, g/L	12	12
$NiSO_4 \cdot 7H_2O$, g/L	—	2
K_2HPO_4, g/L	10	10
Na_1 citrate $\cdot 2H_2O$, g/L	24	28
$NaH_2PO_2 \cdot H_2O$, g/L	5.75	5.75
Ethanolamine, g/L	0.15	—
Diethanolamine, g/L	1.25	1.4
Triethanolamine, g/L	—	0.5
pH (NaOH)	8.3	8.4
Temp., °C	83	83

Table 18.29
Electroless Cobalt Bath
With Tetraborate
And Diethanolamine

Component	Concentration
$CoSO_4 \cdot 7H_2O$	6 g/L
NaH_2PO_2	10 g/L
$Na_2B_4O_7 \cdot 10H_2O$	20 g/L
Na_3 citrate	60 g/L
Glycine*	15 g/L
Diethanolamine	5 g/L
pH (KOH)	9.8
Temp., °C	80

*Or adjusted amounts of α-alanine, serine, threonine, valine, or Na glutamate.

Table 18.30
Horizontal Recording Baths

Component	Bath C	D
$CoSO_4$, g/L	0.1	—
$CoSO_4 + NiSO_4$, g/L	—	0.1
NaH_2PO_2, g/L	0.2	0.2
$Na_2P_2O_7$, g/L	0.4	0.4
$(NH_4)_2SO_4$, g/L	0.5	0.5
pH (NH_4OH)	10.1	9.0-11.0
Temp., °C	70	50-85 (60 opt.)

The claims in a recently issued patent purports the ability to control the coercivity of Co-P, Co-Ni-P, and Co-Ni-Fe-P magnetic layers by means of adding Pb, Si, or Cu to the bath singly or in combination (58). With an exceptionally long roster of possible bath components, magnetic films of the following compositions are claimed (weight percent): $0.5 < P < 10$ percent; $0.05 < Cu + Pb + Si < 1.0$ percent; $0 < Fe < 30$ percent; $0 < Ni < 50$ percent; and Co to 100 percent.

The magnetic and crystalline properties of Co-W-P, Co-Ni-P, and Co-Ni-W-P films generated from an NH_3-malonate-gluconate medium were investigated (59). It was found that the Co-Ni-P forms hcp structures, with the c-axis parallel to the substrate plane, as long as $Ni^{+2}/(Ni^{+2} + Co^{+2})$ ratio is less than 0.7. Otherwise an amorphous fcc structure is formed. Co-Ni-W-P films were likewise oriented in a parallel fashion. As the tungsten content in the film is increased, the orientation of the Co-W-P films becomes increasingly vertical.

Two completely inorganic baths (60) have been reported, which are suitable for Co-P magnetic media, and which can operate at temperatures as low as 50° C. For the cobalt bath (bath C, Table 18.30), the reported maximum Hc of 450 Oe is for a 10 μm film thickness. Under the reported operating conditions, a film of 960 Oe was obtained at 0.1 μm, with a squareness of 0.68. The plate is semibright in appearance under all conditions.

The Co-Ni-P alloy bath (bath D, Table 18.30) shows a coercivity and squareness maximum at a molar ratio of $Co^{+2}/(Co^{+2} + Ni^{+2})$ = 0.9 and a bath temperature of 70° C. The resulting alloy contained 78 mole percent Co, as well as 5.9 mole percent P. In general, the Ni concentration in the film always exceeded that in the bath. When the nickel content in the film was varied from 10 to 24 percent, film structure was observed to change from a needle-like to an aggregate structure. Near the coercivity peak, the grain size was observed to be at its smallest. The phosphorus content of the deposits was not sensitive to the Co/Ni ratio, but was lowered by increasing the bath pH (7.2 percent at pH 9.0, to 2.8 percent at 11.5).

Vertical Recording Baths

Recently, an electroless cobalt plating bath (Bath A, Table 18.31) was developed that yields a quaternary alloy, Co-Ni-Mn-P, which has a vertically oriented crystalline structure (61). The film has a low saturation magnetization (M_s) relative to the anisotropy field (H_k), which makes it suitable for vertical recording. The codeposited manganese was found to enhance the vertical character of the film.

In a follow-up paper (62), the crystalline structure of the quaternary alloy was established, and the evidence was given for its vertical recording capability; i.e., at low bit densities, the appropriate double pulse waveform predicted for a vertical transition interacting with a ring head was observed.

When deposits A and B (Table 18.32) were compared at film thicknesses of 5 and 3 μm respectively, their bit densities were found to be 18.5 and 25 thousand flux reversals per inch (KFRPI) respectively. Their measurements were made at one-half the maximum amplitude (D_{50}).

The M_s values for deposits from the above two baths were reduced still further by the introduction of Re into the films (63), such that complete anisotropy resulted. D_{50} values as high as 68 KFRPI were recorded (Bath C, Table 18.32).

In addition to the malonate, which complexes with Co, tartrate was introduced to complex perrhenate. It is noted in the citation that Mn, in addition to promoting vertical growth, increases phosphorus levels in the film. No such influence was found for Re.

Differences in magnetic behavior between Co-Ni-Re-P and Co-Ni-Re-Mn-P were next investigated (Baths D and E, Table 18.32) (64). Two additional complexors were introduced to the manganese-containing Bath E. For both the vertical and horizontal media, the Mn-containing film increased in coercivity as the film thickness increases, whereas H_c for the Mn-free film decreases with thickness. This is usual behavior for Co-P alloys in their useful range of

Table 18.31
Vertical Recording Baths

Component	Bath							
	A	B	C	D	E	F	G	H
$NiSO \cdot 6H_2O$, g/L	0.01	0.01	0.06-0.16	0.08	0.12	0.08	0.06-0.16	0.04
$CoSO_4 \cdot 7H_2O$, g/L	0.025	0.035	0.06	0.06	0.06	0.06	0.06	0.06
$MnSO_4 \cdot 4\text{-}5H_2O$, g/L	0.04	0.04	0-0.09	—	0.05	—	0.05	0.03
NH_4ReO_4, g/L	—	—	0-0.009	0.003	0.005	0.003	0.005	0-0.02
Na_2 tartrate·$6H_2O$, g/L	—	—	0.20	0.20	0.20	0.20	0.20	0.50
Na_2 malate·H_2O, g/L	0.05	0.05	—	0.20	0.20	—	—	—
Na_2 succinate, g/L	—	—	0.30	—	0.30	—	0.30	—
Na_2 malonate, g/L	0.50	0.50	0.30	0.75	0.30	0.3-0.85	0.30	—
Na_2 tartronic, g/L	—	—	—	0.03	—	0.05	—	—
Na gluconate, g/L	—	—	—	0.30	—	0.30	—	—
$(NH_4)_2SO_4$, g/L	0.50	0.50	0.50	0.50	0.50	0.50	0.50	0.50
$NaH_2PO_2 \cdot H_2O$, g/L	0.20	0.20	0.30	0.20	0.30	0.20	0.30	0.20
pH (NH_4OH)	9.6	9.6	9.2	9.2	9.2	9.2	9.2	9.2
Temp., °C	85	80	80	80	80	80	80	80
Reference	61	61,62	63	64	64	65	66	68

Table 18.32
Baths for Producing
Multilayer Deposits

	Bath	
Component	A	B
$NiSO_4 \cdot 6H_2O$, g/L	—	18
$CoSO_4 \cdot 6H_2O$, g/L	25	30
Na_3 citrate·$2H_2O$, g/L	—	80
Na_2 succinate·$6H_2O$, g/L	25	—
Na_2SO_4, g/L	15	—
NH_4Cl, g/L	—	50
NH_4OH (28%), g/L	—	60
$NaH_2PO_2 \cdot H_2O$, g/L	—	20
Dimethylamine borane, g/L	4	—
pH	5.0	9.3
Temp., °C	60	89

thicknesses. The Mn-free film, because of its high initial coercivity, was proposed for thin vertical films with keeper layers. The Mn-containing film should be useful for thick vertical films without underlayers.

Strikingly, these two films were found to have quite similar compositions, and the difference in magnetic behavior is therefore mainly structural (65). The Co-Ni-Re-P film was found to attain its vertical structure earlier in the course of deposition than Co-Ni-Re-Mn-P. The decline in H_c with thickness for Co-Ni-Re-P films is postulated to be due to a change from single to multiple domains as the columnar crystals lengthen during deposition.

The vertical H_c for the Mn film grows as the vertical orientation grows and becomes dominant. Presumably multiple domains appear at thicknesses beyond those investigated.

The Co-Ni-Re-P film underwent further refinement as a thin vertical recording medium (Bath G, Table 18.31) (66). For films 0.1- to 0.25-μm thick, the vertical H_c was found to increase in a linear fashion with malonate ion concentration. The Co concentration showed a slight increase in the film at the expense of Ni, but with the exception of M_s, no clear relationship between this ratio and magnetic measurements was found.

Recording densities (D_{50}) as high as 64 KFRPI were found for 0.25-μm-thick films, and the dipulse waveforms indicative of a vertical medium were observed. With the addition of a low coercivity underlayer of electroless Ni-Fe-P, a D_{50} of 152 KFRPI was achieved (67).

Refinement of the Mn-Re-Co-Ni-P bath (Bath H, Table 18.31) gave similar gains in density, achieving 68 KFRPI on a 2-μm-thick film (68). On a flexible polyamide film, Co-Ni-Re plated on 7.5 μm of nonmagnetic Ni-P gave densities of 134 KFRPI (69). Films without the underlayer gave poorer results. Here the underlayer is thought to influence the crystal growth of the magnetic layer; clearly it is not acting as a shunt.

A somewhat simpler Co-Ni-Mn-Re-P bath was reported by other authors (70); here, Re showed a strong orienting effect that leveled off with the film strongly vertical. The coercivity continued to climb, as the Re concentration increased, to a maximum of 1100 Oe. The above alloy was plated on Pd-activated polyamide tape, which had been plated with electroless Ni (presumed nonmagnetic), and a recording density of 70 KFRPI was achieved under probable contact recording conditions.

NON-MAGNETIC USES OF COBALT PLATING

While the greatest attention has been given to magnetic uses of cobalt, several other applications have been proposed.

Cobalt, in amounts as small as 0.0005M, allows the use of sodium hypophosphite as the reductant in the plating of copper (71). The system permits copper plating where the usual formaldehyde reductant, or a high pH, is undesirable.

On the other hand, the addition of Co to copper-formaldehyde baths improves ductility, where apparently, the Co is codeposited with the Cu (72). By choosing a complexing agent that binds the Cu more strongly than Co, e.g., triethanolamine or nitrilo-tri-2-propanol [N(-CH$_2$CHOH-CH$_3$)$_3$], sufficient Co is available in depositable form. The codeposited Co is thought to reduce hydrogen generation and thereby reduce hydrogen inclusion in the plate.

Cobalt-nickel-phosphorus and cobalt-boron layers deposited over nickel-phosphorus deposits were found to confer greater corrosion protection to a substrate when compared to an equivalent thickness of nickel-phosphorus alone in salt spray tests (73). Similarly, the Co-B and Co-Ni-P overlayers were found to enhance wear resistance on a Falex wear tester. The baths used in these studies are described in Table 18.32.

In a recent review of electroless nickel plating, Co^{+1} has been cited as a brightener and stabilizer (74). The presence of cobalt in the films reportedly increase corrosion resistance and can be used to control the temperature coefficient of resistance in metal film resistors.

An example of a hypophosphite bath for producing Co-Ni-P metal film resistors is notable for its unusually low operating temperature, and is described in Table 18.33 (29).

Electroless cobalt plate has been proposed as a means of improving the adhesion of rubber to steel (75). In a pull-test, a Co-P-plated steel cord was pulled from rubber vulcanized to it. The adhesion was 30 times greater than steel alone, and was equivalent to a brass-coated cord. After 48 hr at 300 psi O$_2$ and 70° C, the cobalt-treated sample gave twice the adhesion of brass. The percent

Table 18.33
Low Temperature
Ni-Co Bath

Component	Concentration
$NiSO_4 \cdot 6H_2O$	3 g/L
$CoSO_4 \cdot 7H_2O$	30 g/L
Na_2 malate $\cdot \frac{1}{2}H_2O$	30 g/L
Na_3 citrate $\cdot 2H_2O$	180 g/L
$NaH_2PO_2 \cdot H_2O$	50 g/L
pH	10
Temp., °C	30

coverage, i.e., the amount of rubber remaining on the cord after pullout, was equal to or better than the comparison samples.

A Co-Ni-P alloy has been found useful as a barrier layer on brass linotype plates (76), where liquid type-metal (at 280° C) would otherwise dissolve the brass. This bath is shown in Table 18.34.

Samples were annealed for 3 hr at various temperatures and the micro-hardnesses measured. Maxima were found at 260 and 400° C. The brass-alloy interface showed evidence of Cu and Zn migration into the deposited layer, giving a different interfacial hardness in comparison to the bulk plate. The hardness transitions are rationalized in terms of the various Co and Ni phosphorus compounds that have been reported in the literature. The plate was reported to be conformal and wear-resistant.

Table 18.34
Bath Used to Deposit
A Barrier of Brass

Component	Concentration
$CoCO_3$	7 g/L
$NiSO_4$	15 g/L
Na_3 citrate	84 g/L
H_2PO_2	30 g/L
pH	?
Temp., °C	80-90

REFERENCES

1. F.R. Morral, *Plating*, p. 131 (1972).
2. D.E. Speliotis and C.S. Chi, *Plat. and Surf. Fin.*, p. 31 (1976).
3. J. Henry, *Metal Finishing*, p. 17 (Dec. 1984).
4. G.G. Gawrilov, *Chemical (Electroless) Nickel Plating*, Portcullis Press, Surrey, UK, 1979; p. 16.
5. I. Ohno, O. Wakabayashi and S. Haruyama, *J. Electrochem. Soc.*, **132**, 2323 (1985).
6. A.S. Freize, R. Sard and R. Weil, ibid., 115, 586 (1968).
7. O. Takano, T. Shigeta and S. Ishibashi, English abstract to *Metal Surface Technique*, **18**, 299, 466 (1967).
8. N. Feldstein and T.S. Lancsek, *J. Electrochem. Soc.*, **118**, 869 (1971).
9. K.A. Holbrook and P.J. Twist, *J. Chem. Soc. (A)*, p. 890 (1971).
10. Y.H. Chang, C.C. Lin, M.P. Hung and T.S. Chin, *J. Electrochem. Soc.*, **133**, 985 (1986).
11. A. Vaskelis and J. Valsiuniene, *Liet. TSR Moksulu Akad. Darb.*, ser. B, **119** (1972); *Chem. Abstr.*, **77**, 26449j (1973).
12. J. Valsiuniene and A. Prokopcikas, *Liet. TSR Moksulu Akad. Darb.*, ser. B, **119** (1972); *Chem. Abstr.*, **77**, 78564w (1972).
13. M. Lelental, *J. Electrochem. Soc.*, **122**, 486 (1975).
14. P. Josso, P. Lapetit, P. Mazars and M.J. Massard, U.S. patent 4,486,233 (1984).
15. F. Pearlstein and R.F. Weightman, *J. Electrochem. Soc.*, **121**, 1023 (1974).
16. E.I. Saranov, N.K. Bulatov and A.B. Lunden, *Prot. Met.*, **6**, 563 (1969).
17. E.D. Preuter and W.E. Walles, U.S. patent 3.472,665 (1969).
18. K.A. Holbrook and P.J. Twist, *J. Chem. Soc. (A)*, p. 890 (1971).
19. Personal communication from Glenn Mallory.
20. U. George, H. Merk and A.F. Bogenscheutz, *Galvanotechnik*, **72**, 586 (1981).
21. W. vanGoolen and L. Verstraelen, *Plat. and Surf. Fin.*, **65**, 50 (Sep. 1978).
22. E.I. Saranov, N.K. Bulatov and S.G. Mokrushin, *Isz. Vyssh. Ucheb. Zaved., Khim. Khim. Tekhnol.*, **10**, 403 (1967). *Chem. Abst.*, **67**, 93609b (1967).
23. N. Feldstein, U.S. patent 4,318,940 (1982).
24. N. Feldstein, U.S. patent 4,321,285 (1982).
25. N. Feldstein, U.S. patent 4,151,311 (1979).
26. I. Kiflawi and M. Schlesinger, *J. Electrochem. Soc.*, **130**, 872 (1983).
27. B.K.W. Baylis, N.E. Hedgecock and M. Schlesinger, ibid., **124**, 346 (1977).
28. B.K.W. Baylis, A. Busuttil, N.E. Hedgecock and M. Schlesinger, ibid., **123**, 1376 (1976).
29. H. Hamaguchi, U.S. patent 3,932,694 (1976).
30. G.S. Petit and R.R. Wright, U.S. patent 4,125,642 (1978).
31. P.K. Patel, D.H. Johnston and D. Makaeff, U.S. patent 4,124,736 (1978).
32. I. Ohno and S. Haruyama, *Surface Technology*, 13, 1 (1981).
33. M. Paunovic, *Plating*, **55**, 1161 (1968).

34. I. Ohno, T. Tsuru and S. Haruyama, *Metals Australasia,* **14,** 18 (1982).
35. M.C. Lambert, U.S. patent 3,816,075 (1974).
36. M.R. Khan and J.I. Lee, *J. Appl. Phys.,* **57,** 4028 (1985).
37. H. Tanaka, H. Goto, N. Shiota and M. Yanagisawa, ibid., **53,** 2576 (1982).
38. T. Osaka, H. Nagasaka and F. Goto, *J. Electrochem. Soc.,* **128,** 1686 (1981).
39. M.R. Khan and E.L. Nicholson, *J. Mag. and Magnet. Matl.,* **54-57,** 1654 (1986); E.L. Nicholson and M.R. Khan, *J. Electrochem. Soc.,* **133,** 2342 (1986).
40. Tu Chen, D.A. Rogowski and R.M. White, *J. Appl. Phys.,* **49,** 1816 (1978).
41. A. Brenner and G.E. Riddell, *J. Research Nat'l Bur. Std.,* **3,** 31 (1946).
42. V. Morton and R.D. Fisher, *J. Electrochem. Soc.,* **116,** 188 (1969).
43. G.A. Jones and M. Aspland, *Phys. Stat. Sol. (a),* **11,** 637 (1972).
44. S.L. Chow, N.E. Hedgecock, M. Schlesinger and J. Rezek, *J. Electrochem. Soc.,* **119,** 1614 (1972).
45. R.O. Cartijo and M. Schlesinger, ibid., **131,** 2800 (1984).
46. K.M. Gorbunova et al., *Prot. Metals,* **12,** 18 (1976).
47. S.A. Shatsova, S.A. Andreeva and N.V. Koroleva, ibid., **3,** 652 (1967).
48. P. Cavallotti and S. Noer, *J. Mater. Sci.,* **11,** 645 (1976).
49. P. Cavallotti, S. Noer and G. Caironi, ibid., **11,** 1419 (1976).
50. P. Cavallotti and G. Caironi, *Surface Technol.,* **7,** 1 (1978).
51. G. Asti et al., ibid., **10,** 171 (1980).
52. A.R. Baker, R. Carey and B.W.J. Thomas, *Thin Solid Films,* **37,** L8 (1976).
53. A. Amendola, G.W. Brock and R. Travieso, *Lub. Eng.,* p. 22 (1970).
54. H.E. Austen and R.D. Fisher, *J. Electrochem. Soc.,* **116,** 185 (1969).
55. A.M. Lunyatskas, *Prot. Metal,* **6,** 76 (1970).
56. A.J. Kolk Jr., U.S. patent 4,150,1772 (1979).
57. M. Malik and J.L. Livonia, U.S. patent 4,659,605 (1987).
58. R. Shirihata, M. Suzuki and T. Kitamoto, U.S. patent 4,027,781 (1987).
59. T. Osaka and N. Kasai, *Kinzoku Hyomen Gijutsu,* **32,** 309 (1981); *Chem. Abstr.,* **95,** 72158s (1981).
60. O. Takano and H. Matsuda, *Metal Finishing,* p. 63 (Jan. 1986).
61. T. Osaka et al., *J. Electrochem. Soc.,* **130,** 790 (1983).
62. T. Osaka et al., ibid, **130,** 568 (1983).
63. T. Osaka et al., *Bull. Chem. Soc. Jpn.,* **58,** 414 (1985).
64. I. Koiwa, M. Toda and T. Osaka, *J. Electrochem. Soc.,* **133,** 597 (1986).
65. I. Koiwa et al., ibid., **133,** 685 (1986).
66. T. Osaka et al., *IEEE Trans. Mag.,* **MAG-22,** 1149 (1986).
67. H. Goto et al., ibid., **MAG-20,** 803 (1984).
68. T. Osaka et al., ibid., **MAG-23,** 1935 (1987).
69. T. Osaka et al., ibid., **MAG-23,** 2356 (1987).
70. O. Takano and H. Matsuda, *Metal Finishing,* p. 60 (Jun. 1986).
71. R. Goldstein, P.E. Kukanskis and J.J. Grunwald, U.S. patent 4,265,943 (1981).
72. J. Engelbertus et al., U.K. patent application 2,083,080 (1981).
73. L.L. Gruss and F. Pearlstein, *Plat. and Surf. Fin.,* **70,** 47 (Feb. 1983).
74. J. Bielinski, A. Goldon, B. Socko and A. Bielinska, *Metalloberflaeche,* **37,** 300 (1983).

75. G. Rugo, European patent 0159600 (1985).
76. O.P. Stets'kiv, *Prot. Metal.*, **21,** 791 (1986).

ADDITIONAL REFERENCES

A. J.E.A.M. van den Meerakker, *J. Appl. Electrochem.*, **11,** 395 (1981).
B. G.A. Sadakov and Z.K. Slepenkova, *Soviet Electrochemistry*, **9,** 1048 (1972).
C. V.S. Epifanova, Y.V. Prusov and V.N. Flerov, ibid., **13,** 1161 (1977).
D. V.M. Gershov, M. Medne and P.I. Dzyubenko, ibid., **3,** 267 (1974); *Chem. Abstr.*, **81,** 98594s (1984).
E. V.M. Gershov, M. Medne and G. Ozols-Kalnins, *Latv. PSR Zinat. Akad. Vestis, Kim. Ser.*, **3,** 271 (1974); *Chem. Abstr.*, 81, 177135w (1974).
F. R. Shirihata, T. Kitamoto and M. Suzuki, U.S. patent 4,128,691 (1978).

ADDITIONAL REFERENCES

Chapter 19
Chemical Deposition of Metallic Films From Aqueous Solutions

David Kunces

Metallic coatings have been applied since ancient times. Many of the coatings were produced from aqueous solutions, while others were produced from non-aqueous, molten salt baths. One of the earliest known aqueous applications involved the cementation of copper or bronze with arsenic to produce a silvery coating of Cu_3As on art objects. Also, between 1 and 600 A.D., Acudeans plated copper objects by galvanic displacement, by which 0.5-2 μm-thick films of gold or silver were deposited (1).

Chemical plating is a process that requires neither electrodes nor any external source of electricity to deposit a metal. There are several methods of applying a metallic coating to a substrate chemically. Included with these are displacement (immersion), contact, autocatalytic, hot dipping, metal spraying, and chemical vapor deposition (CVD). Of the processes in which a metal is deposited from an aqueous solution, only displacement, contact, and autocatalytic meet the requirements of chemical deposition. Of these three processes, only displacement and contact will be discussed in detail, since the autocatalytic process is covered in more detail in other chapters.

AUTOCATALYTIC (ELECTROLESS) PLATING

The autocatalytic process is the most common of all chemical methods used to deposit metallic films from aqueous solutions. Autocatalytic plating may be defined as "deposition of a metallic coating by controlled chemical reduction that is catalyzed by the metal or alloy being deposited" (ASTM B-374) (2). Autocatalytic plating meets the requirement of not needing any external electrical power. supply by accomplishing the deposition of metal by the chemical reduction of metallic ions in an aqueous solution containing a reducer. The metal salt and reducer react in the presence of a catalyst, i.e., the part to be plated. The chemical reducers most often employed are hydrazine, sodium hypophosphite (3,4), sodium borohydride, amine boranes (5,6), titanium (111) chloride (7), and formaldehyde. The metal salts most often used are those of nickel, copper, tin, and silver. The base material can be itself catalytic or be activated by the use of displacement deposits of palladium compounds (8) or galvanic coupling (contact plating) of the substrate with a catalytic metal. After the reaction is initiated, the metal deposited serves as the catalyst, thus ensuring continuous buildup of the metal. Continuous build-up of the metal is what distinguishes autocatalytic (electroless) plating from the contact and displacement methods.

Autocatalytic, or the commonly used term "electroless", plating is almost as old as electroplating (1). The first description by von Liebig in 1835 was with the reduction of silver salts by reducing aldehydes. Despite its early start, progress in the field remained slow until World War II.

Other chapters will describe autocatalytic (electroless) methods in more detail, including mechanisms, chemistry, components, and applications.

DISPLACEMENT (IMMERSION) PLATING

Displacement plating, sometimes called immersion plating, is the deposition of a more noble metal on a substrate of a less noble, more electronegative metal by chemical replacement from an aqueous solution of a metallic salt of the coating metal. This process differs from the autocatalytic method in both mechanisms and its results (9,1). Displacement plating requires no reducing agents in solution.

The method used is based on the ability of one metal to displace another from a solution as a deposit. The substrate dissolves and is stoichiometrically replaced by the deposit. The electrons are furnished by the dissolution of the metallic substrate. Immersion deposition ceases as soon as the substrate is completely covered by the metal coating, whereas autocatalytic (electroless) plating knows no limit to the thickness of deposit that is obtainable (10). To reiterate, displacement coatings consist of more noble metal ions in solution, which are displaced from their ionic state by the less noble substrate onto the substrate. This process ceases as soon as the substrate is completely covered, thus allowing no further displacement of the metal salts and substrate. Typical thicknesses vary from 10-200 μm. The displaced deposit is formed when the substrate metal is less noble (according to the Electrochemical Series) than the plating metal (11,3) , or is made so by appropriate complexing agents in the solution. The familiar prototype is $Fe^0 + Cu^{+2} \rightarrow Cu^0 + Fe^{+2}$. The displacement process has the advantage of practically unlimited throwing power, like autocatalytic plating. It is limited only by the access and renewal of the solution to the substrate surface. It does differ from all other plating methods by providing very low capital equipment and chemical costs. The major areas of use for immersion methods include:

Base Coat
Nickel sulfate is primarily used as a base coat to provide excellent adhesion for porcelain enameling. In mechanical plating, proprietary salts of copper and tin are frequently used as the base coat (12). The nickel deposition solution for

porcelain enameling (5) on steel consists of: $NiSO_4 \cdot 6H_2O$, 7.5 to 15 g/L; 140 to 160°F; pH 3 to 4; with a 3 to 8 min cycle time.

Zincate Coatings

Although use with electrolytic coatings as a base coat, zincate solutions are primarily used to prevent the rapid formation of oxides of aluminum or magnesium substrates prior to plating by electrolytic or electroless methods. Zinc, nickel, and tin are used as well as alloys of these metals with copper and iron in solution. Several zincate formulations are shown in Table 19.1. (8,6).

Table 19.1
Zincating Formulas

	A	B	C
Sodium hydroxide	120 g/L	500 g/L	10 g/L
Zinc oxide	20 g/L	100 g/L	
Sodium potassium tartrate	50 g/L	10 g/L	
Ferric chloride hexahydrate	2 g/L	2 g/L	2 g/L
Sodium nitrate	1 g/L		
Zinc sulfate			30 g/L
Nickel sulfate			30 g/L
Temperature	23°C	23°C	
Time	30 sec	30-60 sec	
Potassium cyanide			10 g/L
Potassium hydrogen tartrate			40 g/L
Copper sulfate			5 g/L

Metal Fabrication (Liquor Finishing)

Displacement coatings of copper or tin are used for the properties of lubricity and coloring in metal fabrication. Steel wire is often coated with tin or a tin/copper alloy for coloring and as a drawing lubricant. Items such as bobby pins and paper clips are fabricated by this method. A typical formulation (4) is shown in Table 19.2.

Color Identification

Deposits of copper or tin plated by immersion methods are easily applied on small parts to distinguish one size from another when mixed. These coatings offer little corrosion protection but are easily applied and very inexpensive. Substrates used include copper, steel, aluminum and zinc.

Table 19.2
Metal Fabrication Displacement Formulas

Component	Bronze	White
Stannous sulfate	7.5 g/L	0.8-2.5 g/L
Copper sulfate	7.5 g/L	
Sulfuric acid	10-30 g/L	5-15 g/L
Temperature, °C	20	90-100
Time, min	5	5-20

Test for Cleanliness, or Incomplete Stripping, Preece Test (13,14)

Copper sulfate, used at a concentration of 2 oz/gal, with 0.1 oz/gal of H_2SO_4, RT, is used as a dip solution to verify that soils are removed from steel substrates. Another test using the same solution verifies complete stripping of organic, as well as mechanical, coatings from steel substrates (15). The immersed copper coating must be removed prior to any subsequent processing (8).

Another test, ASTM A-239 Preece Test (13), utilizes copper sulfate to locate the thinnest spot in a zinc (galvanized) coating on an iron or steel article.

Decorative Final Finish

Copper, tin (10) or gold are used as the final thin metallic coating for decorative applications. These thin coatings are relatively pore-free. The gold coatings are suitable only for inexpensive and perishable articles. A typical formulation (4) is: gold cyanide, 2.4 g/L; potassium cyanide, 2.1 g/L; at a temperature of 12° C.

Lubricant

Copper or tin coatings such as those used in metal fabrication also provide lubricity during metal drawing. Tin is used on aluminum pistons to prevent scoring of the cylinder wall by the abrasive aluminum oxide. In this case, tin acts as a lubricant during the running-in period. An alkali stannate solution—45 to 70 g/L, 3 to 4 min immersion time, at 50 to 75° C—is often used for this application (8,16).

Purification Process (Cementation)

A widely used method of recovering the last traces of copper from a leaching solution, mill tailing, or the like, is to "cement out" the copper by using scrap iron. Similarly, zinc dust is used to recover cadmium during the zinc refining cycle. Zinc dust is also used to purify electroplating solutions from contamination by unwanted metals and to recover precious metals from waste and stripping solutions. The physical form of the metal deposited is not of interest.

Soldering (17)

Proprietary tin immersion baths are used as a base for soft soldering of some electrical circuits.

Stain Prevention

Copper alloys are coated with immersion tin to prevent or inhibit "green stain", both in the textile industry and on the insides of copper tubing used in water supplies.

Some typical formulations for tinning copper (4) are shown in Table 19.3. The two advantages of immersion plating are:

1. Low capital equipment and chemical costs.
2. The ability to deposit in deep recesses (tubes). Immersion methods are applicable to relatively few situations because the thickness of the deposit is limited. When the substrate is covered, the reaction stops. In those few areas where immersion coatings are suitable, the advantages are evident.

Table 19.3
Copper Tinning Formulations

Component	A	B	C	D
Stannous chloride	4 g/L	8-16 g/L	4 g/L	
Potassium stannate				60 g/L
Sodium hydroxide	5.6 g/L	0.14 g/L		
Potassium hydroxide				7.5 g/L
Sodium cyanide	50 g/L	1 g/L		
Potassium cyanide				120 g/L
Thiourea		80-90 g/L	50 g/L	
Sulfuric acid			20 g/L	
Hydrochloric acid		10-20 mL		
Temperature, °C	20	50-212	25	20-65
Time, min	1-2	5	5-30	2-20

CONTACT PLATING

Contact (galvanic) plating is a process that has lost its identity. There are many reasons for this. Those reasons include:

1. Lack of knowledgeable use.
2. Lack of knowledge of the process.
3. Process has limited use.

Although there is little use of contact plating today, the process is used frequently with autocatalytic nickel to initiate nickel plate on copper and its alloys. This is achieved by coupling the copper or copper alloy workpiece to either aluminum, iron, or nickel. This internal galvanic couple supplies the required flow of electrons so that a thin coating of nickel is produced upon the workpiece. Once the workpiece is coated, the mechanism of the autocatalytic nickel proceeds. The mechanism for contact plating merely replaces the outside source of current with an internal galvanic couple. The galvanic couple supplies the required flow of electrons (4).

To illustrate this mechanism further, a copper sheet is dipped into a solution of potassium-silver cyanide. The copper sheet is touched with an electropositive metal (i.e., zinc) also in the potassium-silver cyanide solution. An electric current is generated that reduces silver ions on the copper sheet, while zinc ions are dissolved into the solution. However, even when the copper sheet has been covered with a coherent deposit of silver, the reduction of silver continues upon the zinc, which will then interrupt contact with the electrolyte. This interruption prevents further migration of zinc ions into the solution unless it has been removed.

The contact process can be applied only to a limited extent (11), because the formation of a uniform, heavy deposit cannot be obtained. This is because the greater current densities appear at the point of contact with the contact metal. A heavier reduction of the metal takes place at this point than on the portions further removed. Another drawback is the constant increase of dissolved contact metal into the solution, which causes short bath life. Finally, the reduction of metal upon the contact metal is not a desirable feature. The contact metal has to be constantly stripped by either mechanical or chemical means.

The chief contact metals used are aluminum, zinc, cadmium, and magnesium. Aluminum possesses three major advantages over the other metals: it is a highly positive metal; it is not reduced into the solution; and it is able to dissolve in nitric acid any of the reduced metal deposited upon the aluminum without attack to the substrate.

The electrolyte used for contact deposition must possess definite properties to yield good results: it must have good conductivity, since the currents that are generated are weak, and the electrolyte must also chemically attack the contact metal.

To increase the conductivity of electrolytes containing potassium cyanide, a greater excess can be used itself or in combination with a chloride such as ammonium or sodium chloride. The attack on the contact metal is promoted by sufficient alkalinity of the electrolyte in connection with chlorides.

Small objects can be plated in bulk in baskets made of the contact metal (14). However, excessive reduction of metal increases the cost of the process. To minimize metal reduction on the baskets, the baskets are coated on the outside only, thereby insuring that the workpieces come in contact with the electrolyte and the contact metal, while decreasing unnecessary metal deposition. The electrolytes themselves become rapidly depleted in metal solutions that require frequent replenishment, which as a rule, is not readily done. This practice makes contact plating quite expensive.

The use of contact plating is also limited because only very thin deposits can be produced. This thin film offers no protection against chemical or atmospheric environments, and is not resistant to mechanical wear. Hence, the contact process is suitable mainly for the coloring of metals or as a process to initiate autocatalytic nickel plating. The metals that have been applied by the contact process include nickel, cobalt, copper, brass, silver, gold, platinum, tin (8), and zinc.

REFERENCES

1. *Kirk-Otlmer Encyclopedia of Chemical Technology*, Third Edition, Vol. 3, pp. 510-512; Vol. 8, pp. 738-750, 826-869; Vol. 15, pp. 241-308; Vol. 20, pp. 38-42; Vol. 10, pp. 247-250; Vol. 11, pp. 988-994; Vol. 14, p. 694.
2. ASTM B-374, *Definition of Terms Relating to Electroplating.*
3. *Modern Electroplating,* The Electrochemical Society, Inc., John Wiley & Sons, Inc., New York (1953), pp. 6, 36-38, 710-747.
4. Frederick Lowenheim, *Electroplating,* McGraw-Hill Incorporated, New York (1987), pp. 389-416.
5. *A.S.M. Metals Handbook,* 9th Edition, Vol. 5, A.S.M., Metals Park, OH, pp. 219-243, 604-607.
6. *Aluminum Finishes Process Manual,* Reynolds Metal Company, Richmond, VA, pp. 123-140.
7. M.E. Warwick and B.J. Shirley, *The Autocatalytic Deposition of Tin,* I.T.R.I. Publication 586.
8. N. Hall, "Immersion Plating," in *Metal Finishing Guidebook and Directory, '86,* Metals and Plastics Publications, Inc., Hackensack, NJ, pp. 393-396.
9. W. Wiederholt, *Chemical Surface Treatment of Metals,* Robert Draper, Ltd., Teddington (1965).
10. J.W. Price, *Tin and Tin-Alloy Plating,* Electrochemical Publication Ltd., Scotland (1983).
11. G. Langbein, *Electrodeposition of Metals,* Henry Carey Baird & Company, Inc., New York (1924), pp. 1-9, 593-622.
12. *Tool and Manufacturing Engineers' Handbook,* 4th Edition, Vol. 3, Chapter 20, pp. 20-50.
13. ASTM A-239-73, *Test for Locating the Thinnest Spot in a Zinc (Galvanized) Coating on Iron or Steel Article by the Preece Test (Copper Sulfate Dip).*
14. W. Blum and G. Hogaboom, *Principles of Electroplating and Electroforming,* McGraw-Hill Book Company, New York (1949), pp. 45, 289, 306, 329, 330.
15. ASTM G-3-74, *Recommended Practice for Conventions Applicable to Electrochemical Measurements in Corrosion Testing.*
16. G.H. Poll Jr., "Phosphating to Retain Lubricant," *Products Finishing,* 7, 64-68 (1986).
17. Z. Kovac, K.N. Tu, "Immersion Tin: Its Chemistry, Metallurgy and Application in Electronic Packing Technology," *IBM J. Res. Develop.,* **28,** 6, 726-734 (November 1984).

Chapter 20
Wastewater Treatment
For Electroless Plating

Robert Capaccio, P.E.

Waste discharges from electroless plating operations are regulated under local, state, and federal discharge regulations. This chapter will cite federal regulations that have been adopted by many state and local sewer authorities (control authorities); however, the reader is cautioned to check with state and local regulatory authorities to ensure that more stringent regulations do not apply.

The federal Environmental Protection Agency (EPA) regulates discharges from electroless plating under either the Electroplating (40 CFR 413) or Metal Finishing (40 CFR 433) point source category. The metal finishing regulations are applicable to all facilities, except indirect-discharging job shops (defined as a facility that owns not more than 50 percent of the basis metal processed annually) that were discharging prior to July 15, 1983. Also exempted were existing indirect-discharging independent printed circuit board manufacturers—including their electroless copper finishing operations. The two exempted facility types are regulated under the federal Electroplating category (see Table 20.1), while all other direct and indirect dischargers must adhere to the federal Metal Finishing regulations. Table 20.2 indicates requirements for existing facilities while Table 20.3 indicates those requirements for new facilities—both under the Metal Finishing guidelines.

Further information on the regulations can be obtained from the literature (1-4). The reader is also cautioned to review other regulations that are applicable to metal finishing operations, including electroless plating. These would include regulations under the Resource Conservation and Recovery Act (RCRA), the Clean Air Act (CAA), the Comprehensive Environmental Response Compensation and Liability Act (CERCLA), the Occupational Safety and Health Act (OSHA), and selected specific standards, such as the Hazard Communication Standard (29 CFR 1910.1200) and the Respiratory Protection Standard (29 CFR 1910.134). A summary of these regulations can be found in the literature (5).

WASTE MINIMIZATION

Waste minimization includes source reduction in which the amount of waste is reduced at the source through changes in industrial processes and resource

Table 20.1
EPA Pretreatment Regulations:
Electroplating Point Source Category
For Plants Discharging More Than 10,000 Gal/Day

Pollutant	Maximum for any 1 day, mg/L	Avg. of daily values for 4 consecutive monitoring days, mg/L
Cyanide (total)	1.9	1.0
Copper	4.5	2.7
Nickel	4.1	2.6
Chromium (total)	7.0	4.0
Zinc	4.2	2.6
Lead	0.6	0.4
Cadmium	1.2	0.7
Total metals*	10.5	6.8
Silver**	1.2	0.7

*Total of Cu, Ni, Zn, and total Cr.
**Applies only to users in the precious metals subcategory.

recovery of metallic contaminants prior to discharge to minimize quantities of hazardous waste (e.g., wastewater treatment sludge) generated (6). Benefits to be derived from source reduction include reduced water and sewer user charges, decreased electroless bath consumption, and minimized hazardous waste sludge production. There are both economic and legal incentives for implementing source reduction at a facility.

If you currently have a conventional end-of-pipe pretreatment system operating at your facility, which is treating non-complexed wastewaters, the addition of electroless finishing solution chelating agents to the treatment system will interfere with conventional precipitation techniques. Therefore, it is also advantageous to treat spent electroless finishing baths and rinsewaters on a segregated basis.

Process Bath Treatment
In the electroless process, unlike the typical electroplating process, the electroless finishing bath itself must be disposed of after a certain number of turnovers.

A turnover is defined as one complete replacement of metal content of the plating solution; therefore, by extending the number of turnovers available from a bath before disposal is required, one can greatly reduce the cost per operating cycle of final bath treatment. By carefully adhering to the bath operating guidelines and techniques described in other parts of this book, the life of

Table 20.2
Federal Effluent Guidelines and Standards:
Metal Finishing Point Source Category[a]
BPT/BAT/PSES Effluent Limitations
And Pretreatment Standards

Pollutant	Maximum for any 1 day, mg/L	Monthly avg. shall not exceed mg/L
Cadmium	0.69	0.26
Chromium	2.77	1.71
Copper	3.38	2.07
Lead	0.69	0.43
Nickel	3.98	2.38
Silver	0.43	0.24
Zinc	2.61	1.48
Cyanide (T)[b]	1.20	0.65
Cyanide (A)[b]	0.86	0.32
TTO[c]	2.13	—
Oil and grease	52.00	26.00
TSS[d]	60.00	31.00
pH[e]	6.0-9.0	6.0-9.0

[a]Source—40 CFR 433 (48 FR 32485, July 15, 1983).
[b]Industrial facilities with cyanide treatment may, upon agreement with the pollution control authority, apply the amenable (A) cyanide limit in place of the total (T) cyanide limit. Cyanide monitoring must be conducted after cyanide treatment and before dilution with other wastewater streams.
[c]Total toxic organics (TTO).
[d]Total suspended solids (TSS).
[e]pH range, standard units.

electroless plating baths can be extended. Remember that improper bath operating techniques not only incur added expense and time replacing the bath, but also the added expense of treating the spent bath materials; therefore, it is crucial that the correct electroless metal formulations are chosen, and that they are operated in the proper manner to enhance stability, optimize bath performance, and thereby reduce hazardous waste generation (7,8).

Of the major electroless finishing processes (e.g., copper and nickel), there is a great deal of interest in regeneration or renewal of spent electroless nickel plating baths using ion-exchange methods. References cited indicate that ion-exchange treatment of spent electroless nickel plating baths with weak-base anionic resins can remove phosphite ions and restore the plating rate, as well as improve the deposit quality on aluminum (deposit quality on steel was not mentioned). Strong alkaline resins treated with sodium hypophosphite removed

Table 20.3
Federal Effluent Guidelines and Standards:
Metal Finishing Point Source Category
NSPS/PSNS Effluent Limitations and Pretreatment Standards[a]

Pollutant	Maximum for any 1 day, mg/L	Monthly avg. shall not exceed mg/L
Cadmium	0.11	0.07
Chromium	2.77	1.71
Copper	3.38	2.07
Lead	0.69	0.43
Nickel	3.98	2.38
Silver	0.43	0.24
Zinc	2.61	1.48
Cyanide (T)[b]	1.20	0.65
Cyanide (A)[b]	0.86	0.32
TTO[c]	2.13	—
Oil and grease	52.00	26.00
TSS[d]	60.00	31.00
pH[e]	6.0-9.0	6.0-9.0

[a]Source—40 CFR 433 (48 FR 32485, July 15, 1983).
[b]Industrial facilities with cyanide treatment may, upon agreement with the pollution control authority, apply the amenable (A) cyanide limit in place of the total (T) cyanide limit. Cyanide monitoring must be conducted after cyanide treatment and before dilution with other wastewater streams.
[c]Total toxic organics (TTO).
[d]Total suspended solids (TSS).
[e]pH range, standard units.

excess phosphite from spent electroless nickel solutions and converted the phosphite to pyrophosphite, which restored the usefulness of the bath. Ion-exchange methods are technically feasible, cost-effective, and preferable to precipitation techniques, which would remove all the nickel from spent solutions (9,10).

Rinsewater Treatment

The other source of waste from electroless finishing processes is the rinsewater used to remove any residual solution from the finished parts. This rinsewater contains metallic contaminants as well as the other bath constituents found in the electroless bath solution, which will require treatment. Dragout from the process bath of contaminants, e.g., by-product phosphite ions, can help extend the life of the bath. Unfortunately, the cost of this means of bath "purification" is to increase the contaminant loading in the subsequent rinsing steps. More research is needed to resolve practical issues relative to removal of bath contaminants from the electroless process bath so that dragout reduction methods can be employed.

Basic rinsewater flow reduction techniques should be utilized on the electroless waste streams because of the complexing agents and chelates they contain. The lower the water flow, the lower the cost will be for implementing treatment or reclamation technologies, such as electrolytic recovery. Significant reductions of up to 90 percent are possible in rinsewater quantities utilized per rinse station by adding two rinse flow tanks in the countercurrent mode instead of one tank. Table 20.4 lists typical rinsewater conservation techniques (11-18).

In summary, it is important to ensure that electroless plating baths are maintained and operated in an optimum fashion according to the manufacturer's recommendations. All feasible wastewater reduction techniques should be instituted to reduce the quantity of rinsewater requiring treatment.

Table 20.4
Rinsewater
Conservation Techniques

- Countercurrent rinsing
- Spray and fog rinsing
- Flood rinsing
- Still rinsing
- Reuse of rinsewater
- Rinsewater agitation
- Conductivity-controlled rinsing
- Flow restriction (regulating) devices
- Higher temperature rinsing
- Drainage boards

CONVENTIONAL TREATMENT TECHNOLOGY

Electroless finishing wastes are typically treated with either conventional techniques, such as chemical precipitation and reduction, or more advanced techniques, such as electrolytic reclamation, ion exchange, reverse osmosis, or electrodialysis (19,20). Conventional chemical treatment of electroless plating baths and rinsewaters involves neutralization and subsequent separation of the precipitated metal hydroxides, and disposal of the sludge. The neutralization/precipitation process is normally conducted in either a continuous-flow or batch system, depending on the volume of wastewater requiring treatment. High flow rates are best handled with a continuous two-stage system to allow for pH adjustment under more controllable circumstances. This aids in reducing the possibility of chemical pump overshoot. The chemistry involved with neutralization involves acid/base reactions. By conventional definition, acid donates hydrogen ions, whereas the base accepts them in a reaction. Acids and bases react to form dissolved salts in water (21).

A plot of the data obtained from monitoring the pH change vs. making incremental additions of an acid-to-base solution results in a neutralization or titration curve. Typical electroless plating wastewaters contain buffering agents, added to resist the pH changes associated with the reduction of the metal ion.

Depending on the pH of the bath and subsequent rinsewaters, the adjustment of the solution to within regulatory requirements, or for precipitation of metallic ions, may require substantial amounts of chemicals to overcome this buffering effect. Laboratory jar tests on the solution to determine titration curves are recommended.

During the neutralization/precipitation reaction, positively-charged metal ions, which exist under acidic conditions, react with negatively-charged base ions. The surface charge of the metal ions are negated. Under proper conditions (e.g., after solution pH is sufficiently raised), this results in the formation of a solid phase known as a *precipitate*. In the precipitation reaction, hydroxides, sulfides, carbonates, carbamates, and starch xanthate all will produce insoluble metallic compounds. These can then be removed after flocculation and clarification (22).

The precipitation reaction is controlled by solubility products of compounds formed, which control the residual concentration of metals that will be in solution at a given pH, in the presence of the complexing agent. Different complexing agents used in the formation of electroless plating baths, such as ammonia, EDTA, citrate, and malic acid, all impact in the ability of conventional technology to meet regulatory standards using various precipitating agents. For most electroless baths, significant precipitation of the metal ions can occur only after the complexing agent is substantially removed or destroyed. Chemical oxidation by ozone, potassium or sodium permanganate, sodium hypochlorite, chlorine, or ozone- and ultraviolet-catalyzed hydrogen peroxide have been reported to be effective pretreatment methods.

Laboratory precipitation tests should be performed to evaluate the impact of the complexing agent contained in the electroless bath on the proposed treatment chemistry and the need for pretreatment to destroy the complexing agent. Each complexing agent offers a varying degree of stability, which inhibits the precipitation of the metallic precipitate. For example, EDTA is much more stable than lactic acid when used as a complexing agent; therefore, baths containing lactic acid will be easier to treat, even at high metallic concentrations (23).

A common method for treating electroless copper rinsewaters utilizes ferrous sulfate as a reducing agent under acidic conditions with subsequent elevations of solution pH for effective copper removal. The acidification step assists in weakening or dissociating the bonds in the copper complex (i.e., between the copper and complexing agent), thus enabling precipitation, as a hydroxide, to occur. It is critical to control this process carefully because of the quantity of sludge that is produced by both copper and iron precipitation (24).

Alternative precipitation chemistries to scavenge heavy metals from complexed or chelated compounds in rinsewater can effectively reduce the quantity of sludge produced in such systems. Reactions of nickel or copper with sulfides, carbonates, and starch xanthate will also produce insoluble precipitates. Hydroxide precipitation can also be enhanced through the use of reducing agents such as sodium borohydride, hydrazine or sodium hydrosulfite. In this case, the reducing agent acts to reduce the metallic ion to its elemental state, e.g.,

$$Cu^{++} \rightarrow Cu^0$$

to aid in precipitation (25).

Another chemical reduction method of treatment of spent electroless baths involves the use of a noble metal catalyzed surface. For example, the bailout or bath growth from electroless copper processing can be successfully "bombed out" onto palladium-activated carbon particles at mass loadings of up to 2.5 pound of copper per pound of carbon (26).

Chemistries such as sodium dimethyl or diethyl dithiocarbamate (DTC) can be used to effectively treat both rinses and spent plating baths of electroless formulations. Studies indicate that DTC effectively precipitates both copper and nickel from rinsewater that contain complexing or chelating agents. Solutions with metal concentrations greater than 10 mg/L were easily treated over the pH range of 3.0 to 10.0, and the low concentrations of residual metals allowed direct discharge of the effluent. Carbon treatment can be used to effect removal of residual DTC, a fungicide, if necessary, but use of a large excess of DTC in a pretreated effluent is not recommended (27).

Various approaches to the design and construction of the actual waste treatment system can be taken once the optimum treatment chemistry has been defined. Whether a company decides to use a vendor or a consulting engineer, its in-house staff should be integrally involved with the project team to provide needed inventory information and quantification of waste and production estimates. The vendor approach taken by managers has the advantage that the vendor knows his particular piece of equipment; however, the inherent disadvantage is that the vendor may sell you a piece of equipment not specific to your needs.

The other route, using the engineer/consultant, has an advantage that the engineer should look at a wide range of equipment approaches available for compliance, and also furnish any detailed installation documents required for bidding. The potential disadvantage is that the consultant may not have knowledge of your particular industry, so selection becomes crucial. Construction documents, either schematic or detailed, should be at the level desired to both install the equipment in a cost-effective manner, as well as meet regulatory requirements. Prior to entering the design phase, it should be determined with state regulatory or local control authorities what type of design documents will be required for submittal for their approval prior to purchase or installation and startup of the equipment (28,29). Further information on conventional technology can be found in the referenced handbooks (30-32).

ADVANCED TREATMENT TECHNOLOGY

Because of the ever-escalating costs of hauling hazardous waste materials, e.g., metallic hydroxide sludge resulting from conventional chemical precipitation of electroless solutions, as well as the need for meeting more stringent discharge requirements, more "advanced" technologies have been installed in electroless finishing shops. These techniques, such as ion exchange and electrolytic recovery, have become more commonly utilized within the finishing industry (33-35).

Ion exchange is a reversible process that interchanges one kind of ion (e.g., hydrogen), present on the insoluble resin, with another ion of like charge (e.g., gold, copper, or nickel), which is present in the wastewater or dragout solution that flows through the ion exchange system. These resins were initially developed to scavenge the gold from the rinsewater because of its high value, but the high cost of conventional treatment of rinsewater from even more common metals baths and disposal of sludge with its inherent liability has subsequently made the use of ion exchange technology more widespread. The ion charge of the resin (e.g., cation or anion), as well as the physical properties of the resin, such as gel type or macroreticular, are important factors in selecting the proper resin. While most heavily chelated metals, such as those found in the electroless bath, are not effectively removed by weakly acidic cation exchange resins, the newer chelating resins can be useful in treatment of electroless finishing spent baths and process rinsewaters. In this case, the ion exchange system acts to concentrate the metallic ions in a more readily treated volume.

The ion exchange resin removes the metals from the more dilute solution, and then, upon subsequent regeneration, the cations (Cu^{+2}, Ni^{+2}, etc.) are stripped from the resin in a concentrated acidic medium, such as 5 percent hydrochloric acid or sulfuric acid. This smaller volume of wastewater can typically be treated in a batch treatment system via chemical precipitation to afford higher levels of chemical efficiency and lower costs in operation, or utilizing electrolytic or evaporative recovery technologies.

Bed loading rates and resin oxidation potential are concerns when utilizing ion exchange. Laboratory-scale testing is recommended to determine whether these critical parameters for particular resins under consideration will enable the use of the desired resin. Various resin manufacturers have information and sample resins available for testing various solutions (36,37).

Electrolytic recovery is also familiar technology to most metal finishers—it is simply reverse plating from less concentrated spent solutions and rinsewaters of metallic ions onto a cathode. As discussed previously, reduction by chemical means results in a precipitate requiring landfilling or incineration as a hazardous waste. Electrolytic recovery and controlled plate-out techniques eliminate the sludge.

Electrolytic recovery has found use in spent electroless plating bath treatment because of the higher content of metallic ions in the spent bath than in the rinsewaters (38,39). Spent electroless baths can also be successfully treated by "controlled plate-out" onto metal scraps that serve as the substrate. Basically, this technique operates the spent bath at normal plating temperatures with steel or aluminum scraps in the bath. With added reducing agents and stabilizers to control the rate of deposition, the metal continues to plate out on the scrap (19).

The major difference between plating from a process bath with a high metal content vs. from rinsewater with lower metal contents, is the role that the cathode polarization layer plays. Cathode polarization is the area near the cathode that tends to become depleted of metallic ions and forms a barrier against migration of metals to the cathode for subsequent plating. In baths with high metal content, this polarization layer is less critical because it can be overcome with minor agitation to present fresh metallic content near the cathode layer. On the other hand, with reduction of metal in rinsewaters that have very low metal content, the polarization layer becomes more difficult to defeat. Most manufacturers have increased the cathode area or agitated the cathode or solution near the cathode to overcome this phenomenon to some degree.

Treatment of spent electroless solutions on a batch scale can also occur by using evaporative recovery technology. In this instance, evaporation of water from the spent electroless bath reduces the volume of electroless solution required to be treated or subsequently hauled for reclamation at an outside facility. Atmospheric evaporators, which are a low-cost means of evaporating water from such solutions, can be used, although variables such as steam and air makeup requirements should be carefully reviewed. Further information on recovery technology, including ion exchange and electrolytic recovery, can be found in the references.

REFERENCES

1. R.S. Capaccio and D. Pierce, *Metal Finishing*, p. 45 (Nov. 1983).
2. U.S. EPA, *Development Document for Existing Source Pretreatment Standards for the Electroplating Point Source Category*, Aug. 1979.
3. U.S. EPA, *Development Document for Effluent Limitations Guidelines Standards for the Metal Finishing Point Source Category*, Jun. 1983.
4. *Federal Register*, 40 CFR 433 (48 FR 32485), Jul. 1983.
5. R.S. Capaccio in *Products Finishing Directory*, Gardner Publications, Inc., Cincinnati, OH, 1990.
6. R.S. Capaccio and S. Greene, paper presented at the 4th Annual Hazardous Waste Source Reduction Conference, Massachusetts Dept. of Environmental Management, Boston, MA 1987.
7. K. Parker, *Plat. and Surf. Fin.*, **74,** 48 (Feb. 1987).
8. H.A. Mackay et al., ibid., **73,** 32 (Jul. 1986).
9. K. Parker, ibid., **67,** 48 (Mar. 1980).

10. F. Levy and S.K. Doss, ibid., **74,** 80 (Sep. 1987).
11. R.S. Capaccio, *Proc. AES 70th Annual Technical Conference* (1983).
12. R.S. Capaccio, *Plat. and Surf. Fin.,* **70,** (Jun. 1983)
13. J.B. Kushner, *Water and Waste Control for the Plating Shop,* Gardner Publications, Inc., Cincinnati, OH, 1982.
14. U.S. EPA, *Control and Treatment Technology for the Metal Finishing Industry: In-Plant Changes,* EPA 625/8-82-002, Cincinnati, OH, 1982.
15. F. Caparelli, P. Kreugar and S. Shepard, *Proc. Circuit Expo '83,* p. 94 (1983).
16. J.B. Mohler, *Metal Finishing,* p. 21 (Aug. 1982); p. 35 (Sep. 1982).
17. J.B. Mohler, *Plat. and Surf. Fin.,* **66,** 48 (Sep. 1979).
18. *Proc. Symposium on Utilization of Water in PCB Production,* European Institute of Printed Circuits, 1981.
19. D. Kunces, *Products Finishing,* p. 85 (Dec. 1987).
20. D. Kunces, "Hope for the Best But be Prepared for the Worst," MacDermid, Inc., Waterbury, CT, 1986.
21. R.S. Capaccio and R.J. Sarnelli, *Plat. and Surf. Fin.,* **73,** 18 (Sep. 1986).
22. R.S. Capaccio and R.J. Sarnelli, ibid., **73,** 20 (Oct. 1986).
23. W-C. Ying and R.R. Bonk, *Products Finishing,* p. 84 (Aug. 1986).
24. R.E. Wing, *Insulation/Circuits,* p. 58 (Mar. 1979).
25. R.E. Wing, ibid., p. 39 (Feb. 1979).
26. T.L. Foecke, *Journal of the Air Pollution Control Association,* p. 283 (Mar. 1988).
27. R.E. Wing and W.E. Rayford, *Plat. and Surf. Fin.,* **69,** 67 (Jan. 1982).
28. R.S. Capaccio, *Proc. 26th Annual Meeting of the Institute for Interconnecting and Packaging of Electronic Circuits,* 1983.
29. R.K. Gingras and R.S. Capaccio, *Printed Circuit Fabrication,* **10**(2), 98 (Feb. 1987).
30. *AESF Environmental Compliance and Control Course,* American Electroplaters and Surface Finishers Society, Orlando, FL.
31. M. Murphy in *Metal Finishing Guidebook-Directory,* Metals and Plastics Publications, Inc., Hackensack, NJ, 1988.
32. L.J. Durney, in *Electroplating Engineering Handbook,* 4th edition, Van Nostrand Reinhold Company, New York, NY, 1984.
33. R.S. Capaccio, *Proc. PC Fab/Expo '85,* 1985.
34. R.S. Capaccio, *Proc. NEPCON West,* 1986.
35. S. Suslik, *Industrial Finishing,* p. 16 (Jun. 1983).
36. "Applications of Ion Exchange: Pollution Abatement," Amber-hi-lites series, Rohm and Haas Company, Philadelphia, PA, 1970.
37. W.H. Waitz, "Ion Exchange for Heavy Metals Removal and Recovery," Rohm and Haas Company, Philadelphia, PA, 1979.
38. E. Hradil, *Proc. Circuit Expo '83,* p. 90 (1983).
39. R.M. Spearot and J.V. Peck, *Environmental Progress,* p. 124 (May 1984).

INDEX

Abrasion resistance, 135, 188, 197, 224, 390
Abrasive wear data for electroless nickel, 274
Abrasive-wear properties, 119
ABS, 377, 378, 381, 383, 384, 392, 393, 399, 433, 436-438
Absorption promoter dip, 244
Accelerators, 295, 388, 389
Activated carbon treatment, 176
Activation, 41, 53, 54, 95, 97, 98, 101, 103, 105, 194, 195, 203-205, 232, 242, 249, 300, 306, 309, 317, 336, 340, 345-349, 351, 364-366, 370, 382, 386, 399, 416, 438, 446, 481, 482, 488, 489, 493-495, 498
Activators, 205, 345, 346, 348, 350, 373, 386, 387
Addition of molybdenum, 119
Adhesion, 61, 62, 103, 105-107, 116, 119-121, 125, 130, 169, 176, 177, 180-182, 189, 190, 193, 195, 197, 198, 202, 203, 212, 231, 233, 240, 243-245, 289, 335-337, 339, 342, 345, 347, 363-365, 369, 377, 378, 381-385, 387, 392, 393, 417, 430, 442, 460, 489, 505, 512
Adhesion measurement, 180
Adhesive wear resistance, 189
Advanced treatment technology, 526
Aerospace
   Landing gear components, 209
Age hardening, 115, 116
Agglomeration of crystallites, 324
Aging, 116, 406
Agitation
   Agitation of the plating solution, 104
   Air agitation, 36, 159, 160, 162, 250, 253, 385
   Work-rod agitation, 104
Alkali metal borohydride, 403
Alkali metal hydroxides, 61, 79
Alkaline deoxidizers, 194, 195
Alkaline hypophosphite plating solutions, 75
Alkaline permanganate, 233, 239
Alkenes, 43
Alkynes, 43
Alloys
   52 alloy, 232, 233
      Co-P, 261, 262, 264, 265, 421, 431, 465, 469, 481, 493-495, 495, 496, 498-502, 505
   Cobalt-nickel-phosphorus films, 469
   Cobalt-phosphorus alloys, 11, 463, 465, 472, 474

   Cobalt-zinc-phosphorus, 472, 477
   Palladium-boron, 422
   Platinum alloy baths, 437
   Ternary alloy deposits, 265, 266, 469, 470, 472, 476
Aluminum, 3, 7, 35, 81, 103, 106, 116, 120, 121, 125-127, 130, 179, 181, 182, 186, 188, 189, 190, 197-199, 200, 201, 208, 209, 225, 229, 232-234, 240, 244, 245, 247, 249, 249-255, 257, 263, 269, 270, 273, 274, 277, 282-284, 364, 404, 406, 409, 417, 435, 436, 439, 471, 494, 513, 514, 516, 517, 521, 527
Aluminum alloys, 106, 116, 120, 190, 197, 198, 200, 249-251
Aluminum nitride, 240
Aluminum-magnesium wrought alloys, 197
Amine borane, 16, 52, 55, 86, 403, 408, 411, 412, 422, 429
Ammonia, 27, 78, 127, 160, 422, 424, 444, 458-460, 499, 500, 524
Ammonium hydroxide, 72, 77, 78, 81, 85, 182, 390, 463, 499
Analysis
   Determination of nickel content, 182
Anodic, 1, 9, 18-22, 22-24, 26, 37-41, 104, 126, 142, 186, 188, 195, 196, 289, 295, 296, 297, 299, 300, 302, 307, 310, 312, 314, 317, 322, 329, 409, 412, 414, 442, 446, 447, 481
Anodic oxidation, 20-22, 22-24, 26, 39, 289, 295, 412, 414, 442, 481
Anodic partial reaction, 299, 300, 302, 307, 314, 329
Anodic polarization, 22, 37-39, 41, 186, 447
Antistiction coating, 255
Appearance, 34, 103, 121, 177, 180, 186, 202, 207, 230, 250, 252, 361, 378, 502
Application
   Chip carriers, 240
   Sucker rod joint, 214
Applications
   Aerospace applications, 202, 208
   Aircraft engines, 208, 209
   Applications and characteristics, 230
   Applications in the chemical processing industry, 211
   Capacitors, 204, 229, 230, 240
   Coating dough kneaders, 217
   Connectors, 229, 230, 235, 238, 245, 253, 259, 421

529

Edge card connectors, 238, 229
EMI shielding, 394, 397
Engineering applications, 125, 207
Food processing industry, 216
Foundry tooling, 225, 227
High-frequency and microwave devices, 244
Hybrid microwave circuits, 240
Magnetic memory disks, 257
Microwave devices, 232, 244
Military applications, 223, 244
Miscellaneous applications, 225
Mo-Mn Application, 242
Printing industry, 225
Shafts of aircraft engines, 209
Sump pumps, 155
TF30 jet engine, 209
Transistor chips, 231
Argon, 36, 184
Arrhenius equation, 53, 316
Arsenic, 105, 461, 511
Autocatalytic, 403, 511, 512
Automatic analyzers and replenishers, 164
Automatic level control device, 161
Barrels, 214, 245, 249-251
Bath chemistry, 101, 103, 106, 256
Beryllium, 121, 190, 202, 206, 209, 240
Bismuth, 104, 105, 169, 171
Blue ammonium (amine) complex, 30
Borogluconate, 78, 79
Borogluconate solutions, 79
Borohydride, 4, 5, 8, 12-15, 24, 52, 81-85, 93,
    111, 133, 139, 201, 243, 261, 403-405, 407,
    408, 411-414, 416, 417, 432, 436, 438, 439,
    445, 474, 483, 484, 489, 511, 525
Boron, 5, 6, 11-16, 85, 93, 111, 112, 119, 122-
    124, 139, 207-209, 229, 231, 232, 238, 240,
    247, 261, 264, 265, 273-275, 278, 282, 422,
    463, 474, 484, 505
    Boron contents, 111
Bright dip solution, 233
Buffers, 33, 60, 61, 102, 293
Cadmium, 104, 105, 152, 169, 171, 224, 225,
    264, 436, 514, 516, 520-522
Calcium, 105, 275, 382
Cannizzaro reaction, 360, 361, 371, 486
Carbon and low-alloy steels, 194
Carboxylic acid, 48, 424
Cast iron, 194-196, 217, 224, 225
Catalyst, 16, 31, 35, 45, 47, 83, 239, 243, 317,
    318, 321, 322, 324, 342, 347, 364, 365, 371,
    375, 382, 388, 409, 412, 414, 421, 443, 511, 512
Catalyst predip, 239
Catalysts
    DMAB, 4, 15, 16, 18, 21-26, 86-91, 93, 98,
        101, 133, 139, 261, 264, 265, 291, 429, 430,

443, 458, 482, 484-486, 489, 491
Catalytic, 2-4, 6-10, 13, 18, 20, 22, 35, 36, 42, 43,
    45-49, 53, 62, 69, 71, 83, 89, 102, 105, 108, 193,
    195, 201, 204, 205, 238, 249, 266, 269, 282-
    284, 297, 309, 312, 313, 315, 317-319, 321,
    322, 324, 327, 347, 348, 351, 357, 364, 371,
    387, 404, 412, 414, 421, 423, 424, 434, 483,
    489, 491, 495, 511, 512
Cathodic, 1, 9, 18-21, 24, 37-41, 83, 188, 195,
    196, 238, 249, 289, 290, 295-297, 299-302,
    311, 313, 314, 329, 374, 409, 442, 446, 447,
    454, 455, 481
Cathodic partial reaction, 300, 302, 311, 313,
    314, 329
Cathodic polarization, 20, 37-41, 446, 447
Ceramics, 144, 204, 207, 230, 239, 240, 243,
    245, 405, 411, 417, 423, 434, 436, 438, 441,
    460, 461
    Alumina, 111, 204, 240, 243, 269, 335
    Barium titanate, 229, 240
    Beryllium oxide, 240
Chelating agent, 359, 361
Chemical kinetics, 94
Chemical passivation, 142
Chemical plating, 1, 401, 441, 456, 511
Chemical vapor deposition, 244, 441, 511
Chemistry imbalance, 101
Chlorinated polyvinyl chloride, 141
Chromic acid, 127, 202, 239, 336, 364, 373, 382
Cleaning
    Alkaline soak cleaning, 194
    Desmutters, 251
    Non-etch aluminum cleaner, 250
    Poor surface preparation, 103, 107
    Preplating 2000 and 6000 series, 249
    Pretreatments, 103, 194
    Removal of surface oxidation, 193
Cobalt, 2, 7, 11, 124, 257, 258, 261, 262, 264,
    266, 429, 431, 443, 463-467, 469-477, 479,
    481-495, 498, 499, 502, 505, 517
    Cobalt boranes, 464
    Horizontal recording baths, 499-501
    Non-magnetic uses of cobalt plating, 505
    Plating electroless cobalt on electroless
        nickel deposits, 257
    Process developments for cobalt plating,
        488
Coercivity, 124, 463-467, 469, 471, 472, 474-
    476, 481, 483, 494, 495, 498, 499, 501, 502,
    504, 505
Complexing agents, 3, 26, 27, 29-31, 33, 34, 60,
    61, 69-71, 78, 82, 97, 256, 424, 486, 512, 523,
    524
Composites
    Codeposition of silicon carbide, 269

Composite electroless plating, 269, 270
Mechanics of composite electroless plating, 269
Multilayer composite, 332, 364
Particle-solution interactions, 280
Sputtered chrome layer, 254
Typical composites used in circuitry manufacturing, 331
Concentration of dissolved oxygen, 329
Concentration of HCHO, 329
Controlled mass transfer, 299
Copper, 201, 265, 267, 414, 415, 515
Bulk concentration of copper ions, 329
Catalysts for electroless copper plating, 317
Components of electroless copper baths, 291
Composition of electroless copper plating solutions, 291
Copper, 1, 7, 10, 11, 21, 22, 104-106, 108, 119, 125, 130, 151, 179, 181, 186, 193, 198, 201, 202-204, 229, 232, 234, 235, 237-239, 243, 244, 249, 251, 261, 263, 265, 267, 289-293, 295-297, 299, 300, 302-307, 309, 311, 313-318, 320, 322, 324, 325, 327, 329, 331-333, 335-337, 339-342, 344-375, 384, 389, 390, 392, 393, 395, 396, 401, 404, 413-415, 417, 423, 426, 429, 432, 435-437, 439, 442, 459-461, 471, 494, 505, 511-517, 519-522, 524-526
Copper heat sink, 232
Copper oxide, 238, 345, 350
Copper strike, 108, 203, 239, 249, 251
Electroless copper-based alloys, 265
Electroless copper plating 21, 289-291, 293, 313, 314, 317, 357, 359, 366, 373, 374, 414, 461
Formaldehyde-based electroless copper baths, 324
Kinetics of electroless copper deposition, 314
Measurements in the complete copper bath, 309
Microetch copper, 239
Migration and diffusion of the copper, 237
OFHC, 232
Post-electroless copper consider ations, 362
Properties of electroless copper deposits, 322, 325
Summary of reaction orders for electroless copper baths, 315
Typical electroless copper baths, 351

Corrosion
Comparison of the corrosion rates, 214
Corrosion resistance, 59, 125-127, 129, 130, 177, 186, 187, 197, 207, 209, 210, 212, 216, 217, 223, 224, 227, 230, 231, 245, 263, 264, 269, 272, 276, 391, 393, 505
Cupric chloride, 239, 344, 365
Current density, 37-39, 41, 126, 143, 196, 251, 295, 297, 309, 314, 329, 367, 379, 447, 482
Cyanide, 106, 194, 202, 203, 251, 291, 307, 325, 405-408, 411, 412, 414, 488, 513-516, 520-522
Data
Coulometry data, 322
Current potential curves, 26, 296
Diffusion coefficient of HCHO, 329
Polarization curves, 19, 20, 22, 23, 37, 39-41, 295, 297, 298, 302, 446, 447, 454, 481, 490, 491
Dehydrogenation, 4, 7, 8, 18, 35, 42, 46
Deionized water, 104, 105, 155, 186, 404
Density, 37-39, 41, 122-126, 143, 179, 188, 196, 208, 213, 248, 251, 254, 295, 297, 309, 314, 325, 329, 353, 366, 367, 379, 423, 447, 464-467, 470, 472, 482, 488, 493, 494, 505
Deoxidizers, 194, 195
Depolarizers, 405
Deposition on metals for engineering use, 247
Deposition potential, 36, 176, 177
Deposits
Cloudy deposits, 102, 105
Compressively stressed deposits, 126
Dark deposits, 103, 105, 108
Deposit properties, 73, 169, 171, 176, 177, 327, 425, 439
Phosphorus content of deposits, 39, 72, 75
Zinc immersion deposits, 197, 198, 202
Diallyl phthalate, 378
Diamond, 204, 227, 254, 270, 273-275, 277-284, 286, 287
Dies, 225
Diffusion parameter, 299, 329
Dilute zincate, 198
Dimethylamine borane, 4, 5, 15, 24, 86, 261, 412, 429, 463, 484, 489, 504
Dimethylformamide, 382
Direct plating on ceramic, 3, 245
Displacement, 1, 7, 27, 31, 35, 159, 197, 198, 401, 402, 404, 405, 408-411, 413, 417, 421, 423, 460, 511-514
Economics
Economics of electroless nickel solutions, 215, 216, 227
Plating costs, 104, 252
EDTA, 78, 79, 82, 85, 265, 291, 293, 309, 315, 316, 351, 405, 414, 422, 425, 432, 436, 482,

485, 524
Effect of bath temperature and pH, 264
Effect of reaction products, 93
Effects of pH, 77
Effects of stabilizers, 93
Electrical resistivity, 123, 208, 229
Electrochemical potential, 125
Electrochemical reaction, 26
Electroless cobalt, 257, 258, 264, 463, 464, 467, 472, 473, 476, 481, 482, 484, 489, 502, 505
Electroless plating of platinum group metals, 421, 439
Electroless platinum, 432, 433, 435-437
Electroless rhodium bath, 439
Electroless ruthenium bath, 437
Electroless silver plating, 22, 441, 443, 446, 449, 457, 460
Electron donors, 33, 49
Electronic sensors, 162
Elongation, 118, 220, 223, 325, 367
Empirical rate equation, 64, 95, 316
Energy, 3, 29, 53, 54, 95, 97, 98, 150, 202, 206, 230, 232, 269, 279, 306, 317, 461, 464, 465, 482, 498
Environmental Response Compensation and Liability Act, 519
Equilibrium constant, 32, 34
Equilibrium potential, 20, 295, 297
Equipment
  Applied anodic current to tank, 143
  Bag filters, 152, 155
  Bag liners, 143
  Coated tanks, 144, 146
  CPVC, 141, 154, 155, 248
  Dipping baskets, 249, 250
  Double boiler, 149
  Equipment, 101, 103, 104, 109, 125, 129, 135, 139, 151, 155, 157, 158, 167, 180, 186, 187, 213, 214, 217, 226, 245, 247, 248, 394, 401, 512, 515, 525, 526
  Equipment/mechanical problems, 101
  Exhaust, 140, 212
  Float control valves, 161
  Fuse link heaters, 163
  Heat exchangers, 146, 148, 149
  Immersion heaters, 141, 144-146, 159, 163, 358
  Magnetic impellers, 155
  Mechanical pump, 155
  Mechanical stirrer mixing, 160
  Panel coils, 146
  Plating facility, 139, 226
  Polypropylene felt sleeves, 153
  Pumping systems, 154, 160
  Reservoir tanks, 248

  Special plumbing, 140
  Stainless tanks, 104, 142, 143
  Steam heaters, 146
  Steel valves, 213
  Submersible pumping equipment, 214
  Tank construction, 104, 140, 357
  Teflon coils, 146, 147
  Water deionizers, 160
Equivalent weight, 188
Erosion resistance, 217, 224
ESD, 397
Etch-back, 335, 340, 342
Etched tanks, 104
Etching, 106, 143, 197, 200, 239, 243, 244, 250, 257, 340-342, 344, 346, 353, 364, 365, 367, 368, 380, 382, 384, 385, 416, 417, 456, 461
Ethanolamine, 405, 500
Ethylene, 49, 474
Ethylenediamine, 28, 82-85, 291, 425, 436, 453, 454
Ethylenediaminetetraacetic acid, 293
Exaltants, 61, 291, 295
Extrinsic stresses, 121
Faradaic process, 295
Faraday's law, 20
FB-1, 257
Ferric chloride, 68, 69, 239, 243, 365, 513
Ferrous alloys, 3, 194, 197
Field performance, 127
Fill speed, 381
Film structures, 244
Filtration, 69, 104, 107, 151-155, 160, 171, 248, 253, 256, 278, 356, 358, 427
Fine line patterning of ceramic substrates, 239
Fluoride predips, 202
Fluoride solutions 243
Formaldehyde, 21, 261, 265, 290-292, 295, 299, 307-309, 311, 313, 314, 316-322, 324, 327, 351, 354, 356, 357, 359-361, 372, 374, 378, 390, 412, 414, 422, 431, 458, 485, 486, 505, 511
  Formaldehyde oxidation data, 309
Formic acid, 290, 431
Fracture stress values, 118
Free metal ions, 33
Galvanic displacement, 401, 402, 404, 409, 410, 413, 511
Glass and ceramics, 204, 423, 441
Glycine, 327, 374, 499, 501
Glycolate strike, 196
Gold
  Applications of electroless gold and gold alloys, 416
  Applying a preplate of displacement gold, 408
  Au(I) sulfite baths, 412

Au(III) chloride with weakly reducing amine boranes, 411
Au-Ag alloy, 414
Au-Ag alloys, 414
Au-Cu alloy, 414
Au-Sn alloy, 415
Electroless gold bath with hydrazine, 410
Electroless plating of gold and gold alloys, 401
Gold 11, 229, 230, 232, 237, 238, 240, 244, 261, 317, 322, 366, 368, 386, 392, 401, 402, 403-417, 419, 421, 423, 426, 429, 430, 460, 488, 511, 514, 517, 526
Gold alloys, 401, 414, 416
Gold-copper alloys, 414, 415
Gold-tin alloy, 415
Hypophosphite electroless gold bath, 409
Non-Cyanide baths, 411
Plating electroless gold on nickel substrates, 408
Pure gold, 403, 414
Trivalent gold cyanide complex, 406, 407
Guidelines, 139, 253, 373, 378, 386, 519-522,
Hastelloy, 214
Heat sinks, 210, 212, 229-231
Helmholtz double layer, 312
Hermetic, 241, 245
Hexamine ion, 29
Hexavalent chromium, 385, 386, 388, 389
High alloy steels, 194, 195
High-strength steels, 120
Hydrazine, 4, 5, 17, 18, 139, 261, 410, 412, 413, 422-424, 428, 432-435, 437, 439, 443, 445, 446, 458, 459, 463, 464, 482, 484, 511, 525
Hydride ion, 9, 19
Hydride transfer, 8
Hydrogen, 2, 6-14, 17-25, 33, 34, 36, 39, 49, 50, 58, 73, 75, 93, 104, 113, 120, 129, 136, 143, 159, 160, 176, 177, 182, 190, 194, 239, 242, 293, 295, 313, 315, 317, 319, 320-322, 324, 325, 327, 351, 355-357, 371, 403, 404, 417, 426, 428, 434, 436, 465, 481, 483, 486, 487, 505, 513, 524, 526
Hydrogen embrittlement, 120, 177, 182, 190, 195
Hydrogen evolution reaction, 39
Hydrogenation, 7, 35, 43, 45, 49
Hydrolysis of DMAB, 90
Hydroxyacetate anion, 30
Hypophosphite, 1, 3-11, 13, 17-20, 22, 24, 26, 29, 37, 40, 42, 46, 52, 55, 57, 61-63, 65, 68, 71, 72, 75, 82, 86, 91, 93, 96, 106, 111, 133, 139, 171, 201, 239, 243, 261, 263, 264-266, 286, 291, 409, 410, 412, 422, 424-429, 431, 463-467, 469, 471-473, 481, 485, 487, 493, 499, 505, 511, 521

Index, 529
Inhibitors, 3, 35, 104, 194, 485
Inorganic ions, 105
Instability constant, 32
Intermetallic compounds, 103, 111, 115, 123
Iron, 7, 35, 69, 125, 131, 171, 194-196, 198, 213, 217, 224, 225, 232, 254, 264, 265, 289, 389, 413, 470-472, 476, 513, 514, 516, 517, 525
Isomerism, 43, 49, 50
Kinetics, 31, 42, 53, 92, 94, 300, 302, 314-316, 328, 373, 486
Lead, 5, 11, 14, 18, 21, 38, 40, 41, 43, 47, 52, 81, 84, 86, 102-106, 143, 171, 207, 230, 232, 238, 239, 257, 286, 291, 294, 331, 348, 365, 366, 369, 382, 384, 386, 407, 432, 442, 471, 520-522
Lead chloride, 84
Ligand, 27, 28, 30-32, 34, 69, 70, 91, 92, 98, 315
Line definition, 241
Magnesium, 105, 197, 201, 202, 206, 257, 513, 516
Magnesium alloys, 201, 206
Magnetic layer deposition, 254
Magnetic properties, 124, 255, 264, 265, 463-466, 469, 472, 483, 493-495, 498, 499
Magnetic thin film coating, 254
Maleic acid, 43, 44, 49, 50
Maleic anhydride, 43, 44
Mannitol/boric acid complex, 78, 79
Masking, 105, 176, 205, 238, 250
Mass transfer, 37, 41, 299, 300, 304, 315, 316
Mechanical techniques, 103
2-mercaptobenzothiazole, 84, 126, 265, 294, 430
Mercaptoformazan, 423, 424, 428
Metal turnovers, 67, 68, 93, 256, 278
Metallic contaminants, 104, 151, 205, 520, 522
Metallic optics, 209
Metallized conductor patterns, 239
Metallizing plastics, 204
Methanol, 210, 265, 356, 360
Methylene glycol, 295
Microstrain, 473, 474
Mineral-reinforced nylon, 378, 382, 384
Mining and associated materials handling, 219
Miscellaneous applications
  Miscellaneous ions, 73
Modulus of elasticity, 118
Molds, 225
Molten inorganic compounds, 243
Molybdenum, 119, 203, 230, 241, 244, 262, 263, 471
Molybdenum content, 119
Molybdenum-manganese (Mo-Mn) metallization system, 241
Monocarboxylate anions, 30
Monocrystalline-type diamond, 283

Nernst equation, 295, 299
Neutralizers, 385, 386
Nickel
  Applications of electroless nickel, 135, 207,
    212, 213, 217, 223, 224, 227, 229
  Brittle electroless nickel coating, 120
  Classes of electroless nickel coatings, 190
  Corrosion resistance of electroless nickel,
    125, 127
  DMAB-nickel plating solutions, 91
  Electroless nickel for memory disks, 254
  Electroless nickel-tungsten alloys, 262
  Free nickel ion, 31, 32, 34, 71
  Major uses of electroless nickel, 209, 210
  Mechanical properties of electroless nickel,
    118
  Nickel acetate, 3, 196
  Nickel alloys, 5, 123, 133, 196, 266, 482
  Nickel boride, 34, 81, 82, 207
  Nickel chelate compounds, 30
  Nickel chloride, 3, 73, 196, 201, 513
  Nickel concentration, 6, 62, 63, 72, 75, 78, 88,
    89, 96, 167
  Nickel hypophosphite, 3
  Nickel phosphite 65-67, 69, 96
  Nickel plating 1, 3, 7, 9, 11, 15-17, 19, 20, 24,
    26, 28, 31, 33, 34, 36, 39, 45, 52, 53, 55, 57,
    58, 60-62, 64, 68, 69, 71, 72, 75, 76, 78-80,
    82, 86, 87, 89, 91, 93, 94, 99, 101, 104-106,
    121, 139, 193, 194, 197-199, 202-204, 207,
    217, 224, 226, 229, 232, 234, 238, 241, 243,
    245, 247-249, 251, 252, 256-258, 264, 269,
    390, 391, 446, 479, 489, 505, 507, 517, 521
  Nickel sulfate, 3, 73, 239, 265, 286, 512, 513
  Nickel sulfate hexahydrate, 265
  Non-magnetic, electroless nickel deposits,
    125
  Photomicrographs of cross-sectional cuts
    of electroless nickel, 270
  Plating electroless nickel on 5086 aluminum
    alloy, 257
  Properties of electroless nickel coatings,
    127, 208
  Sulfur-free Ni deposits 39
Nickel-200, 213, 215
Nickel composites
  Electroless nickel containing alumina, 269
  Electroless Nickel-PTFE composite, 276-278
  Electroless nickel/teflon, 210
Nitrile compounds, 325
Nitrogen, 6, 15, 30, 78, 140, 187, 432, 433, 473
Noble metals, 289, 404, 432
Non-conductive surfaces, 207
Nonmetallic surfaces, 203, 204
Octahedral, 27
On-line quality control, 170

Optimum plating rate, 252
Organic contaminants, 104, 105, 160, 176, 197,
  404
Organic stabilizers, 36, 45, 407
Orthophosphite, 8, 61, 127, 490
Oxidation 1, 2, 4, 13, 19-26, 37, 39, 65, 75, 183,
  190, 193, 237, 289, 290, 293, 295, 307, 308,
  309, 311, 313, 314, 317-322, 329, 336, 347,
  348, 362, 363, 375, 384, 412, 414, 416, 417,
  442, 447, 481, 483, 487-489, 491, 493, 524, 526
Oxyanion, 262
2-oxazolidinone, 464, 486
Oxygen, 27, 30, 35, 36, 232, 238, 291, 300-302,
  318, 329, 474
Palladium, 7, 10, 11, 36, 105, 201, 203, 205, 239,
  243, 244, 261, 266, 267, 317, 336, 344, 346,
  347-349, 351, 361, 366, 370, 373, 375, 386-
  390, 399, 401, 416, 421-432, 437, 488, 511, 525
Partial reactions, 19, 20, 38, 296, 298, 299, 302,
  309, 314, 409
Particles, 34, 35, 48, 105, 111, 115, 121, 129,
  130, 139, 144, 147, 152, 155, 156, 158, 160,
  204, 269, 270, 273-275, 277-284, 286, 294,
  324, 349, 356, 357, 382, 387, 442, 455, 474,
  495, 525
Perchloroethylene, 250
Persulfate or hydrogen peroxide-sulfuric acid
  etchant, 239
Phase diagrams, 111, 135
Phenolic, 331, 378
Phosphite anion, 65
Phosphorus contents, 111, 112, 126, 262, 469,
  473, 475
Phosphorus reduction, 39, 71, 74
Physical properties, 3, 12, 122, 124, 318, 324,
  325, 327, 366-368, 374, 401, 495, 526
Pickling, 103, 107, 126, 193-195, 202
Pits, 103, 177, 380
Pitting, 102, 104, 105, 107, 125, 126, 160, 171,
  213, 257
Plated through-hole, 238, 243, 332, 375
Plating electronic components, 252
Plating magnetic media, 499
Plating of non-conductors, 377
Plating on plastics, 204, 377, 378, 394, 399
Plating rate, 5, 6, 32, 33, 39, 42, 46, 47, 55, 58, 61,
  62, 65, 68-73, 75, 77-79, 83-85, 88, 89, 91,
  94-96, 105, 171, 176, 252, 256, 269, 277, 295,
  297, 299, 316, 325, 327, 361, 368, 403-406,
  408-411, 413, 415, 416, 422-427, 429, 431,
  433, 436, 437, 447, 453, 454, 458, 483, 491,
  492, 495, 499, 500, 521
Plating rates, 53, 69, 70, 77, 98, 102, 171, 316,
  325, 361, 403, 405, 407, 408, 490
Plating silicon wafers with p-n junctions, 79
Plating with amine boranes, 86, 99

Plating with sodium borohydride, 81
Platinum
  Chloroplatinic acid, 432
Polarization, 19, 20, 22-24, 37-41, 126, 186-188, 295, 297, 298, 300, 302, 304, 317, 406, 446, 447, 454, 481, 490-492, 527
Polarography, 72, 488
Polishing and texturing, 254
Polyacetal, 378
Polyarylether, 378
Polycarbonate, 378, 381, 384, 436, 438, 474
Polycrystalline diamond, 273, 275, 279, 282-284
Polyetherimide, 377
Polyhydric materials, 79
Polyphenylene oxide (modified), 378
Polypropylene, 104, 140, 141, 151-153, 248, 249, 356, 357, 377, 383, 386, 404, 434
Polysulfone, 377, 382, 383, 434
Porosity, 59, 103, 122, 125, 146, 169, 171, 176, 177, 182, 185, 186, 194, 207, 212, 243, 249, 366, 428
Post-activation, 349
Post-treatments, 189, 190
Postplating treatments, 125
Potassium perrhenate, 264, 265
Pourbaix diagrams, 290
Printed wiring boards, 229, 235, 240, 331, 370
Process
  Additive processing, 332, 363, 365-368, 374, 375
  Alternative processes, 370
  Burnishing, 255
  Contact process, 516, 517
  Fully-additive processing, 366-368
  Glass removal, 242
  Handling procedures, 248
  Heat treatment 107, 116, 119, 121, 123-125, 127, 130, 132, 134, 135, 177, 190, 207-209, 227, 255, 256, 273, 474, 480, 497
  Immersion plating, 289, 512, 515, 517
  Immersion zinc, 234, 235, 254, 257
  Microetching, 242
  Passivation method, 143
  Photoresist and etching techniques, 244
  Plating procedures and processes, 247
  Plating servo valves, 209
  Semi-Additive process, 365-368, 374, 375
  Sensitization, 243, 244, 438, 489, 495
  Smear removal, 335, 336, 348
  Step plating, 171, 176
  Subtractive processing, 332, 363, 365-367
  Typical process cycle, 232, 238, 244, 257
Process bath treatment, 520
Process steps for plating, 238
Process variations, 170

Properties, 111, 130, 270, 274, 276-278
  Mechanical properties of parts, 118, 120, 181, 331, 374, 472
Quality control, 160, 169, 170, 189-191, 342, 392
Quality method, 169
Quality of the substrate, 103
Quaternary alloy, 263, 502
Racking, 248, 352, 358, 359, 391
Rate constant, 7, 26, 53, 54, 94, 95, 98
Rate of deposition, 6, 9, 20, 33, 36, 47, 58, 71, 75, 78, 79, 88, 89, 91, 94, 293, 353, 355, 463, 476, 488, 527
Rates of chemical reactions, 94
Reaction mechanisms, 5, 6, 18, 94, 307
Redox potential, 1, 12, 289, 290, 292
Redox reaction, 295
Reducing agents, 4, 5, 11, 19, 22-24, 55, 102, 139, 171, 290, 293, 350, 403, 407, 411-413, 422, 429, 441, 443-445, 459, 463, 464, 481, 512, 525, 527
Reduction, 1, 2, 4-9, 12-20, 25-29, 35, 38, 39, 41, 55, 61, 62, 64, 71, 74, 78, 82, 84, 89, 90, 92, 96, 101, 106, 121, 169, 171, 176, 177, 193, 199, 201, 290, 296, 301, 313, 317, 324, 329, 336, 371, 374, 377, 384, 385, 389, 390, 406, 426, 432, 433, 439, 442, 446, 489, 511, 512, 516, 519, 520, 523-527
Regulations
  Clean Air Act, 519
  Regulations, 519, 520
Residual contamination sources, 158
Resistors, 240, 489, 505
Reverse etch-back, 342
RFI, 397
Rhodium, 7, 421, 439, 473
Rinsing, 35, 105, 107, 108, 159, 185, 205, 242-244, 250, 336, 339, 340, 342, 344, 345, 348, 350, 359, 362, 373, 385, 386, 388, 389, 522, 523
Rinsing after activation, 348
RMA fluxes, 229
Rochelle salt, 265, 291
Sacrificial layer, 127
Safety, 139, 140, 145, 149, 151, 211, 251, 403, 519
  Respiratory Protection Standard, 519
Salt spray resistance, 127
Screened conductive patterning, 244
Shop traveler, 170, 190
Shot peening of parts, 121
Silanes, 325
Silicon carbide, 111, 269, 270, 273-275, 277, 278, 282-284, 286
Silicon devices, 204
Silicon nitride, 240
Silicon wafers, 79, 204, 231

Silver, 10, 21, 22, 229, 232, 244, 261, 289, 290, 392, 414, 417, 441-446, 449, 453-461, 471, 511, 512, 516, 517, 520-522
  DIT bath, 446, 449, 454, 455
  Electroless plating of silver, 441
  Practice of electroless silver plating, 457
  Silvering reaction in a non-aqueous solution, 449
Skinning, 381
Sodium bisulfate, 336
Solution contamination, 101, 104
Solution replenishment, 163
Specifications
  2404B, 247
  2405A, 247
  5086, 257
  5252, 257
  5586, 257
  AMS 2399, 247
  ASTM A307, 215
  ASTM A193 B8M, 215
  ASTM B568, 180
  ASTM B733, 177, 190
  ASTM G5, 187
  ASTM standards. 186
  B733-86, 247
  CW66, 257
  MIL-C-26074, 247
  MIL-C-26074C, 177, 190
  OFHC, 232
  Specifications, 177, 187, 190, 191, 247, 372
Stabilizer concentration, 35, 36, 45-47, 71, 72, 108, 429
Stabilizers, 3, 34-38, 43, 45, 47-50, 52, 58, 71, 72, 75, 84, 85, 93, 102, 106, 126, 265, 291, 294, 295, 351, 359, 361, 374, 405, 407, 411, 425, 456, 527
Stainless steel, 104, 108, 125, 126, 129, 133, 142-144, 148, 152-155, 179, 182, 195, 196, 213, 248, 249, 277, 357, 384
Steel, 7, 35, 61, 62, 104, 106, 108, 116, 120, 121, 125-127, 129-133, 135, 142-144, 148, 151-155, 179-182, 185, 187, 188, 193-197, 199, 201, 210, 212-217, 223-226, 232, 248, 249, 262-265, 272, 275-278, 357, 384, 423, 426, 463, 473, 505, 513, 514, 517, 521, 527
Stop-off materials, 205
Stress
  Internal stress, 67, 71, 93, 121, 125, 177, 428
  Intrinsic stress, 59, 67, 96, 121, 189, 256
  Tensile internal stresses, 127
Stripping, 105, 106, 141, 142, 144, 238, 248, 322, 366, 368, 514
Stripping from steel, 106
Structure
  Amorphous microstructure, 126

Banded structure, 115, 116, 127
Crystal growth, 324, 498, 505
Crystal structure, 112, 123
Diffusion barrier, 230, 232, 235, 237, 238
Fibrous-appearing microstructure, 126
Formation of micelles, 48, 50
Lamellar structure, 113
Lattice defects, 169
Microcracking, 127
Nodules, 103
Photomicrographs of cross-sectional cuts of electroless nickel, 270
Roughness 102-107, 119, 160, 171, 248, 278-284, 384, 385, 493, 498
Structure of electroless nickel, 111
Structures
  Grain sizes, 112, 324
Substrate material, 179, 331
Substrate preparation/activation, 101
Substrate-catalyzed processes, 401, 402
Succinate anion, 30
Sulfamate strike, 196
Sulfate ions, 127
Sulfur
  Divalent sulfur compound, 93
Sulfur-bearing substance, 38
Superheated water, 146, 149
Supplemental part agitation, 160
Surface concentration of HCHO, 329
Surface contaminants, 103
Surface preparation, 103, 107, 193-195, 199, 202, 370, 373, 490
Teflon, 139, 144, 146, 147, 156, 187, 210, 216, 248, 277, 357, 358, 377, 404
Tensile strength, 118, 119, 169, 182, 190, 325, 327
Tests
  Accelerated yarnline wear test, 272, 273
  Alfa wear test, 272
  Alloy composition, 59, 63, 177, 198, 415, 437
  Atomic adsorption, 72, 171
  Bend test, 180, 254
  Beta backscatter method, 180
  C.A.S.S., 393
  Coefficient of thermal expansion, 121, 124, 208, 378
  Coulometric method, 180
  Ductility, 118, 245, 253, 254, 263, 324, 325, 327, 375, 505
  Ferroxyl test, 185
  File test, 181
  Friction coefficients and wear data, 277
  Friction coefficients for miscellaneous composites, 279
  Hardness, 116, 119, 120, 127, 129-131, 133-136, 147, 169, 181, 190, 197, 207, 209, 210,

213, 223, 225, 230, 231, 235, 240, 245, 264, 266, 274, 325, 423, 426, 428, 430, 473, 474, 506

Hot chloride porosity test, 185

Hot hardness, 119, 136

Hot water test, 185

Jacquet test, 392

Lubricity, 210, 212, 217, 224, 225, 245, 269, 272, 513, 514

Magnetic method, 180

Measuring corrosion rate, 186

Metallographic sectioning, 179

Microhardness, 177, 181

Micrometer method, 180

Microporosity, 46, 242, 243

Porosity measurement, 185

Porosity test for aluminum substrates, 186

Porosity test for copper substrates, 186

Preece test, 514, 517

Preparation of test specimens, 182

Process analysis, 171

Process capability index, 170

Pull-away, 337, 339

Punch test, 181

Quench test, 180, 181

Rotating disk electrode, 297, 299, 302, 309, 329

Salt spray test, 125-127, 393

Service conditions, 129, 130, 391

Service tests, 127

Solderability, 197, 209, 212, 229-232, 245

Spectra analysis, 184

Stability constant, 28, 32, 33, 70, 71, 293, 327

Stress analysis, 189

Substrate densities for weigh-plate-weigh method, 179

Substrate heat-treatment temperatures for quench test, 181

Surface roughness, 119, 248, 278-281, 283, 284, 498

Taber Abrader, 188

Taber wear test data, 275

Tafel parameter, 297

Tafel slopes, 297, 307, 312, 314

Tape test, 342

Test methods, 177, 189, 247

Testing of deposit properties, 177

Thermal conductivity, 123, 124, 208, 240, 463

Thermal cycling, 377, 393

Wear of steel pins in V Blocks, 276

Wear resistance, 111, 133, 177, 189, 190, 207, 209-212, 217, 219, 223, 224, 231, 235, 237, 245, 264, 266, 269, 272-276, 463, 505

Weigh-plate-weigh method, 179

Weldability, 245

Wire bonding, 231, 232, 240, 406

X-ray diffraction studies, 263

X-ray spectrometry, 180

Textiles, 226

Thallium nitrate, 84, 85

Thallium salts, 52

Theory of mixed potentials, 295, 297

Thin film evaporation and sputtering, 244

Thiodiglycolic acid, 52, 93, 428

Thiosulfate, 427, 460

Thiourea concentration, 39

Through-hole interconnections, 238

Tin-palladium activating solutions, 348

Titanium, 121, 127, 144, 148, 202, 206, 209, 226, 244, 274, 382, 384, 416, 511

Toxic chemicals, 139

Treating electroless copper rinsewaters, 524

Treatment

Annealing, 116, 120-122, 125, 127, 254, 325, 469, 471, 476, 483, 495

Cementation, 511, 514

Double zincate, 251, 417

Treatments

Carbon treatment, 105, 176, 525

Ceramic conditioning, 243

Conditioning, 197, 201, 243, 340, 350, 386

Ion exchange, 69, 436, 438, 523, 526, 527, 528

Removal of foreign contaminants, 193

Trichloroethane, 250

Troubleshooting, 101, 106, 107, 372, 382, 383, 385-387, 390

Tungsten, 203, 262, 265, 275, 417, 432, 470, 472, 498, 502

Type I reagent water, 188

Ultrasonics, 243

Urea formaldehyde, 378

Vertical recording baths, 502, 503

Voltammetry, 72, 317, 322

Waste minimization, 519

Waste treatment

Conventional techniques, 194, 250, 523

Electrolytic recovery, 523, 526, 527

Hazard Communication Standard, 519

Heavy metals, 105, 345, 525,

Occupational Safety and Health Act, 519

Resource Conservation and Recovery Act, 519

Rinsewater conservation techniques, 523

Rinsewater treatment, 522

Wastewater treatment

Wastewater treatment for electroless plating, 519

Water molecules, 14, 18, 27, 30-32, 82, 91, 92

Wiring, 229, 230, 235, 240, 249, 331, 370, 375

Woods strike, 195, 196
Zinc, 1, 79, 103, 145, 152, 171, 181, 197, 198,
  202, 203, 206, 209, 234, 235, 251, 254, 257,
  264-266, 295, 432, 442, 471, 472, 477, 513,
  514, 516, 517, 520-522
Zinc die castings, 79, 203
Zincate, 182, 197, 198, 200-202, 235, 251, 417,
  513

ABOUT THE EDITORS

Mallory

Hajdu

Glenn O. Mallory

Glenn O. Mallory has more than 30 years' experience in the surface finishing industry, with specialization in electroless deposition processes. In 1986, Mr. Mallory formed the Electroless Technologies Corporation (ETC) in Los Angeles, CA. ETC is an independent R&D organization which licenses its technical developments. Prior to that he was vice president of R&D for electroless nickel development for the Allied-Kelite Division of Witco Chemical Corporation in Los Angeles.

Mr. Mallory is the author of many papers and patents on electroless deposition. He was elected a fellow of the Institute of Metal Finishing in 1974, and received the AES Silver Medal for a paper published in *Plating and Surface Finishing* in 1975. Mr. Mallory received his BS degree from UCLA and his MS from California State University, Los Angeles.

Juan B. Hajdu

Juan B. Hajdu has extensive experience in surface treatments and electroless plating. During his 30-year career he has been affiliated with ENTHONE, Inc., in New Haven, CT. He joined ENTHONE in 1961 as a research chemist and is currently Vice President, Technology, for ENTHONE—OMI, Inc., a subsidiary of Asarco Inc.

Mr. Hajdu obtained his PhD at the University of Buenos Aires, Argentina, then joined Pantoquimica, S.A., an affiliate of ENTHONE. He has written many papers on electroless deposition and electrodeposition, and is the inventor or co-inventor on some 20 patents. In 1966, Mr. Hajdu received the AES Gold Medal for work published in *Plating* concerning plating on plastics, and in 1970, the Eugene Chapdelaine Memorial Award for his work in zinc plating.

Printed and bound by CPI Group (UK) Ltd, Croydon, CR0 4YY

03/10/2024

01040434-0012